Studies in Vertebrate Evolution

List of Contributors

H. P. Whiting / *Department of Zoology, Bristol University*

C. C. D. Shute / *Department of Physiology, Cambridge University*

S. M. Andrews / *Royal Scottish Museum, Edinburgh*

A. A. Howie / *Department of Geology, Melbourne University, Australia*

A. L. Panchen / *Department of Zoology, Newcastle University*

A. R. I. Cruickshank / *Bernard Price Institute, Johannesburg, S. Africa*

A. J. Charig / *Department of Palaeontology, British Museum (Nat. Hist.)*

A. d'A. Bellairs / *St Mary's Hospital Medical School, London*

C. B. Cox / *Department of Zoology, King's College, London*

Ch. H. Mendrez / *Institut de Paléontologie, Muséum National d'Histoire Naturelle, Paris*

T. S. Kemp (Editor) / *University Museum of Zoology, Cambridge*

A. W. Crompton / *Museum of Comparative Zoology, Harvard, U.S.A.*

P. M. Butler / *Department of Zoology, Royal Holloway College, London*

K. A. Joysey (Editor) / *University Museum of Zoology, Cambridge*

W. G. Kühne / *Geologisch-Paläontologisches Institut, Freie Universität Berlin*

5675
JOY 1

Studies in Vertebrate Evolution

Edited by
K. A. Joysey
T. S. Kemp

Essays presented to
Dr F. R. Parrington, FRS

Oliver & Boyd
Edinburgh

OLIVER & BOYD
Tweeddale Court
14 High Street
Edinburgh EH1 1YL
A Division of Longman Group Limited

ISBN 0 05 002131 1

First published 1972

© 1972 The Contributors
All rights reserved
*No part of this publication may be
reproduced, stored in a retrieval system,
or transmitted in any form or by any means
—electronic, mechanical, photocopying, recording
or otherwise—without the prior permission of the
Copyright Owners and the Publisher. The request
should be addressed to the Publisher in the first instance.*

Set in 10/12 pt. Monotype Plantin 110
Printed in Great Britain by
T. & A. Constable Ltd, Edinburgh.

15672
15666

Contents

Francis Rex Parrington, ScD, FRS
Preface

H. P. Whiting	Cranial anatomy of the ostracoderms in relation to the organisation of larval lampreys	1
C. C. D. Shute	The composition of vertebrae and the occipital region of the skull	21
S. M. Andrews	The shoulder girdle of '*Eogyrinus*'	35
A. A. Howie	On a Queensland labyrinthodont	51
A. L. Panchen	The interrelationships of the earliest tetrapods	65
A. R. I. Cruickshank	The proterosuchian thecodonts	89
A. J. Charig	The evolution of the archosaur pelvis and hind-limb: an explanation in functional terms	121
A. d'A. Bellairs	Comments on the evolution and affinities of snakes	157
C. B. Cox	The pectoral region and habits of a new digging dicynodont	173
Ch. H. Mendrez	On the skull of *Regisaurus jacobi*, a new genus and species of Bauriamorpha Watson and Romer 1956 (= Scaloposauria Boonstra 1953), from the *Lystrosaurus* zone of South Africa	191
T. S. Kemp	The jaw articulation and musculature of the Whaitsiid Therocephalia	213
A. W. Crompton	The evolution of the jaw articulation of cynodonts	231
P. M. Butler	The problem of insectivore classification	253
K. A. Joysey	The fossil species in space and time: some problems of evolutionary interpretation among Pleistocene mammals	267
W. G. Kühne	Progress in biological evolution	281

Dr F. R. Parrington, FRS

Francis Rex Parrington, ScD, FRS

Dr Rex Parrington retired to Scotland in 1970 after more than forty years of teaching and research in Cambridge, where he was Reader in Vertebrate Zoology and Director of the University Museum of Zoology.

The young Mr Parrington came up to Sidney Sussex College in 1924 and attended lectures on vertebrates, given at that time by Hans Gadow and Clive Forster-Cooper, which captured his interest and gave direction to the rest of his academic career. After graduation in 1927 he was appointed as Assistant to the Director in the Museum of Zoology and, determined to supplement the collections for teaching purposes, he went on an expedition to East Africa in 1933 in order to collect fossil vertebrates. The fruits of this expedition not only fulfilled the intended purpose but also formed the basis of many of his own subsequent publications and provided material for all of his research students.

As a junior worker Rex Parrington had frequent contact with Professor D. M. S. Watson in London and was privileged to accompany him on a tour of the principal Museums and Universities of North America. In so doing they developed a mutual respect and established a personal relationship which, in later years, transcended any difference of opinion on scientific matters.

When Clive Forster-Cooper was appointed to the Directorship of the British Museum (Natural History) in 1938, Rex Parrington succeeded him as Director of the University Museum of Zoology in Cambridge. Only one year later his career was interrupted by the war and he served in the Royal Artillery until 1945; during this period Major Parrington visited every continent except Antarctica and seized the opportunity to make a series of surprise calls upon his international colleagues.

After the war he returned to teaching and research in Cambridge and his elementary courses on Vertebrate Morphology and Evolution are remembered by generations of students as some of the most inspiring, best presented lectures which they attended. For him teaching and research were inseparable, much of his teaching arising from his research and much of his research arising from his teaching. In his advanced courses on Vertebrate Palaeontology each problem was reviewed in its historical perspective and then the currently accepted interpretation was brought to a climax in a spirit of personal involvement.

Rex Parrington was awarded the degree of Doctor of Science in 1958, elected as a Fellow of the Royal Society in 1962 and appointed Reader in Vertebrate Zoology in 1963. Those who have been privileged to work closely with Rex Parrington are deeply grateful for the stimulation, devotion and high scientific ideal that he has engendered, and wish him a happy retirement among the trout and the heather.

The Publications of Dr F. R. Parrington, FRS

1933   On the cynodont reptile *Thrinaxodon liorhinus* Seeley. *Ann. Mag. nat. Hist.*, (10), **11**, 16-24.

1934   On the cynodont genus *Galesaurus*, with a note on the functional significance of the changes in the evolution of the theriodont skull. *Ann. Mag. nat. Hist.*, (10), **13**, 38-67.

1935   On *Prolacerta broomi*, gen. et sp. n., and the origin of lizards. *Ann. Mag. nat. Hist.*, (10), **16**, 197-205.

1935   A note on the parasphenoid of the cynodont *Thrinaxodon liorhinus* Seeley. *Ann. Mag. nat. Hist.*, (10), **16**, 399-401.

1936   On the tooth-replacement in theriodont reptiles. *Phil. Trans. R. Soc.*, B, **226**, 121-142.

1936   Further notes on tooth replacement. *Ann. Mag. nat. Hist.*, (10), **18**, 109-116.

1937   A note on the supratemporal and tabular bones in reptiles. *Ann. Mag. nat. Hist.*, (10), **20**, 69-76.

1939   On the digital formulæ of theriodont reptiles. *Ann. Mag. nat. Hist.*, (11), **3**, 209-214.

1940   On the evolution of the mammalian palate: (with T. S. Westoll). *Phil. Trans. R. Soc.*, B, **230**, 305-355.

1941   On two mammalian teeth from the Lower Rhætic of Somerset. *Ann. Mag. nat. Hist.*, (11), **8**, 140-144.

1945   On the middle ear of the Anomodontia. *Ann. Mag. nat. Hist.*, (11), **12**, 625-631.

1946   On the cranial anatomy of cynodonts. *Proc. zool. Soc. Lond.*, **116**, 181-197.

1946   On a collection of Rhætic mammalian teeth. *Proc. zool. Soc. Lond.*, **116**, 707-728.

1946   On the quadratojugal bone of synapsid reptiles. *Ann. Mag. nat. Hist.*, (11), **13**, 780-786.

1948   Labyrinthodonts from South Africa. *Proc. zool. Soc. Lond.*, **118**, 426-445.

1949   Remarks on a theory of the evolution of the tetrapod middle ear. *J. Laryng. Otol*, **63**, 580-595.

1949   A theory of the relations of lateral lines to dermal bones. *Proc. zool. Soc. Lond.*, **119**, 65-78.

1950   The skull of *Dipterus*. *Ann. Mag. nat. Hist.*, (12), **3**, 534-547.

1953   On *Aenigmasaurus grallator*, gen. et. sp. nov., a problematic reptile from the L. Trias. *Ann. Mag. nat. Hist.*, (12), **6**, 721-738.

1955   On the cranial anatomy of some gorgonopsids and the synapsid middle ear. *Proc. zool. Soc. Lond.*, **125**, 1-40.

1956   A problematic reptile from the Upper Permian. *Ann. Mag. nat. Hist.*, (12), **9**, 333-336.

1956   The patterns of dermal bones in primitive vertebrates. *Proc. zool. Soc. Lond.*, **127**, 389-411.

1958   On the nature of the Anapsida. pp. 108-128 in *Studies on Fossil Vertebrates*. ed. T. S. Westoll, Athlone Press. Univ. of London.

1958   The problem of the classification of reptiles. *J. Linn. Soc. (Zool.)*, **44**, 99-115.

1959   A note on the labyrinthodont middle ear. *Ann. Mag. nat. Hist.*, (13), **2**, 24-28.

1960   The angular process of the dentary. *Ann. Mag. nat. Hist.*, (13), **2**, 505-512.

1961   The evolution of the mammalian femur. *Proc. zool. Soc. Lond.*, **137**, 285-298.

1962   Les relations des cotylosaurs diadectomorphes. *Problèmes actuels de Paléontologie. Colloques Internationaux du Centre National de la Recherche Scientifique.* Paris. No. **104**, 175-185.

1967   The vertebrae of early tetrapods. *Problèmes actuels de Paléontologie. Colloques Internationaux du Centre National de la Recherche Scientifique.* Paris No. **163**, 269-279.

1967   The identification of the dermal bones of the head. *J. Linn. Soc. (Zool.)*, **47**, 231-239.

1967   The origins of mammals. *Advancement of Science*, **24**, 165-173.

1971   On the Upper Triassic mammals. *Phil. Trans. R. Soc.*, B, **261**, 231-272.

Preface

This volume is presented as a tribute to Dr F. R. Parrington, FRS, and we trust that it will give him great pleasure. In 1968 a number of his former research students and colleagues discussed ways of expressing their appreciation to him, and it was agreed that the publication of a volume of essays and original work would be the ideal. We felt that such a volume would give the greatest pleasure to Rex Parrington if the contributions came from those who had been closely associated with him, and so it was determined that invitations to contribute would be confined to his own former students, and those who had either worked in the same department or published jointly with him.

We have subsequently heard from others who would have been glad to join us and we wish to assure them that their exclusion arose only from our intention of making a personal rather than a general tribute. We deeply regret that Professor Westoll was seriously ill during part of the production programme and we are grateful to him for volunteering to withdraw rather than delaying the rest of the volume. All the manuscripts were in our hands before the end of 1970.

So far as possible we have arranged the contributions in their systematic order because this reflects the pageant of vertebrate evolution on which Rex Parrington based his own lecture courses. The width of interest represented in the current work of his former students and immediate colleagues reflects the exceptional span of his own research interests. Each author was invited to contribute an article in a particular field, and all agreed to work within this framework except Professor Kühne who preferred to write on a philosophical topic rather than Jurassic mammals. We have been persuaded to include his contribution on the basis of his own claim that it is just what Rex would expect from him!

We should like to express our gratitude to the contributors and to the publishers for all the hard work which made this volume possible, and for their willing co-operation which has greatly eased the editorial load. We have done our best, but we should like to apologise to the other contributors for those errors and omissions which have escaped our notice.

Cambridge 1971

K. A. J.
T. S. K.

H. P. WHITING

Cranial anatomy of the ostracoderms in relation to the organisation of larval lampreys

INTRODUCTION

Knowledge of jawless vertebrates has recently been increasing rapidly, and a closer integration between what is known about the modern and the palaeozoic forms should be possible. But the new information tends to be more specialised, and so the biologist, 'palaeo-' or otherwise, finds himself non-expert over a greater sector of the picture.

For an integration of this sort, 'Occam's Razor' must be used—*Frustra fit per plura quod potest fieri per pauciora*, Occam, W. (1324-7): there is no need to do by many things (hypotheses) what can be done by fewer. But hypotheses must be set out in sufficient detail for them to be re-examined when new facts become available, as Parrington (1958) has done in a brief discussion of anaspid agnathans. There, structure and function, habit and habitat are joined together to form a hypothesis that can be tested.

Here, the head of the larval lamprey, especially in its youngest free-living stage, is compared with what is known of ostracoderm heads. This proammocoete head is likely to provide the simplest picture of structural and functional features common to the ancient and the modern. Hagfish, though perhaps more directly related to the earliest ostracoderms known, are very specialised and have a direct development into their adult form.

Plates, rather than diagrams, are used so that there should be less of an 'author's selection' of the facts, and alternative views may more easily be considered.

LARVAL LAMPREYS CONSIDERED IN RELATION TO ANASPID HEAD-STRUCTURE

The lamprey embryo hatches into a larval form known as the proammocoete which is less specialised than the older ammocoete stage. It is also easier to understand. Fig. 1 is intended as an introduction to a description of Plate I. The figure was traced from photographs of a living *Lampetra* proammocoete 9 mm long. A similar figure emphasising different features is given by Young (1962). A painting by Shipley (1887) shows the beauty of the proammocoete; modern colour-photographs show similar tones.

At this stage of development the almost transparent larva lies in the stream-bed, usually just buried between the pebbles or gravel downstream of the 'nest' in which the eggs were laid (cf. Young, 1962). The figure does not show the median pineal eye between the paired eyes nor the naso-hypophysial opening in front of the pineal. It shows the gill-openings of the left side only, though both sides are really visible. The larva draws in a respiratory and feeding current by means of a pair of muscular folds called the velum. Food such as diatoms, *Paramecium*, *Arcella* and other Protozoa is trapped in mucus strands and so passes back along the pharynx while the exhalant respiratory current passes out through the minute gill-openings. There is a vertical band of cilia at one place on each gill-bar that probably keeps the mucus strands in position. The intestine contains the mucus column

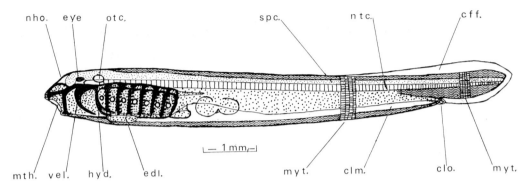

Fig. 1 Diagram of *Lampetra planeri*, at a late Stage 14 of Damas: from a photograph of a living proammocoete.
nho., opening of nasohypophysis; *otc.*, otic chamber; *spc.*, spinal cord; *ntc.*, notochord; *cff.*, caudal finfold; *mth.*, mouth; *vel.*, velar fold; *hyd.*, hyoidean arch and front of pharynx; *edl.*, endostyle; *myt.*, myotomes at trunk and tail; *clm.*, coelom; *clo.*, cloaca.

and is long and straight. The heart, arteries and veins are fully functional. The kidneys are pronephric with widely open ciliated coelomic funnels. There is no separation of pericardial and peritoneal coelom. The animal swims well but for brief periods.

Plate I shows frames from a cine film of a proammocoete of *Petromyzon marinus* L., 8·5 mm long. No structural differences are known between proammocoetes of different species. The eggs were collected from nests where there were several spawning adults, by Dr Q. Bone and myself. The film was taken by Dr M. A. Sleigh and myself.

Plate I, 1 shows the body proportions with the long straight notochord, very caudally placed cloaca, the myocommata separating the myotomes and the large gill-pouches with their small gill-openings. I, 2 shows the eyes, and it can be seen that there are seven pairs of gill-openings. I, 3 and 4 show the endostyle and the auditory or 'otic' capsules shining refringently above the hyoid arches. In the film, the blood-corpuscles can be seen coursing along the blood-vessels. There is an impressive supply to the velum and to the cerebral arteries. I, 5 shows a *Paramecium* being drawn toward the mouth.

Looking at I, 1-5 together, we can follow the brain back into the spinal cord. The latter is seen to leave only a small space dorsally between itself and the meningeal membrane, because the spinal cord is round in T.S., not flattened as in the ammocoete and adult. Neuronal details of the spinal cord and its relations to the myotomes have been given by Whiting (1948). The notochord becomes slightly thinner as it runs forward as far as the front of the midbrain floor, but is obscured from view by the upper part of the pharynx.

The feeding and respiratory mechanism can now be looked at, with the possibility in mind that creatures like the proammocoete might have developed ontogenetically into the various palaeozoic anaspids.

The pharynx extends from the visceral arch behind the last gill-opening, forwards to the hyoid arch. The branchial bars 1 to 6 reach inwards, bearing gill-filaments on their anterior and posterior

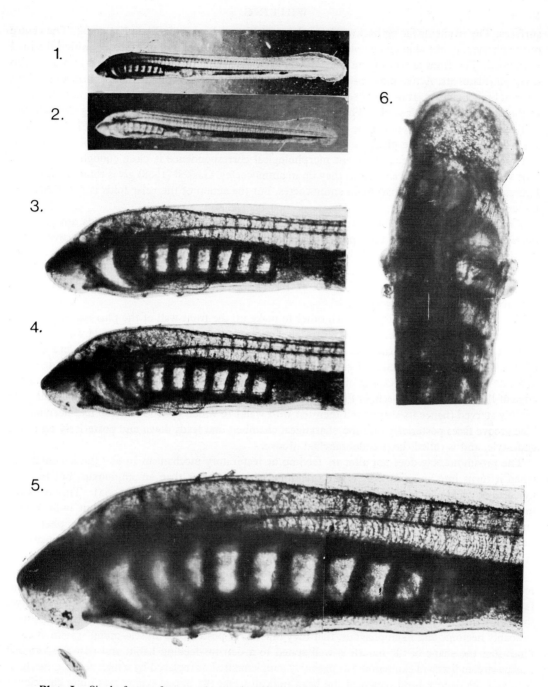

Plate I. Single frames from a 16 mm cine film of an active, 8·5 mm long, proammocoete of *Petromyzon marinus* L. figs. 1-5 lateral view, fig. 6 dorsal view. They show the velar movements, the gill-ports (fig. 2), and the mobility of the snout. The film was taken during one hour, in August 1967, by M. A. Sleigh and H. P. W.

surfaces. The filaments facing backwards are longer; this can be seen in Plate I, 4 and 5. The centre of the pharynx is free along its whole length, at this time, to the column of mucus in which the food is trapped. The front aspect of the last visceral arch and the hinder aspect of the hyoid arch also carry gill-filaments. Active movement of the pharynx can be seen, with rhythmic contraction of the pairs of branchial bars. It does not seem very powerful at this stage but it does alter the external form of the pharynx with each stroke. This is presumably a respiratory more than a feeding mechanism.

The buccal cavity in front can be altered in shape by the visceral muscle visible in Plate I, 6. The velar folds in front of the pharynx represent the mandibular pair of arches of gnathostomes: this statement may be a generalisation but morphological correspondence is close enough to justify it. The velar folds are shaped and act as they do in ammocoetes. Gaskell (1908) gives three-dimensional figures of the mouth and pharynx in ammocoetes, but the action of the velar folds is not difficult to understand.

The left and right velar folds are attached laterally to the buccal wall. The folds are powerful muscular organs, concave on their posterior surface. Each reaches forward in the buccal cavity, conforming to its shape, and then pushes back a body of water into the pharynx. The movement is reminiscent of a human hand, the shape changing during the driving stroke like that of the hand during a crawl-stroke in swimming (Plate I, 4 and 3). In this way, the ventral tip of each velar fold extends far back underneath the hyoid arch.

The hyoid arches extend toward each other to make up the front wall of the pharyngeal chamber. They are not mobile, but their medial ventral part curves back, so that the lower hyoid wall conforms, without changing shape during the velar stroke, to the form of the velar folds at the end of their stroke. There is, in fact, a ciliated groove on the medial edge of the hyoid wall, on each side of the aperture of the diaphragm leading into the pharynx. This groove is shown up, and with it the medial edge of the hyoid visceral arch, in Plate II, 2a and 2b, in a *Lampetra* pro-ammocoete that had been slightly starved (hence its shape) and then fed with water to which opaque material had been added. The groove faces posteriorly into the pharyngeal chamber, and leads down and posteriorly on to the endostyle, and is called the pseudobranchial groove.

The proammocoete does not alter its feeding or respiratory mechanism in any fundamental way when it turns into an ammocoete. The proammocoete is smaller than an amphioxus, but the ammocoete or pride may reach a weight of 10 g and a length of 17 cm (Young, 1962). This is of the same order of size as that of anaspids, several genera being 10 or 12 cm long, but bulkier than ammocoetes. The larval lamprey feeding system, with muscular velar folds and mucus pharyngeal sieve, limits the animal to very small branchial openings which also are found consistently in anaspids. It is therefore justifiable to pursue the comparison.

Many authorities have discussed the way anaspids may have lived, including their respiration and feeding. Among recent workers, Parrington (1958) and Ritchie (1964) have added some important facts, and Ritchie has summarised earlier work. Ritchie concludes his summary by stating that two forms have infillings of the intestine which indicate they ingested organic-rich mud; that most were probably nectonic and microphagous; that they must have possessed a muscle-pump system of some kind; that the shape of the mouth is well suited to a detritus-feeding habit; and that the 'original mucus stream for food extraction was probably supplemented or replaced by a filter feeding mechanism, and the deep ventral region of the head indicates that the buccal region was quite capacious'. This implies that the fish lived in midwater but acquired its food from the bed of the stream or lake.

The ammocoete pharynx is expanded and contracted by a most elaborate and diagnostic muscle-system. There is a valve and a muscle rhythmically closing the small gill-openings. There is an elaborate and again highly diagnostic folding of the branchial bar which is 'trefoil' shaped in horizontal

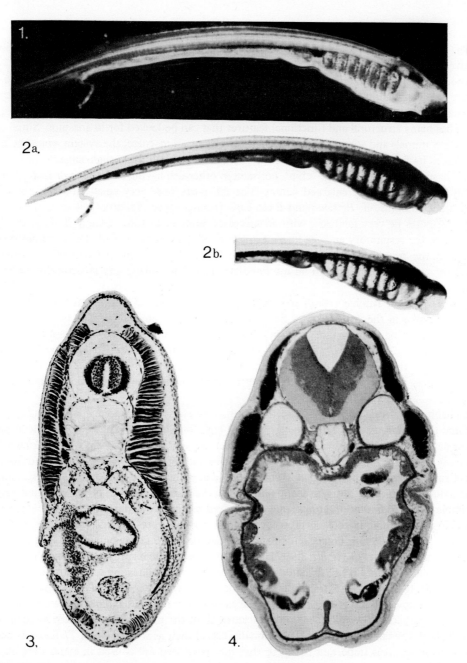

Plate II.
fig. 1 A feeding *Lampetra* proammocoete, recently underfed, about 10 mm long.
fig. 2 The same individual under different illumination. Photographs by G. L. E. Wing and H. P. W.
fig. 3 Transverse section of a 7 mm *Lampetra* proammocoete at the posterior pharyngeal level: silver-impregnated preparation; intermyotomal on left, mid-myotomal on right. H. P. Whiting Collection (H. P. W. Coll.), Slide No. L55A, 4 (1947). England Finder Graticule (E. F. G.) position J20-1. × 144 linear.
fig. 4. T.S. 8 mm *Lampetra* proammocoete at otic level, stained with phosphotungstic acid haematoxylin. Slide L4′, 2. E. F. G. position K32-3. × 135.

section. Alcock (1898), and Gaskell, in papers summarised in his book of 1908, have described this system.

Although these diagnostic features need not be described here, it is important that they are already present in the proammocoete stage, e.g. the special 'tubular' type of muscle in part of the visceral constrictor muscle (Séwertzoff, 1916). The proammocoete to ammocoete feeding and respiratory system has many structural and functional features that can be looked for in anaspids. Since anaspids were active animals and some were larger than the largest ammocoete, the system would be tested in evolution to its functional limits and these special features should be recognisable.

The anaspids have a very large 'chin', not seen in other fish having either jaws or rasps, that would suitably house a relatively enlarged velum. The gill-ports were very small, and remained so even in the larger anaspids, cf. *Birkenia* and *Endeiolepis* (Romer, 1966, fig. 20).

The gill-ports lie very far back, with an increased number in some genera. This gives a greater surface area for respiratory and feeding mechanisms if the medial part of the branchial system lay forward in the primitive normal position in the head. The slanting position of the gill-ports of ammocoetes would then represent an intermediate stage functionally and structurally between pro-ammocoete and anaspid.

In *Pharyngolepis*, Ritchie (1964) describes a pattern of ventral scaling that seems to separate a myotomal area in the trunk from a flexible pharyngeal one anteriorly. The pattern shown, with the pharynx running back further on each side of the ventral midline than it does along the midline itself, would match the shape of gill-bars extending caudally each under the one behind, in the same way that the velar pair extend under the hyoid pair in the lamprey larva. The absence of larger anaspids is then accounted for because the apparatus for respiration and feeding has reached its limit.

The habitat of larval lampreys has been described by Hardisty (1944). All larval stages require to draw into the mouth well oxygenated water with a high food-content. They could not feed by sifting mud or detritus from the bottom or they would block the mucus flow and asphyxiate. Anaspids, from their body form, are to be considered midwater or surface fish. They might have swum to aggregations of small animals at the surface and sucked these in with the velar apparatus. The size and shape of the mouth agrees with this possibility. There are, therefore, a number of features that agree with the simple hypothesis that anaspids fed and respired as the most simply organised vertebrates do today. No other simple hypothesis would fit so many of the facts, for example filter-feeders using a straining mechanism may be expected to have especially large gill-openings.

The characid teleost *Thayeria*, Plate III, 2, provides an interesting comparison with anaspids. The lower lobe of the tail is lengthened and strengthened. It feeds near or at the surface. When swimming forward, the body-axis is horizontal, but when stationary near the surface it holds itself poised with the head upward, at a fairly constant angle as shown in the illustration, where the white line is one of some parallel horizontal lines in the background of the photograph. The fish frequently moves to the surface; the tail-shape may be functionally useful then as well as when *Thayeria* is stationary below the surface. It is important to notice that the specialised tail is used in manoeuvre not merely in forward swimming, a point that may often be relevant in discussing the locomotion of extant and extinct fishes. Details of *Thayeria* species are given by Sterba (1962). An example of *Birkenia elegans*, Plate III, 1 has been placed with the dorsal outline and tail in a similar position to that of a poised *Thayeria*. The similarity in ventro-lateral outline is fortuitous. When *Birkenia* swam forward, it would carry itself as shown in Romer's fig. 20 (1966). But it is suggested that *Birkenia* might flex its tail down, as it lies in the specimen of *Birkenia* illustrated, at other times such as when poising at an angle or manoeuvring at the surface. This comparison is some further support for considering

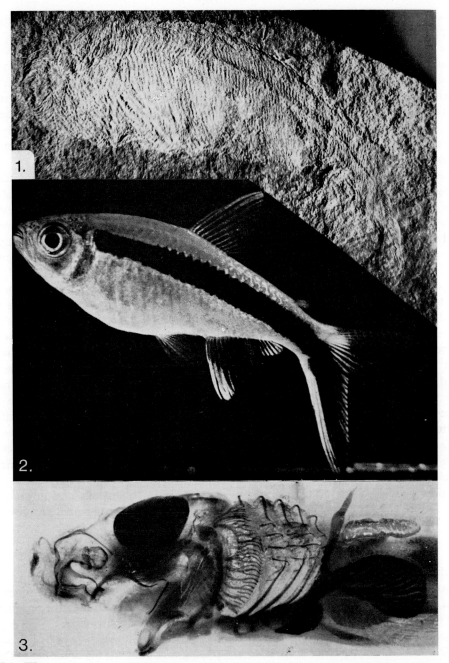

Plate III.
fig. 1 *Birkenia elegans* Traquair. Univ. of Bristol Geol. Dept. No. 7935. Specimen 62 mm long.
fig. 2 *Thayeria obliqua* Eigenmann. A characid teleost poising itself in a typical position: horizontal line shows in lower background. 30 mm long.
fig. 3 *Scyliorhinus caniculus* (L.). Cartilage stained with methylene blue; extrabranchial elements show clearly. Univ. of Bristol Zoology Dept.: prepared by E. Etemad, J. T. Ablett and H.P.W. ×4·5

anaspids to be surface-feeding forms which could have used the ammocoete feeding and respiratory cranial apparatus.

On the simplest hypothesis, the anaspid would develop directly from a proammocoete, moving from a bottom-living habitat without passing through a burrowing phase where the eyes were reduced. In visceral anatomy the various anaspids would come to resemble ammocoetes and then 'super-ammocoetes' as described already. This hypermorphosis would involve support for the pharyngeal structure as they grew larger (whatever their respiratory mechanism). Where the internal skeleton is weak, the external one will require to be more developed: the stage of *Scyliorhinus* shown in Plate III, 3 may be an example of this. In any case, the form of its extra-branchial elements makes an interesting mechanical comparison.

As Watson (1954) remarks, regarding anaspids such as *Birkenia* and *Lasanius*, the equivalent of a shoulder-girdle is required. Even though it has now been shown that these forms probably had a lateral fin or fin-fold behind the last branchial rod, the gill-ports anterior to the last, being at different vertical levels, would each have to take the pull of the lateral body-muscles. They would require a branchial skeleton to do so because there is a large water-cavity anterior and ahead of each port. In this respect the anaspids probably had an ammocoete structure in the splanchnocranium, with additional support externally because they were more advanced along the same lines.

Some conjectures can be made about the non-visceral part of the head if the anaspids developed from a proammocoete. The paired eyes would develop from those of the proammocoete with its retina and lens (Studnička, 1898), and with vertebrate extrinsic eye-muscles (Koltzoff, 1901), without the recession of the visual system occurring in prides. The brain would resemble the adult lamprey's, with a layered optic tectum and functionally compartmented hindbrain. The diencephalon would be different; the epithalamus would be larger than in lampreys because the pineal system was more developed; it would be far more asymmetrical, this being an ancestral vertebrate feature. The hypothalamus would be different because there would be a well-developed endostyle and no thyroid gland; the effects of this would be shown in the ventral diencephalon just as much as any differences in the pituitary gland.

The body-scaling of anaspids indicates the siting of the myotomes (Westoll, 1958). Hence, the scaling in such forms as *Pharyngolepis* implies that the myotomes of the postorbital and otic region did not extend far forward over the eyes, with oblique myocommata as in lampreys, but retained their original position, with vertical myocommata. Furthermore, there was evidently no ventral extension forwards of hypobranchial myotomal muscle, under the maximally-sized pharynx.

LARVAL LAMPREYS AND HETEROSTRACAN HEAD-STRUCTURE

The development of a heterostracan proammocoete would in the simplest hypothesis be like that of a petromyzont or anaspid one. Its minimal differences would perhaps be two: in early forms in the geological series even the young larva might be more primitively organised than the proammocoete we know today, and secondly, this proammocoete might be divergently organised.

Divergently, the paired olfactory bulbs at the front of the endbrain would here be in contact with a nasal system that remained anterior in the head, not passing back on to the top of the head. The nasal openings would be single or paired according to the condition in the adult, a matter recently discussed by Romer (1968). They might or might not be joined to the hypophysial opening: presumably not, in any heterostracan line leading to the gnathostomes.

In more primitive organisation, the third and second, and possibly the first, pairs of pre-otic myotomes would remain in position alongside the notochord, not moving out to become extrinsic eye-muscles as they do in petromyzont proammocoetes (Koltzoff, 1901; Damas, 1944, fig. 76). The first pair of myotomes could in heterostracans have provided adequate movement of the eyeball in the orbit, for it divides in the advanced embryo into three parts differently inserted on to the eyeball. The second and third pairs we see apparently in their original position, in the adult, as indicated in Plate IV, 1 and 2, as proposed by Tarlo and Whiting (1965), and supported by Carter (1967). But this view is not taken by all (cf. Jollie, 1968, fig. 9).

Such a position of the pre-otic myotomes implies a more primitive, and very different, neuronal system in the midbrain and hindbrain in larva and adult.

Certainly heterostracans, as we find them as fossils, provide direct evidence on the relation between somatic and visceral organisation in the ostracoderm head. In Plate II, 3 and 4, the pharyngeal and buccal arrangement of the petromyzont proammocoete is shown for comparison. Fig. 3 traverses the posterior pharynx well behind the endostyle; at mid-myotomal level, on the right, successive muscle-plates lie one above another, their myofibrils contracting in a direction parallel to the longitudinal axis. The axons of somatic motor neurons leave the spinal cord in the ventral root which comes out at each mid-myotome: a ventral root is visible in the silver-impregnated section. Peripherally, the axons turn up or down on to the inner face of the muscle-plates but remain very much in the same mid-myotomal plane. Part of the ventral branch of the ventral root descends past the outside of the visceral muscle to reach the hypobranchial part of the myotomal muscle; at the posterior end of the pharynx the axons remain in the same vertical plane, as in this section. The central organisation of the somatic neurons, and their relation to myotomes and skin, have been described by Whiting (1948).

On the left, an inter-myotomal level is transected. The complexity of even the early stage of the visceral muscle and its trefoil-shaped branchial bar can be imagined. The main fact is that the three types of visceral muscle, which produce the pharyngeal movements by their integrated action, are aligned in the transverse plane (Séwertzoff, 1916, text fig. 8) and are supplied by two visceral motor nerve components between them. There are, therefore, clear-cut differences in the orientation of muscle and in the origin and course of nerve-supply between the somatic and visceral pharyngeal organisations, which become even more marked as the animal matures.

The myotomes are an important morphological and anatomical barrier between the visceral branchial system and the dorsal surface at this level of the pharynx. This barrier continues to the front of the pharynx; then at the level of the otic capsule and the buccal cavity it becomes more slender.

In Plate II, 4, at this level, the myotomal muscle is stained red in the section and shows as a darker tone in the plate. There is ventral somatic muscle as in the posterior pharynx, while the dorsal part of the myotomes has divided, as it extends forward of the otic capsule, into two lateral parts and a medial one between the notochord and the otic capsule. A section of this level of the head is shown in a coloured diagram by Séwertzoff (1916, plate-fig. 2), and sections of this region in other planes are shown by Whiting (1957). Successive stages in the development of the somatic and visceral muscle and the early branchial skeleton of the proammocoete and ammocoete can be seen in Séwertzoff's elegant illustrations.

So the morphological data given here and elsewhere are a strong basis for identifying the medial pairs of ridge-and-furrow folds under the dorsal shield of cyathaspids such as *Anglaspis* as myotomal muscle of successive somites. The complicated hypotheses which would have to be made if these medial ridges were regarded as visceral can be realised if these two illustrations on Plate II are compared with the reconstruction made by Watson (1954, fig. 4) for a cyathaspid head.

The hypothetical heterostracan proammocoete would presumably pass through a second larval stage, when it was larger and with more complicated gill-structure as in ammocoetes, but developing the exhalant aperture systems of adult heterostracans at some stage. At all events, there would be a larval stage when the animal would retain effective eyes and be free-living but would be unarmoured, unlike the adult heterostracan. The myotomal head-musculature would be functional just as in larval lampreys.

When the various heterostracans became adult, was their armour flexible, or made of rigid units that could move on each other like the scales of *Polypterus*, or was it strong and rigid? Extensive evidence is available, but it is a specialist's field. Great variation in structure existed in this great and long-lived order (or higher taxon); this seems to have included variation in strength and in degree of flexibility. The degree of flexibility is important regarding cyathaspids and pteraspids where good data for the internal structure are available. Denison (1961) stated that these heterostracans had gills "enclosed within an essentially immovable shield".

Were the armour at all flexible, even in one direction only, there would remain an important function for the somatic muscle of the head that seems on morphological grounds to have been well developed there, not in directly assisting forward locomotion but in counteracting the rhythmic changes in tension from the vertical component in the underlying visceral muscle.

Anglaspis may have possessed slightly flexible armour. On the left side of the specimen in Plate IV, 1, the outline of successive muscles can be seen on the outer surface of the armour: trunk-muscles posteriorly, two metameric series anteriorly; the medial one being the apparent myotomic muscle, and continuous with the ridge-and-furrow system exposed at otic level on the right; and a lateral series that is branchial and should in part represent visceral muscle. Plate IV, 4 is of a section of the armour of the Ordovician heterostracan *Astraspis*. This seems almost entirely to lack any lower laminated layer and it can hardly have been strongly rigid (cf. Ørvig, 1967, fig. 3). Myotomic muscle would be needed to support its notochord in the absence of a strong neurocranium.

Three further items in larval lamprey stages are relevant to heterostracan head structure. First, the swelling lying caudal and lateral to the posterior semi-circular canal in *Anglaspis*, Plate IV, 3a, may have been the 'vagal' *lateralis* ganglion: the line leading medially back from it may have been the lateral-line nerve, with the first few sensory branches rising to the shield just behind the supposed ganglion. Plate IV, 3b shows these structures in a parasagittal section of a proammocoete. The two animals portrayed are of entirely different size, of course, but the relative proportions seem to be correct, the eye in the petromyzont lying at about the level of the pineal, not further forward as in (adult) *Anglaspis*. Plate IV, 3b is shown on a larger scale on Plate VI. The lateral-line nerve runs back toward the midline in the proammocoete but not so obliquely. There it runs back along the trunk in a dorsolateral part of the *meninx primitiva* around the spinal cord (Plate II, 3).

If the identification of the vagal *lateralis* ganglion and nerve is accepted, the occurrence of a proammocoete stage in the life-history of a heterostracan becomes more probable. Watson (1954) thought it feasible that the myxinoids, lampreys, cephalaspids and anaspids are to be grouped together by

Plate IV.
fig. 1 *Anglaspis macculloughi* Woodward Univ. of Bristol Geol. Dept. No. 7939. Photograph by E. W. Seavill. ×3·4.
fig. 2 Diagram of fig. 1:—*orb* = orbit; *pin* = pineal complex; *mdb* = midbrain; *myt* = myotome; *vsa* = visceral arch; *ps-cc* = posterior semi-circular canal.
fig. 3a *Anglaspis macculloughi* Woodward. Univ. of Bristol Geol. Dept. No. 10315. Photograph by E. W. Seavill. ×3·4.
fig. 3b *Lampetra*, parasagittal section through the otic capsule. For explanation see Plate VI, 5d. The otic capsule is here aligned with those of fig. 3a. The dorsal surface is at *d*, and the more ventral part at *v*. The lateral-line nerve and ganglion can be seen caudal to the otic capsule.
fig. 4 *Astraspis* sp., section through the dermal armour. Material provided by F. R. Parrington, sectioned by M. White, photographed by K. Wood. H. P. W. Coll., Slide No. *Astraspis* 2, E.F.G. position G39-4. ×90.

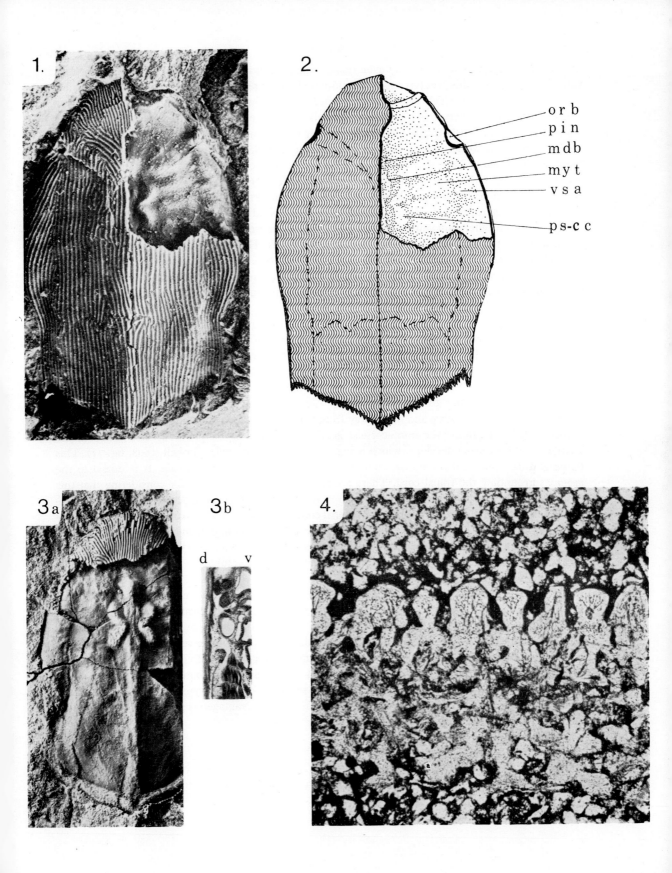

their common possession at an early evolutionary stage of an ammocoete larva. "The Heterostraci, whose relationship to this group seems to be supported by the probable character of their gill-pouches form a different group which branched off before the introduction of the ammocoete". This view would now seem to involve a number of difficult conjectures, especially if the proammocoete is also excluded from heterostracan life-history on that hypothesis.

The second item relates to the head in front of the mouth; in the proammocoete, this region is highly flexible due to local visceral muscle. The snout can be raised or lowered, altering the head profile. The cast of the internal surface of the head-shield varies within one species of some heterostracans. Alteration in head profile would be of great importance in the forward locomotion and manœuvre of heterostracans, especially such as lacked paired fins, as the illustrations of Halstead (1969) show. The alteration could be achieved by visceral muscle acting on a slightly flexible armour or on joints in the armour of such an animal as *Nektaspis* (Heintz, 1968). The possible alteration in head profile of heterostracans when swimming seems to have been little discussed. Probably it has seemed too conjectural a topic. But all stages of lampreys, which lack paired fins, also have mobile head profiles, no doubt using myotomal and visceral snout-muscle in co-ordination; and this aspect of head structure is linked with the last item.

If the lateral-line is correctly identified then we have strengthened the validity of identifying the *meninx* where it leaves the posterior edge of the lower edge of the head-shield, as shown in the impression of that shield in *Anglaspis*, Plate IV, 3a. The *meninx* appears to pass under a rim formed by the head-shield impression, and at the same time to expand laterally. This marked expansion needs explanation. It seems to have no relation to any external trunk feature of the trunk of *Anglaspis*, (cf. Romer, 1966, fig. 22), nor to any feature of the proammocoete, where the nervous system and its *meninx* are of constant calibre and draw away from the dorsal surface in maintaining contact with the notochord (Plate I). But in the ammocoete stage, the spinal cord at this point becomes flattened, and a large *dorsal fat-column* develops, which is biggest at mid-trunk levels (Gaskell, 1908, fig. 73). This tissue is usually outlined, in ammocoetes, by black pigment on its dorsal side. It is dorsal to the *meninx* surrounding the spinal cord. There are two morphological reasons for considering that this tissue may have occupied a larger volume still in the ancestor of the ammocoete. Expansion downward would explain the flattened form of the spinal cord as compared with proammocoetes and gnathostomes. A like expansion upwards would explain the condition visible in many ammocoetes, where the black pigment forms an inverted Y, as if in modern forms the dorsal part of the *dorsal fat-column* had been pressed together when the column became reduced in volume.

If *Anglaspis* possessed a fat-column like that of the ammocoete it would lie in this position, and would begin its expansion at the beginning of the trunk. This inference about the trunk of *Anglaspis*, from its posterior head-structure, is very much conjecture, but it could indicate that in that genus, and perhaps in cyathaspids generally, there was a dorsal organ in the trunk, of low density, similar to, but larger than, that in ammocoetes. This tissue would hardly achieve the volume to provide neutral buoyancy but might be an important factor in keeping the dorsal surface uppermost—an important aid to manœuvrability in a fish lacking paired fins and lacking also the smooth outlines and flexibility of the ammocoete's trunk. This tissue is absent over the ammocoete pharynx, and tapers away again towards the tail; such effect as it has is produced in the most advantageous position.

LARVAL LAMPREYS AND CEPHALASPID HEAD-STRUCTURE

So much of the cranial anatomy of cephalaspids is known, and it resembles that of lampreys so closely, that they are commonly grouped together in systematic position, for example, as orders of

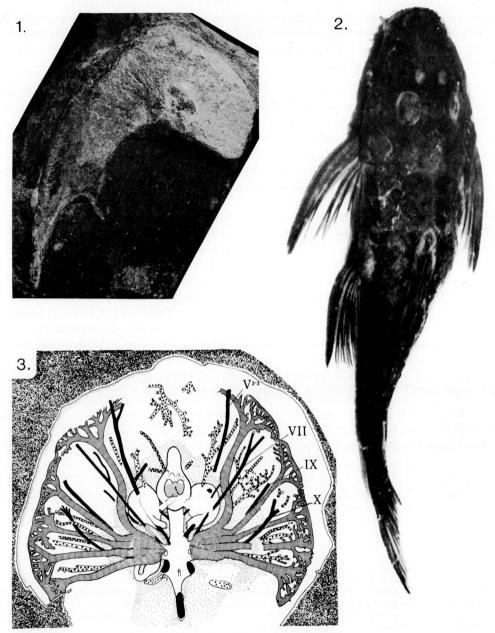

Plate V.

fig. 1 *Cephalaspis* sp. Univ. of Bristol. Geol. Dept. No. 17149. ×1. Photograph by K. Wood.

fig. 2 *Lithoxus lithoides*, Eigenmann. 4·85 cm long. A loricariid teleost: the anterior and posterior nostrils are adjacent on each side; the tail is hypocercal, with the upper and lower margins both supported by a spine.

fig. 3 *Kiaraspis auchaspidenoides*, Stensiö. ×7. Diagram after Stensiö (1927), plate 49. *Black* = nerves; *wavy lines* = 'sensory electric system'; *white spots on black* = arteries; *black spots on white* = veins. The labelling does not follow that of Stensiö. Note that the specimen is viewed from the ventral aspect.

the same sub-class (Romer, 1966). Stratigraphically there is no great gap, a lamprey having been described from Carboniferous rocks (Bardach and Zangerl, 1968). Ontogenetically, the cephalaspid may well have developed through a proammocoete and an ammocoete stage, though the latter may not have been adapted to burrowing.

External form of the head. Cephalaspids are clearly bottom-living fish. Some modern bottom-living fish have a similar shape to them, as Sanderson (1939) observed in his travels. The loricariid, *Lithoxus lithoides* (Plate V, 2) has a flattened head, paired eyes close together on the top of its head, large pectoral spines, and nostrils noticeably far back, close to the eyes. These fish live in fast, turbulent water. The olfactory sampling mechanism probably works better on top of the head than it would on the snout, being subject to less of a direct onrush of water and less change in water pressure from moment to moment. There is an obvious advantage in having the eyes high in the head, as in a skate.

The profile and rigidity of the cephalaspid head were no doubt significant in locomotion. The elasmobranch *Torpedo* has a head of similar contours. If the tail of a *Torpedo* is prodded, the fish swims violently forward, the head rises relative to the tail and continues to do so until the fish comes to be upside down, in an 'uncontrollable' loop-the-loop. Another elasmobranch with a similar profile, *Squatina*, prodded on the tail, rushes forward with the same result. (Dr Q. Bone and Mr J. Tayless, respectively, pointed out these examples to me.) It is the head profile, and not the shape of the tail, that is the more important in controlling forward swimming. Three related *Loricaria* species are illustrated by Boulenger (1896, plate VII); one with the upper margin of the tail strengthened, one with the lower, and one with the tail symmetrically strengthened above and below. The first and second appear to be mechanically equivalent to heterocercal and hypocercal fish. The heterocercal tail of cephalaspids was perhaps important in manœuvre, as in turning, not in the continuous correction of some dynamic imbalance during horizontal forward swimming.

Internal structure of the head. The dorsal part of the cephalaspid head contained a solid tissue, broadly equivalent in position to the neurocranium and dorsal musculature. Its anatomy can be described by the courses taken by tubes, identifiable as those of nerves, arteries and veins, as they pass through the

Plate VI.
Silver-impregnated sections of the head of *Lampetra* proammocoetes. In the longitudinal sections, the head faces to the left (as in Plate 1). The small arrow at the bottom of the plate indicates the position of the IXth nerve in the four sections above it. The otic capsule can be seen anterior to the glossopharyngeal ganglion in these sections. Sections were cut at 10μ or at 12μ.

fig. 1 Parasagittal section through a late embryonic stage, passing lateral to the brain. Nerves and ganglia of VII, IX and X^1, three myotomes and two gill-pouches are visible. 12 days after fertilisation, 4 mm long, (stage XI of Damas). H. P. W. Coll. L25, 1. E.F.G. position N47-2. $\times 168$.

fig. 2 Similar section through a newly-hatched stage. Below the brain are the eye, maxillo-mandibular ganglion, otic capsule, IXth ganglion and three myotomes. 15 days after fertilisation, 6 mm long (stage XIV of Damas). H. P. W. Coll. L12, 1. E.F.G. position P42-3. $\times 168$.

fig. 3. Transverse section across a late embryonic stage, at the level of the IXth ganglia; the area on the left is two sections (20μ) caudal to that on the right, along the diagonal line indicated. Myotomal tissue can be seen medial and lateral to the ganglia. 13 days after fertilisation, 4 mm long (stage XI of Damas). H. P. W. Coll. L21, 1. E.F.G. positions K41-2 and K37-2. $\times 173$.

fig. 4 Medial sagittal section through a proammocoete. Note the naso-hypophysial complex, the pineal complex, and the infolding between midbrain and hindbrain. 15 days after fertilisation, $7\frac{1}{2}$ mm long (stage XIV of Damas). H. P. W. Coll. LX2. E.F.G. position F51-2. $\times 110$.

fig. 5 Five successive sections cut at 10μ, 5a being the most medial and separated by 70μ from that shown in fig. 4, through the same animal. 5c shows ganglia V^1, V^{2-3}, VII, IX, X, and X-lateralis. The peripheral course of V^1, above the eye-position, can be seen in 5e. E.F.G. positions E26-1, P54-2, P51-1, P48-1 and P44-2. $\times 110$.

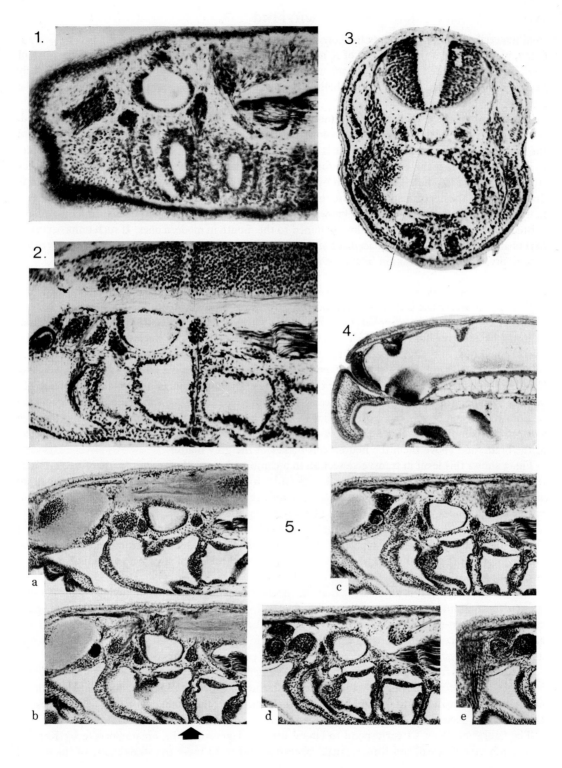

solid material. Plate V, 3 attempts to show in diagrammatic form the structure described by Stensiö (1927, plates 49 and 50) in the head of *Kiaeraspis auchenaspidoides*, as seen from below. This work of 1927 is rightly regarded as the classic one upon the subject. The figure shows the tubes identified by Stensiö as nerve-paths, in black. The labelling does not follow Stensiö and should for the present be ignored.

The lower part of the head, the splanchnocranium, did not contain this matrix and is not preserved. But the lower side of the preserved part shows foramina, and foldings corresponding to gill-bars and the branchial cavities preceding them. It is clear that the branchial openings were on the lower side, as in a skate, and that these openings lay further forward than the corresponding openings of gnathostome fish or cyclostomes. This is shown by the courses of nerves and blood-vessels. The problem is first to identify the nerves and other structures found, and then to determine, if possible, the visceral morphology underneath. It has been known for a long time that anterior branchial units may have existed in early vertebrates, but are lost or joined to the mouth in modern ones. If such units occurred in cephalaspids, the forward displacement of the visceral system may not be so great as it appears topographically. This is the position in abrupt and brief generalisation.

Nervous organisation in the proammocoete pharynx. The visceral units that have to be identified in a cephalaspid head are the branchial bars, and the best means of doing so are the dorsal cranial nerves. Plate VI, 1-3, shows the glossopharyngeal nerve IX, which runs very much in one transverse plane, as might be expected from Plate I. This cranial nerve innervates a bar that is not distorted by the buccal apparatus, nor pushed back by the expansion backwards of the pharynx relative to the overlying somatic structure. Fig. 1 shows a *Lampetra* embryo, oriented as the *Petromyzon* in Plate I, 1-5, The *IXth ganglion* lies behind (to the right of) the otic capsule. Its central root to the brain dorsally and its peripheral innervation of the visceral bar ventrally, can be seen in the silver-impregnated section. Its visceral pouch, the most anterior to have a gill-opening, lies in front of it. Fig. 2 shows a *Lampetra* proammocoete; above the *IXth ganglion* lie the longitudinal tracts of the 'white matter' of the brain, and above them the neuron cell-bodies of the grey matter.

Fig. 3 shows this level in transverse section in a combined microphotograph (parts of two sections from one individual). The *IXth ganglion* is seen on both sides, with its central root entering the brain and its peripheral root descending into the splanchnic region. There a main branch, less obvious but visible on both sides, runs inwards, some fibres reaching the pharyngeal epithelium.

The animal is still very young and simply organised. Fig. 4 is a medial sagittal section through a slightly older animal, Stage 14 of Damas (1944) as staged by the position of the nasohypophysial opening in the first place. Fig. 5 a-d shows the glossopharyngeal ganglion of this individual in successive parasagittal sections.

Returning to Fig. 3, it is noteworthy that this dorsal cranial nerve is already divided centrally into a dorsal part, containing sensory fibres with their cell-bodies in the ganglion, and a ventral part, containing viscero-motor fibres with cell-bodies in the brain that run out into and through the ganglion. One viscero-motor neuron can be identified in Fig. 2. So the brain is already dividing into its *longitudinal functional components*, as described, for example, by Herrick (1934, fig. 68).

One conclusion from these figures is that the IXth nerve is entirely post-trematic at this time, that it divides into lateral and medial branches after entering the splanchocranium, and that some fibres of the medial branch run to the pharyngeal epithelium. The other branchial nerves, the successive divisions of X, have a similar splanchnic arrangement.

The figures on Plate VI correspond to lateral aspect diagrams of the proammocoete by Koltzoff (1901), Séwertzoff (1916) and Damas (1944, plate-figs 11-14). In these the straightness of the ventral path of the IXth nerve is apparent but the peripheral division of branchial nerves into lateral and medial branches is not visible.

Nervous organisation in the ammocoete pharynx. The medial and lateral divisions of each branchial nerve are further developed in the ammocoete. Just after passing through the dorsal wall of the splanchnic skeleton, the division occurs, the medial part sending a branch to the pharyngeal wall, where it innervates five very large taste-buds that lie one above another on the inner wall of each pharyngeal bar. Both divisions include a viscero-motor component.

This is portrayed in the superb histological work of Alcock (1898). She followed the nerve-branches through the complex visceral muscle of the pharynx without the aid of metallic impregnation. Later work has shown her to be accurate. Johnston's account (1905) is based chiefly on one 6 cm ammocoete, fixed whole in Zenker's fluid. He also used a non-metallic stain, but he was meticulous and he had Golgi preparations, no doubt from his important study of the lamprey brain (1902), to which he could refer. His fixation was not so good (cf. Alcock's text-fig. and plate II fig. 4 with Johnston's figs 17 and 18) and he could not follow nerve-paths so far. Both authors describe only post-trematic branches of branchial nerves, although this is not obvious from their lateral aspect projections because the taste-buds on the inner wall, in most gill-bars, lie anterior to the corresponding gill-opening, the gill-bar running obliquely forward as well as inwards. However, in the case of the medial-visceral branch of X for example; "It is post-trematic as in the case of IX", (Johnston, p. 173). Later research extended this knowledge but it was not fully confirmed until the account of Lindström (1949). Lindström made silver-impregnated preparations of the ammocoete; his illustrations show that he could rely on his material. Incidentally to a study of pre-otic cranial nerves, Lindström checked much of Johnston's account of the branchial region and reviewed the work of the intervening years, including the studies of Tretjakoff (1910-1929), not discussed here.

Dorsal cranial nerves in relation to other cranial structures in proammocoetes. Nerves V^1, V^2, VII, IX and the successive 'branches' of X resemble each other in their central relations as they do, in general, peripherally. In spite of this metamerism, each has its distinctive position in the head. What is surprising is the extent of these distinctions. These can probably be applied more than has been done in the past to the identification of nerves in the cephalaspid head. Other vertebrate material could be used for the comparison, but the ammocoete, with a reduced visual system, is probably less like cephalaspids.

Plate VI can now be looked on as a whole. A section similar to Fig. 4 is drawn and annotated by Whiting (1957). Laterally to this position, other sections pass through the brain, and some of its structure is also described by Whiting (1957). Outside the brain come the cranial ganglia shown in Fig. 5 (cf. Séwertzoff, plate VI).

The *IXth ganglion* lies tucked against the otic capsule: its other special features have been seen already. The *Xth ganglion* is bigger and spreads laterally and caudally in close relation to the cardinal sinus; its central fibres sweep backwards as well as upwards as they enter the hindbrain, Fig. 5, b. The *Xth Lateralis* or *Posterior Lateralis* lies far dorsally and more laterally; it is very large; its central root enters the brain at a very dorsal level, Fig. 5, d and c.

The *facial ganglion, VII*, last in the pre-otic series, is tucked under the front edge of the otic capsule, Figs. 1, 2 and 5. Its central root is flattened by the otic capsule. Peripherally the facial nerve seems to have two partly motor divisions, one running anterior to the pseudobranchial groove, *N. hyoideus* of Alcock, apparently described, but not drawn, by Johnston, and several sensory branches, quite large and joined by anterior *lateralis* branches. These peripheral facial branches have no value in comparisons with cephalaspid material because there might be a breakdown of the tissues in this region in metamorphosis into adult cephalaspids as there is in petromyzont metamorphosis (cf. Lindström, 1949), due to changes in feeding mechanism. But the facial nerve is also diagnosed by a small nerve that runs caudally under the otic capsule and then to the *IXth ganglion*, just visible in Fig. 2. This runs far back over the pharynx rather in the position of a sympathetic chain, along the whole length of the pharynx in ammocoetes, and remains in the adult. It is figured by Séwertzoff, Alcock, Johnston, and other authors up to Lindström under different names. Its terminations in the pharynx are described by Alcock, and Gaskell (1908), on the basis of her work, ascribes importance to it in the origin of vertebrates. It seems not to be identical with any sympathetic chain in the trunk (Sigmund Freud and later workers, as summarised by Lindström, p. 444). This facial branch should be observable in cephalaspids if their pharynx resembled that of any stage of lamprey. There is also known to be a facial branch running along the 'pseudo-branchial' groove (cf. Plate II) and apparently reaching the endostyle.

The V^2 ganglion, *maxillo-mandibular*, V^{2-3}, of gnathostomes is very large and occupies most of the space between eye and ear, Figs 2 and 5, (cf. Damas, fig. 74 and Séwertzoff). The central root runs backward and inward to reach the hindbrain. Besides cutaneous branches there are two large branches with visceral-motor components (cf. Johnston, 1905), the velum and anterior prestomial visceral muscle being innervated by this nerve. Its organisation in ammocoete and adult has been carefully described by Lindström. There seems to be no homology between the branches here and in gnathostomes, for the maxillary nerve of the latter, like all their pre-trematic branches, carries no motor component.

The *Profundus* nerve, V^1, lies above the retina and is large and elongate, Fig. 5, c, d and e, and its central root, running back to the hindbrain above V^2, is distinct from any lateral-line fibres in this part, Fig. 5, b, c. Its peripheral fibres are seen beginning to run forward above the retina and, of course, quite dorsal to the optic tract which reaches the retina from its antero-ventral origin at the optic chiasma, Fig. 5, d, e, and more laterally. This may be compared with Damas (1944, text-fig. 74 and plate-figs 10-14), Séwertzoff (plate-fig. 16) and Koltzoff (1901). There is a ventral angle on the lower side of the *Profundus* ganglion, but no fibres from the ganglion run down at this point. A few somatic motor axons pass close to the ganglion, from the trochlear nerve, but they do not run along with the *Profundus* nerves and cannot be confused with them. This accords with descriptions of older lamprey stages, e.g. Lindström (1949, p. 392) and Cords (1929, fig. 1).

There remain the *Nervus Terminalis* and the anterior *Lateralis* nerves. The former is prominent in many primitive gnathostomes, e.g. *Chimaera*. It is distinctive in position and origin, but seems not to have been observed in petromyzonts of any age nor in cephalaspids. The anterior branches of the *Lateralis* system, while not part of the metameric dorsal cranial nervous system, can confuse the morphological picture in gnathostomes. In proammocoetes the anterior *Lateralis* system is distinct from the proximal part of the trigeminal complex. It remains separate in the ammocoete stage (Johnston, 1905, plate 5), and remains distinct from the V^2 ganglion in the adult, and does not join *Profundus* fibres even in the adult until well beyond the *Profundus* ganglion, as Marinelli and Strenger show (1954, fig. 53).

The dorsal cranial nerves do differ from each other in a diagnostic fashion in the proammocoete, both in their position relative to the optic tract and the otic capsule, and in their shape and central relations, and in their branches. The same distinctions can be seen in the ammocoete stage, though not all to the same extent.

Comparison of cranial nerves in cyclostomes and cephalaspids. When the information from Plate VI is applied to the study of the cephalaspid shown on Plate V, an identification of the nerves, in the way Lindström in particular proposed, seems logical.

On this identification the V^2 nerve runs below the orbit and the optic tract. All available evidence about the *Profundus* nerve shows that it and its ganglion lie entirely above the optic tract and the orbit, with no branch below at all. This is a diagnostic difference which could hardly change.

Secondly, the nerve now identified as the facial runs proximally towards where the facial ganglion is known to be. This identification is also much the more suitable in the case of Zych's figure (1937, plate II) of *Cephalaspis kozlowski*. The glossopharyngeal nerve is directed proximally towards the hinder part of the otic capsule instead of to a position in front of it. Further evidence could be given using other examples, e.g. from Zych (1937), in support of the view of Lindström and his predecessors in that view.

On the other hand, Stensiö and others have published a great deal of work on cephalaspids, and somewhat qualified the identifications given by him in 1927. But as the matter stands, the identification used by Stensiö is still accepted in text-book accounts such as those of Romer (1966), Parker and Haswell (1962), Young (1962) and Carter (1967). The onus is on those who accept the earlier identification to support it. Until the identification of cephalaspid cranial nerves is agreed, the identification of the successive visceral bars cannot be made with any reliability.

The identification of cephalaspid head-structure does not depend only on the cranial nerves, but evidence from other aspects of embryology and morphology, such as that of Smith (1963), Claydon (1938), and Damas (1954), also casts doubt on the morphology of cephalaspid head-structure as proposed by Stensiö.

Finally, it may be useful to state that evidence from the proammocoete does not support the existence of a pretrematic branch in agnathan dorsal nerves. It does support the importance, even in the early development, of the longitudinal functional components in the nervous system. It is now considered doubtful whether the trunks running from the lower side of the otic capsules were nerve-trunks (Romer, 1968), but if they were, then their ventral origin shows that they were motor nerves, presumably viscero-motor nerves, and not sensory nerves, going to the tessellated plate areas of the cephalaspid dorsal shield.

ACKNOWLEDGEMENTS

I have pleasure in thanking Dr Q. Bone, Dr M. A. Sleigh, Dr J. C. Brown and also the editors of this volume, for their help, and I very gratefully acknowledge the technical assistance of Mr N. Ablett, Mr K. Wood, Mr E. Seavill and Mr M. White.

REFERENCES

ALCOCK, R. 1898. The peripheral distribution of the cranial nerves of Ammocoetes. *J. Anat. Physiol*, **13**, 131-153.
BARDACH, D. and ZANGERL, R. 1968. First fossil lamprey: a record from the Pennsylvanian of Illinois. *Science*, **162**, 1265.
BOULENGER, G. A. 1896. On a collection of fishes from the Rio Paraguay. *Trans. zool. Soc. Lond.*, **14**, ii, 25-39.
CARTER, G. S. 1967. *Structure and habit in vertebrate evolution*. Sidgwick and Jackson, London.
CLAYDON, G. J. 1938. The premandibular region of *Petromyzon planeri*.—Part I. *Proc. zool. Soc. Lond. (Ser. B)*, **108**, 1-16.
CORDS, E. 1929. Die Kopfnerven der Petromyzonten. (Untersuchungen an *Petromyzon marinus*.) *Z. Anat. EntwGesch.*, **89**, 201-249.
DAMAS, H. 1944. Recherches sur le développement de *Lampetra fluviatilis L.*—contribution à l'étude de la céphalogenèse des vertébrés. *Archs Biol., Liège*, **55**, 1-284.
—— 1954. La branchie préspiraculaire des céphalaspidés. *Annals Soc. r. zool. Belg.*, **85**, 89-102.
DENISON, R. H. 1961. Feeding of agnatha and early gnathostomes. *Am. Zool.*, **1**, 177-181.
GASKELL, W. H. 1908. *The origin of vertebrates*. Longmans, Green, London.

HALSTEAD, L. B. *The pattern of vertebrate evolution*. Oliver and Boyd, Edinburgh.
HARDISTY, M. W. 1944. Life history and growth of *Lampetra planeri. J. Anim. Ecol.*, **13**, 110-122.
HEINTZ, N. 1968. The pteraspid *Lyktaspis* n.g. from the Devonian of Vestspitzbergen. In *Current problems of lower vertebrate phylogeny* (ed. T. Ørvig), 73-80. Alqvist & Viksell, Stockholm.
HERRICK, C. J. 1934. *An introduction to neurology*. 5th edition. Saunders, Philadelphia and London.
JOHNSTON, J. B. 1902. The brain of *Petromyzon. J. comp. Neurol.*, **12**, 1-86.
—— 1905. The cranial nerve components of *Petromyzon. Morph. Jb.*, **34**, 149-203.
JOLLIE, M. 1968. Some implications of the acceptance of a delamination principle. In *Current problems of lower vertebrate phylogeny* (ed. T. Ørvig), 89-107. Alqvist & Viksell, Stockholm.
KOLTZOFF, N. K. 1901. Entwicklungsgeschichte des Kopfes von *Petromyzon planeri. Bull. Soc. Nat. Moscou*, **15**, 259-589.
LINDSTRÖM, T. 1949. On the cranial nerves of the cyclostomes with special reference to *N. Trigeminus. Acta zool., Stockh.*, **30**, 315-458.
MARINELLI, W. and STRENGER, A. 1954. *Vergleichende anatomie und morphologie der wirbeltiere*, Lfg. l. Vienna.
OCCAM, W. 1324-7. *Summa logicae pars prima* (ed. P. Bochner, 1951). New York and Louvain, Franciscan Institute Publications.
ØRVIG, T. 1967. Phylogeny of tooth tissues: Evolution of some calcified tissues in early vertebrates. In *Structural and chemical organisation of teeth*, Vol. I (ed. A. E. W. Miles): 45-110. Academic Press, New York and London.
PARKER, T. J. and HASWELL, W. A. 1962. *A text-book of zoology*, Vol. II (7th edition, revised A. J. Marshall). Macmillan, London.
PARRINGTON, F. R. 1958. On the nature of the anaspida. In *Studies on fossil vertebrates; essays presented to D. M. S. Watson* (ed. T. S. Westoll): 108-128. Athlone Press, London.
RITCHIE, A. 1964. New light on the morphology of the Norwegian anaspida. *Skr. norsk. videnski Akad. i Oslo., Mat.-Nat. Kl.*, N.S. 14, 1-35.
ROMER, A. S. 1966. *Vertebrate paleontology*. 3rd Edition. Chicago University Press, Chicago.
—— 1968. *Notes and comments on vertebrate paleontology*. Chicago University Press, Chicago.
SANDERSON, I. T. 1939. *Caribbean treasure*. Viking Press, New York.
SÉWERTZOFF, A. N. 1916. Etudes sur l'évolution des vertébrés inférieurs. 1. Morphologie du squellette et de la musculature de la tête des Cyclostomes. *Archs russ. Anat. Histol. Embryol.*, **1**, fasc. 1, 1-104.
SHIPLEY, A. E. 1887. On some points in the development of *Petromyzon fluviatilis. Q. Jl microsc. Sci.*, **27**, 325-370.
SMITH, S. 1963. The origin of the vertebrate head. *Proc. XVIth. Int. Congr. Zool. Washington*, **155**.
STENSIÖ, E. A. 1927. The Downtonian and Devonian vertebrates of Spitzbergen I. Family Cephalaspidae. *Skr. Svalbard. Ishavet.*, **12**, 1-391.
—— 1964. Les cyclostomes fossiles ou ostracodermes. In *Traité de Zoologie*, **13** (i), 173-425.
STERBA, G. 1962. *Freshwater fishes of the world* (translated and revised by D. W. Tucker). Vista Press. London.
STUDNIĆKA, F. K. 1898. Untersuchungen über den bau der sehnerven der wirbeltieren. *Zeitschr. Naturw.*, **31**, 1-25.
TARLO, L. B. H. and WHITING, H. P. 1965. A new interpretation of the internal anatomy of the Heterostraci (Agnatha). *Nature*, **206**, 148-150.
TRETJAKOFF, D. 1910. Das Nervensystem von Ammocoetes. II. Gehirn. *Arch. mikrosk. Anat.*, **74**, 636-779.
WATSON, D. M. S. 1954. A consideration of ostracoderms. *Phil. Trans. R. Soc.*, B., **238**, 1-25.
WESTOLL, T. S. 1958. The lateral fin-fold theory and the pectoral fins of ostracoderms and early fishes. In *Studies on fossil vertebrates; essays presented to D. M. S. Watson*. (ed. T. S. Westoll): 180-211.
—— 1960. Recent advances in the palaeontology of fishes. *Liverpool & Manchester Geol. J.*, **2**, 568-596.
WHITING, H. P. 1948. Nervous structure of the spinal cord of the young larval brook-lamprey. *Q. Jl microsc. Sci.*, **89**, 359-383.
—— 1957. Mauthner neurones in young larval lampreys (*Lampetra* spp.). *Q. Jl microsc. Sci.*, **98**, 163-178.
WHITING, H. P. and TARLO, L. B. H. 1965. The brain of the Heterostraci (Agnatha). *Nature*, **207**, 829-831.
YOUNG, J. Z. 1962. *The life of vertebrates*. Clarendon Press, Oxford.
ZYCH, W. 1937. *Cephalaspis kozlowskii*, n.sp., from the Downtonian of Podole, Poland. *Arch. Tow. nauk. Lwów*, (Sect. III), **9**, 49-96.

C. C. D. SHUTE

The composition of vertebrae and the occipital region of the skull

INTRODUCTION

The analysis of vertebrae raises a number of problems relating to the formation of the centrum (C), the status of arch elements, their relation to the embryonic sclerotomes, and the nature of the occipito-spinal articulation which are still not satisfactorily solved. Are the Cs, for instance, strictly homologous structures with a similar embryological origin in all vertebrates? Is their development autocentral, i.e. independent of the vertebral arches? To what extent are Cs and arches intersegmental (Remak, 1855), in the sense that the definitive vertebra is formed by the recombination of caudal and cranial halves (sclerotomites) of successive sclerotomes? Arch elements or arcualia have been known since Gadow (1895, 1896) as the basidorsal (BD) and the interdorsal (ID), matched ventrally by a basiventral (BV) and an inconstant interventral (IV). The BD may be surmounted by a supradorsal (SD). Yet another element the 'dorso-interdorsal' (DID) has been described in amniotes (Piiper, 1928; Dawes, 1930). What contribution do these elements make to the definitive vertebra?

Many attempts have been made to answer these questions with widely discrepant results, so—as Williams (1959) has pointed out—the present situation is very confused. This study is based on a re-examination of adult, larval and embryonic stages of representative vertebrate types. I am happy to acknowledge my indebtedness to Dr F. R. Parrington for making available to me some of the material. Space does not permit me to detail all the ways in which my observations and conclusions differ from those of other workers, or, unfortunately, to refer to fossil forms.

NOTOCHORDAL SHEATHS AND THE FORMATION OF CENTRA

In adult and larval lampreys, the notochord is surrounded by three distinct sheaths: an inner epithelial sheath formed by peripheral notochordal cells, a middle fibrous sheath derived from these cells which is sufficiently thick to maintain the turgor of the notochord, and an outer elastica sheath (the elastica externa) which is thin, contains elastic tissue, and is mesodermal in origin. In ammocoetes a thin covering of perichordal mesenchyme surrounds the elastic sheath and intervenes between it and the developing neural arch cartilages, but after metamorphosis the cartilages become closely applied to the sheath. A similar triple arrangement of the notochordal sheaths is seen in bony fish. In *Amia* the Cs ossify in the perichordal mesenchymatous layer around the bases of arcual elements. The caudal region exhibits diplospondyly: between each true arch-bearing C an extra C is formed from a ring of perichordal skeletogenic mesenchyme. Diplospondyly also occurs in the caudal region of elasmobranchs, with complete reduplication of arcualia as well as of Cs.

Understanding of how the C is formed in elasmobranchs has, it seems, been hampered by a false identification of the various notochordal sheaths. In orthodox accounts the thick fibrous ring which chondrifies to form the C is regarded as an expanded equivalent of the fibrous notochordal sheath of other vertebrates. The prominent elastic sheath which lies on the inner surface of the C-forming ring has, therefore, to be called an 'elastica interna'—a structure without parallel elsewhere, although some attempts have been made to recognise its equivalent in bony fish. The elasmobranch 'elastica externa' has to be found in the tenuous external limiting membrane of the C-forming ring, although it bears no resemblance to a typical elastic notochordal sheath, and is properly regarded as a perichondrium. The C-forming ring is the equivalent of the perichordal skeletogenic layer of other vertebrates, which is particularly thick in elasmobranchs, as it is also in amniotes. The cartilage cells of the developing C differentiate *in situ* from the mesenchyme of this layer. The so-called 'elastica interna' is the true elastic sheath of the notochord, and structurally resembles that of other vertebrates. It remains in apposition with the inner epithelial sheath, and no fibrous sheath is ever developed.

The formation of the Cs in elasmobranchs is, therefore, not fundamentally different from that of other vertebrates. In all cases the skeletogenic material surrounds the notochord and proceeds to constrict it to a greater or lesser extent. Because of its situation it is convenient to speak of the skeletogenic region as the perichorda. The perichorda surrounds the bases of the arcualia, and these become incorporated into the C in many fish, e.g. *Lamna*, *Amia*, *Esox*. In teleosts and amphibia the C is thin, and ossifies from the outer part of the perichorda. In the intervertebral region the inner perichordal membrane may form a ring of secondary cartilage constricting the notochord, as in the holostean *Lepidosteus*, and in urodeles and anurans. These rings subsequently break down to form the intervertebral joint, which is opisthocoelous or procoelous according to which articulating surface is hollowed out and which is convex. The intervertebral rings are not derived from arcualia.

In considering whether the C should be regarded as intersegmental it is necessary to distinguish between the primitive embryonic segment, bounded by the somatic arteries, and the definitive vertebral segment bounded by the myosepta. The somatic arteries are paired branches of the dorsal aorta which run laterally to the junctions between successive myotomes and the overlying dermatomes or cutis plates. The primitive myosepta consist of densely aggregated mesenchymal cells deep to the junctions between the embryonic myotomes. Later these cells form definitive fibrous septa. Their relationship to the C is determined by their attachment to the dorsal and ventral arcualia, as will be described later.

In amniotes the C develops on either side of the somatic artery from adjacent half-segments or sclerotomites. At a certain stage of development the mesenchyme of the caudal sclerotomite forming the cranial half of the C appears relatively dense as compared with that of the cranial sclerotomite forming its caudal half. This dense appearance is caused by cells proliferating in the bases of the neural arches, which contrast with the already chondrified caudal half of the definitive C. The precocious development of the caudal half of the C which is a feature of reptiles and mammals may be due to the regression of the arcualia (IDs and IVs) originally associated with the cranial sclerotomite. The two components of the amniote C normally fuse rapidly but in whales they begin to ossify separately, so that in embryos the C may appear bipartite on X-ray.

VERTEBRAL ARCHES, RIBS AND THE OCCIPITO-SPINAL REGION IN FISHES

Cyclostomes, elasmobranchs, holocephalians

In metamorphosed lampreys the spinal notochord is surmounted throughout its length by paired cartilages, typically two for each segment, one on each side of the ventral nerve root. Of these cartilages, the one in front of the ventral root has been regarded as a BD and the one between the ventral and dorsal roots as an ID. Tretjakoff (1926), however, pointed out that at the anterior end of the series the two cartilages unite to form a single cartilage pierced by the ventral root, and another smaller intervening cartilage is present behind the fused larger ones. Serial sections through the postcranial region in *Lampetra* show that some of these smaller cartilages are pierced by dorsal roots. Comparison with *Squalus* suggests that only the small cartilages supporting the dorsal roots in the postcranial region are IDs, and that the cartilages on either side of each ventral root form two parts of the BD. The more posterior cartilage, i.e. the one behind the ventral root, will be called the postbasidorsal (PoBD). In the postcranial region the PoBDs are reduced and fused to the bases of the more anterior cartilages, which form the main part of the neural arch. The more anterior or prebasidorsal (PrBD) cartilages lie immediately in front of the dorsal branch of the somatic artery, and are presumably derived from the caudal sclerotomite of the primitive embryonic segment. The myosepta are attached to the anterior borders of the PrBDs. These cartilages, therefore, occupy the front part of the definitive segment, as defined by the myosepta.

There are no ventral arcualia in cyclostomes. Anteriorly, their place is taken by the dorsal ends of branchial arch cartilages which unite to form the subchordal rod. Cyclostomes also lack an occipital region of the skull; that is to say, the most anterior dorsal arch cartilages corresponding to the occipital arches of jawed vertebrates are not fused to the backs of the parachordal cartilages or the otic capsules. The first BD is pierced by a dorsal root carrying a rudimentary sensory ganglion and two ventral roots, of which the more anterior innervates the first two myotomes and the more posterior innervates the third. The dorsal root associated with the ventral root supplying the third myotome is supported by the first ID, and passes out in the septum behind the third myotome. The second BD is pierced by the ventral root supplying the fourth myotome. It would appear that the ventral root belonging to the first metotic segment has been lost, and that the BD associated with that of the second metotic segment has fused with that of the third.

Typical elasmobranchs (Squaloidea, Notidanoidea, Batoidea) show a basic arrangement of dorsal arcualia similar to that found in the postcranial region of cyclostomes; i.e. the BD is pierced by the ventral spinal nerve root and the ID by the dorsal root. The BDs develop as squat elements seated on the perichorda on the upper surface of the future C, whereas the IDs are taller and make contact through their narrow lower ends with the perichordal mesenchyme in the intercentral region. The vault of the neural arch, therefore, in these groups is derived from the IDs and occupies an intervertebral position: this may have the effect of increasing the flexibility of the vertebral column. In Scylloidea the dorsal root passes caudal to the ID and an SD fuses with the BD.

Ventral arcualia are present in elasmobranchs. BVs develop in the same vertical plane as the BDs, and give articulation to the ribs. Although the ribs pass into the intermuscular septum between the epaxial and hypaxial musculature, and so occupy the position of dorsal ribs of higher forms, they are regarded as homologues of the pleural ribs of bony fish (Emelianov, 1935). IVs develop intervertebrally in the same vertical plane as the IDs, and in *Squalus* often help to give articulation to the ribs belonging to the BVs immediately behind.

In the trunk region, the somatic artery arises from the dorsal aorta immediately behind the BV. Its dorsal branch passes upwards in front of the roots of the spinal nerve. The relationship of the BVs to this artery indicates that they develop from the caudal sclerotomite of the primitive embryonic segment. In the caudal region of *Squalus* the vertebral elements are reduplicated, so that for every BD and ID pierced respectively by ventral and dorsal nerve roots, there is another intervening pair of non-neural arch elements which have no spinal nerves associated with them. The somatic artery arises behind the primary (neural) BV; behind each secondary (non-neural) BV lies a tributary of the caudal vein. The resulting diplospondyly may serve to increase the flexibility of the tail. At the junctional region behind the mesonephros and in front of the first non-neural vertebra there is a compound vertebra due to a BV having moved forwards under the BD of the segment in front (Fig. 1A). In consequence the somatic arteries of the tail run obliquely upwards and backwards—an arrangement which is haemodynamically advantageous, since the direction of blood flow is no longer opposed to that in the dorsal aorta.

The relationship of the vertebral arches to the definitive segment as defined by the myosepta can be seen in *Scyllium* and *Squalus* larvae. The myosepta pass in front of the BDs and behind the corresponding BVs. They are also attached to the backs of the pleural ribs. In the tail the main dorsal attachment of the myoseptum is to the non-neural BD (Fig. 1A).

In the occipito-spinal region, the arcualia are modified to form the cranio-vertebral articulation. In *Squalus*, the occipital condyles are formed from the BDs of an occipital arch fused to the backs of the parachordal cartilages. The first free neural arch is formed from IDs, each pierced by a dorsal root as in more caudal segments. The corresponding ventral root passes behind (not through) the condylar BD and supplies the fifth metotic = first spinal myotome. Three hypoglossal roots (XIIb, c and d) are enclosed in foramina in the occipital cartilage, which is formed from the arcualia of preoccipital segments. The hypoglossal root (XIIa) supplying the first metotic myotome degenerates (de Beer, 1922). In *Scyllium*, only XIIb and XIIc are enclosed in the occipital region of the skull. XIId, supplying the fourth metotic myotome, passes out directly behind the skull, and the first free neural arch is formed by the BD of the occipital arch, which is not attached to the back of the skull as in *Squalus*. In some elasmobranchs, e.g. *Hexanchus*, and in holocephalians (Furbringer, 1897), the 'hypoglossal nerve' has five roots, presumably as a result of the incorporation of an anterior cervical ('occipito-spinal') arch into the occipital region of the skull.

In the holocephalian fish *Chimaera*, the notochord is strengthened by multiple annular calcifications but no true Cs are formed. Comparison with *Squalus* supports the view (Goodrich, 1909) that the low triangular element containing the ventral nerve root opposite the BV is a BD, while the ID forms the main neural arch. The dorsal nerve root emerges in front of the ID, as in higher forms. There are no IVs and no ribs.

Holosteans, teleosts

In the holostean *Amia* each bony C in the trunk region ossifies around a pair of squat dorsal cartilaginous elements immediately opposite the paired ventral elements which support the pleural ribs, while the cartilages forming the main neural arch are situated intervertebrally. The ventral rib-bearing cartilages lie immediately in front of the somatic arteries and the ventral ends of the myosepta, and are, therefore, BVs. The intervertebral dorsal element also lies immediately in front of the somatic artery and the myoseptum is attached to its anterior border. Consequently, this element is not an ID like the tall dorsal arch cartilage of *Squalus*, but a BD, and the dorsal cartilages immediately above the BVs helping with them to form the body of the C, are IDs (Fig. 1B). The primitive segments as defined by the arteries have become secondarily oblique. The myosepta are attached to the posterior borders of the IDs and BVs of each C. Each BD is continued upwards and backwards in a long spine, between the bases of which are small paired SDs below the longitudinal ligament.

No IVs are formed in the trunk region of *Amia*. A downgrowth from the BV, in the trunk region only, forms a haemal process (Goodrich, 1930) ventrally on either side of the dorsal aorta. Another process, the basapophysis (Remane, 1936), grows out laterally from the BV in the trunk region to articulate with the pleural rib. The basapophyses form haemal arches in the tail.

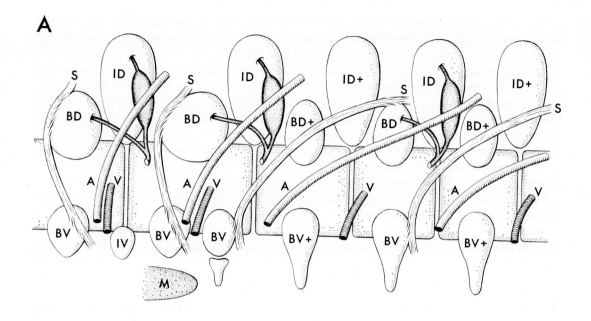

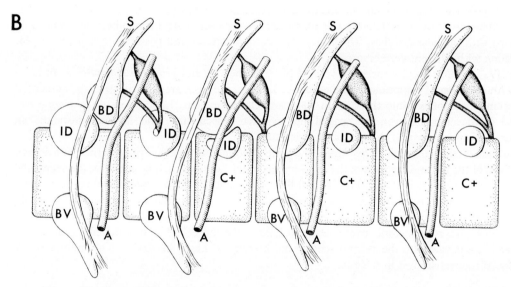

Fig. 1 Vertebral column, left-sided diagrammatic view of junctional region between trunk and tail in **A**, *Squalus* and **B**, *Amia*.

Abbreviations: A, somatic artery; BD, primary (neural) basidorsal; BD+, secondary (non-neural) basidorsal; BV, primary basiventral; BV+, secondary basiventral; C+, secondary centrum (pleurocentrum); ID, primary (neural) interdorsal; ID+, secondary (non-neural) interdorsal; IV, primary interventral; M, caudal end of mesonephros; S, myoseptum; V, somatic vein.

The posterior caudal region of *Amia* is diplospondylous, and the segments lose their obliquity As in elasmobranchs, the reduplication is effected by the anterior sclerotomites, but only extra Cs are produced. The primary Cs are formed in relation to the BVs as in the trunk, but the BDs move forwards on to their dorsal aspects. The secondary Cs form in relation to the IDs. Such Cs are termed pleurocentra. The myosepta are attached to the front of the BDs and the back of the BVs (Fig. 1B).

Amia differs from *Squalus* in the occipito-spinal region in that the arcual elements fused to the back of the parachordal cartilages are not BDs but IDs belonging to the hindmost preoccipital arch, while the first free neural arch is formed from BDs of the occipital arch. The ventral root (XIId) supplying the fourth metotic myotome lies in front of the first free neural arch, as in *Scyllium*. Two hypoglossal roots emerge in front of the occipital cartilage, the smaller (XIIb) supplying the first and second metotic myotomes, and the larger (XIIc) the third myotome.

Vertebral development in *Lepidosteus* (Balfour and Parker, 1882) is recognisably holostean in type. The neural arch of the definitive vertebra ossifies on each side around a low ID anteriorly, lying vertically above the BV as in *Amia*, and a tall BD posteriorly. The BVs form basapophyses. The joint between adjacent Cs develops in an intervertebral ring of cartilage (derived from perichordal mesenchyme and not of arcual origin) which constricts the notochord.

Although at first sight the vertebrae of teleosts, with their dorsal and ventral arches apparently symmetrically attached to the Cs, seem to be very different from those of holostean fish, a careful study of trout larvae shows that the construction of the vertebra is easily derived from the holostean model. The ID forms the base only of the neural arch, and the BD is fused to the back of the ID to give an arch which is firmly attached, through the ID, to the C. This arrangement probably makes for a more powerful vertebral column, as compared with the more flexible column in which the neural arches remain free. In some teleosts the spinal nerves are caught up in membranous ossification associated with the neural arches, which may also develop anterior and posterior 'zygapophyses' for articulation with adjacent vertebrae.

The BVs in *Salmo* occupy the same vertical plane as the IDs and grow basapophyses as in holosteans. An ill-defined IV appears behind each BV. In *Esox*, the basal arcualia constrict the notochord and become incorporated into the C as in *Amia* and elasmobranchs. In addition to the pleural rib, a dorsal rib develops between the epaxial and hypaxial musculature, and in some instances an epipleural rib appears within the epaxial muscle bundles.

Horizontal sections through *Salmo* larvae show the obliquity of the myosepta and their attachment to the front of the BD and the back of the ID elements of the neural arch. The course of the segmental artery is also oblique, as in *Amia*. The ID and BD are separate in the occipito-spinal region, where, also as in *Amia*, the IDs of the hindmost preoccipital arch are incorporated in the back of the skull, and the BDs of the occipital arch form the first free neural arch. The first spinal nerve supplies, through its ventral root, metotic myotomes as far back as the fourth. This root, therefore, represents XIId and is equivalent to the ventral root of the first spinal nerve in *Amia*. The more anterior hypoglossal rootlets are lost in teleosts.

Chondrosteans: Acipenser, Polypterus

Acipenser does not form Cs. The BVs are large and in the trunk region develop basapophyses and haemal processes. The neural arch is composed of three pairs of cartilages. The largest of these lie in front of the ventral nerve root. They extend upwards on either side of the longitudinal ligament, where they are surmounted by a separate neural spine which may represent fused SDs. Behind the large cartilages of the neural arch are two smaller cartilages which have traditionally been interpreted as secondarily subdivided IDs, the large cartilages being BDs. It is probably better to regard the more anterior of the smaller neural arch cartilages as PoBDs. They lie immediately behind the ventral roots, and have a broad base in contact with the notochord which, together with that of the large PrBDs of the neural arch, is coextensive with the base of the BVs. The more posterior of the smaller neural arch cartilages, which may or may not make contact with the notochord, are then IDs. The subdivision of the BDs can be regarded as a primitive character in sturgeons, resembling the condition found in lampreys. In *Polyodon* (Schauinsland, 1906) there are no separate IDs and the PrBDs grow backwards to enclose the dorsal nerve roots.

The composition of the occipito-spinal region is complicated by the fact that in the adult a number of anterior spinal vertebrae and their related ribs become secondarily attached to the back of the skull, so that, as in *Amia*, its true posterior limit becomes obscured. In larval stages of *Acipenser* the occipital cartilage is pierced by three hypoglossal roots (XIIb, c and d) supplying the first four metotic myotomes; the first metotic myotome appears only transitorily (Sewertzoff, 1899, 1928). The BDs of the occipital arch form the back of the larval skull as in *Squalus*, and the dorsal and ventral roots of the nerve which, through its ventral root, supplies the fifth metotic myotome pass out behind them.

In *Polypterus* larvae the arcualia consist of basal elements only (Budgett, 1902); no IDs or IVs are present. The dorsal myoseptum is attached to the anterior border of the BD, as in other forms. Each spinal nerve root emerges behind the BD of the same definitive segment, but in the trunk region the roots become secondarily included in membranous ossification which extends forward from the BD of the segment behind. In this way the spinal nerves come to be associated with the neural arch of the vertebra behind their proper segment.

In the trunk region of *Polypterus* the BVs are subdivided on each side into two separate cartilages, one ventral and one lateral. The ventral portions of the BVs are basapophyses, which in the tail fuse with pleural ribs to form haemal arches. In the trunk the basapophyses scarcely chondrify, but membranous 'aortic supports' form in their place on either side of the aorta. The free pleural ribs are pushed laterally by the developing mesonephros on each side, and so come to be loosely slung by ligaments from the under surface of the lateral portions of the BVs. In the adult vertebra these ossify to form the bony transverse processes (synapophyses; Remane, 1936) which support dorsal ribs. The myosepta are attached to their anterior borders.

Owing to lack of information on early larval stages of development, it is impossible to be sure what modifications have befallen the segmental pattern of arcualia and nerve roots in the occipito-spinal region in *Polypterus*. At the 70 mm stage there is one hypoglossal root, possibly XIIc, which passes in front of the occipital cartilage and supplies the first two metotic myotomes which are then present. Later another ventral root (? XIId) becomes caught up in the occipital ossification (Lehn, 1918).

Dipnoans

In *Ceratodus* (Klaatsch, 1895; K. Furbringer, 1904; Goodrich, 1909) no bony centra are laid down, but the perichorda thickens to produce a tough semicartilaginous sheath, and the cartilaginous bases of the arches expand on its surface. These bases remain cartilaginous, but the narrow distal parts of the arches ossify.

In larval stages of *Protopterus* the myosepta attach to the backs of the ventral arcualia and to the anterior margins of the dorsal arcualia—the typical arrangement for BVs and BDs. The cartilaginous bases of the BDs lie slightly behind the BVs. The main part of the arch extends in an upwards and backwards direction to enclose the canal for the spinal cord and extends as a spine on either side of the longitudinal ligament, like the BDs of *Amia*. BVs articulate with pleural ribs in the trunk region and form haemal arches in the tail. IVs appear to be absent. The course of the segmental arteries is as in other bony fish: they arise behind the BVs and ascend obliquely backwards to pass in front of the nerve roots. One may conclude that the apparent alternation of ventral and dorsal arcualia in lungfish is due to the obliquity of the segments, the BDs being carried backwards as in *Amia*. Dipnoans differ, however, from holostean fish in that they do not form IDs.

The occipito-spinal region of *Protopterus* is characterised by a reduction of the BDs immediately behind the occipital cartilage and of the nerve roots which lie in front of them. The ventral root innervates the fourth metotic myotome, and is, therefore, XIId. Consequently, the vestigial BDs belong to the occipital arch and correspond to the first free dorsal arch of *Amia*. In later stages they come to be included in the ossification of the exoccipital bone. In very early larvae three hypoglossal roots supplying the first three metotic myotomes are present in front of the occipital cartilage (Agar, 1906; Fox, 1962), but these are soon reduced to a single root, XIIc. The occipital cartilage, then, is formed from one or more pre-occipital arches. A large cranial rib articulates with its base.

VERTEBRAL ARCHES, RIBS AND THE OCCIPITO-SPINAL REGION IN AMPHIBIA

Modern Amphibia

Larval stages of the apodan *Geotrypetes* show the myoseptum attached to the anterior borders of the neural arch cartilage and rib, and in *Hypogeophis* (Marcus and Blume, 1926) cellular myosepta, like those found in early stages of *Scyllium*, pass in front of the neural arch primordia. One may conclude that apodan neural arches are derived from BDs arising primarily from the posterior sclerotomite which come, as a result of the mode of attachment of the dorsal myosepta, to occupy the anterior part of the definitive segment. The spinal nerve roots in each definitive segment originally emerge behind the BDs, but in the anterior trunk region they become enclosed in membranous ossification associated with the back of the neural arch. The zygapophyseal articulation between adjacent arches is achieved through contributions to the BDs from the anterior sclerotomite, forming a DID. This process, so essential for the stability of the vertebral column in tetrapods, may have been facilitated by the regression of the IDs.

The adult apodan vertebra is characterised by prominent paired antero-inferior processes, each of which carries on its upper surface near the base a facet or parapophysis for articulation with the ventral head of the rib. The dorsal head of the rib (probably a secondary upgrowth) articulates with another process situated just below and behind the prezygapophysis on the side of the neural arch. The antero-inferior process of the vertebra arises embryologically from a separate cartilaginous element called the 'parachordal' by Marcus and Blume (1926). Since this terminology leads to confusion with the parachordal cartilages of the basis cranii, an alternative name 'paracentral' (PC) is suggested. The PCs are not BVs (lacking in apodans); they occupy a more dorsal position and, unlike typical basal arcualia, develop outside the skeletogenic perichordal tissue at the inner ends of the horizontal intermuscular septa separating epaxial and hypaxial musculature. PCs possibly represent separate medial elements of the dorsal ribs which fuse secondarily with the neural arches to form transverse processes. The myosepta are attached to their anterior borders.

The most anterior member of the apodan vertebral series is specialised to provide a type of cranio-vertebral articulation peculiar to modern Amphibia. The nature of the modification which has taken place can be seen in larval specimens of *Geotrypetes*. There are no free ribs related to the first vertebra, but the PCs are applied to the sides of the BDs and are prolonged forwards so as to pass inside the divergent posterior ends of the parachordal cartilages of the basis cranii. These anterior prolongations (responsible for the apodan 'spinal tube', Gadow, 1933) form condylar processes which articulate with the parachordals. There is no BD element included in the occipital condyle which is, therefore, formed solely from the parachordal. The hypoglossal nerve roots are all lost, and the first five metotic myotomes are supplied by the first spinal nerve.

The vertebral column of urodeles differs from that of apodans in possessing BVs in the caudal region which form haemal arches lying in the same vertical plane as the neural arches. The attachment of the myosepta, seen most clearly in the trunk region, is to the anterior borders of the neural arches, which is the typical relationship to BDs. The spinal nerve roots of each definitive segment pass out behind the BDs, and no IDs are formed. Ventral arch elements are usually lacking in the trunk, but *Necturus* possesses BVs in this region (Gamble, 1922), each partly or wholly separated into two components which can be compared to the synapophysis and basapophysis of *Polypterus*. The basapophyses form the complete haemal arch in the tail region.

The transverse processes of the vertebrae of adult urodeles are formed from separate elements, the so-called 'rib-bearers', which eventually fuse to the neural arch. The rib-bearer can be recognised as the urodele equivalent of the apodan PC, and like it provides a process for articulation with the ventral head of the rib. In some instances it later grows upwards between the neural arch and the dorsal head of the rib (itself a secondary extension from the rib up into an intermuscular septum) and in this way acquires another process for articulation with the dorsal head. The urodele PC does not project freely forwards as in apodans, but instead grows down in front of the segmental artery to make contact with the side of the bony C or, in *Necturus*, with the synapophyseal element of the BV. A longitudinal anastomosis (the 'vertebral artery') joins successive segmental arteries and becomes enclosed between the PC and the C.

The region of the cranio-vertebral joint, as seen in urodele larvae, is constructed in a fashion which in a number of respects resembles that of apodans. The occipital condyles are formed by the caudal ends of the parachordal cartilages, which develop facets on their inner aspects. The bony C of the first cervical vertebra is prolonged forwards as a pseudo-odontoid process, and modified PCs fuse with its sides to form condylar processes. Caudally, the cartilage of each condylar process grows up under the base of the BD and encloses a small foramen.

The segmentation of the occipito-spinal region in urodeles is complicated by the fact that the first metotic somite fails to form muscle, and the second and third metotic myotomes fuse (Platt, 1897; Goodrich, 1911). In typical urodeles the first free spinal nerve, which passes out in front of the first spinal arch (that of the so-called 'atlas' vertebra), consists of a ventral root only which innervates the fused second and third metotic myotomes and the fourth metotic myotome. This root can be identified, therefore, as XIId, and the neural arch of the first spinal vertebra is formed by the BDs of the occipital arch. In *Cryptobranchus* the occipital arch fuses with the occipital cartilage, so that the hypoglossal root XIId is enclosed in a hypoglossal foramen. The condition then resembles that described for *Squalus*.

In anurans the dorsal myoseptum is attached to the anterior border of the neural arch, as in urodeles. A large transverse process grows out from the side of the neural arch, passes deep to a 'vertebral artery', which is much more dorsally placed than that of urodeles, and articulates with a short single-headed rib. The relationships suggest that the single head of the reduced rib of anurans may correspond to the dorsal head of the urodele rib, and that the transverse process of the anuran vertebra is an enlarged articular process, corresponding to the process high on the side of the neural arch which supports the dorsal head of the rib in urodeles if the PC does not intervene.

The cranio-vertebral articulation of anurans, as seen in larval stages of *Rana* and *Bufo*, is in general similar to that of urodeles. Condylar processes are formed from cartilages which are largely distinct from the occipital arch BDs of the first cervical vertebra, although they fuse with them posteriorly.

These condylar processes can be regarded as rudimentary PCs, lost in more caudal regions of the vertebral column. The anterior surfaces are hollowed out to receive the caudal ends of the parachordal cartilages, which form the occipital condyles. The articular surfaces of the condyles, therefore, face posteriorly, rather than medially as in apodans and urodeles, and the anuran first cervical vertebra differs from that of urodeles in not possessing an odontoid-like process. The hypoglossal nerve XIId is lost in *Rana*, as in apodans, but it may be retained in *Bufo*.

The evidence of the vertebral column, and especially of the occipito-spinal region, definitely favours the view advocated on dental grounds by Parsons and Williams (1962) that the three living amphibian Orders have a common origin. The Anura have departed furthest from the ancestral condition, but there would seem to be no overriding objection to deriving all three groups from a primitive lepospondylous amphibian stock of nectridian type.

VERTEBRAL ARCHES, RIBS AND THE ATLANTO-AXIAL COMPLEX IN AMNIOTES

Lepidosauria: rhynchocephalians, lizards. Autotomy

The vertebral column of *Sphenodon* possesses discrete hypocentra (HCs) formed from cartilaginous BVs which ossify as chevrons in the tail and are united by a mid-ventral membranous ossification in the trunk. In some lizards, e.g. geckoes, the trunk HCs form median crescents; in others, e.g. *Lacerta*, they are absent from the trunk and form V-shaped chevrons in the tail. The chevrons and HCs may maintain an intercentral position, but in some lizards, e.g. *Varanus*, the chevrons gain attachment to the back of the C in front. In the teiid lizard *Tupinambis*, on the other hand, the chevrons bend backwards to connect with the C behind (Boulanger, 1891).

In *Sphenodon* (Howes and Swinnerton, 1901) and certain lizards the more posterior caudal vertebrae are subdivided by a septum or fissure across which the vertebra fractures during autotomy or spontaneous shedding of the tail. In lizards where the caudal vertebrae possess two spines, the anterior spine lies near the vertical plane of the autotomy fissure (Gadow, 1896; Pratt, 1946).

In embryonic stages of *Lacerta* the autotomy fissure can be seen to arise bilaterally in two parts, which later become continuous. The fissure for the neural arch forms in early embryonic life as a foramen in the BD. This foramen is invaded by blood vessels derived from the segmental artery and becomes extended into a slit. The fissure for the C appears in late embryonic life on its ventro-lateral aspect, just anterior to the transverse process. The notochord in this region is unconstricted, and the notochordal cells undergo a cartilaginous change to form the so-called 'chordal cartilage'. Foramina are formed as a result of proliferating mesenchymatous cells and blood-vessels derived from the segmental artery actively invading the shell of perichordal bone which ossifies, in membrane, around the chordal cartilage. These foramina are extended into slits, which unite with each other and with the slits for the neural arch to form the definitive fissure, and the chordal cartilage degenerates.

It is clear from the above account that the autotomy fissure is a new formation, achieved as a result of two distinct processsses occurring at different stages of embryonic development. Nevertheless, its relationship to the segmental artery indicates that, particularly at its base, the fissure represents approximately the junction between two primitive segments. The position of the fissure with respect to the definitive septum is determined by the attachments of the myoseptum. The dorsal part of the myoseptum is attached in front of the fissure for the neural arch, and its ventral part behind the fissure for the C. It follows that the myoseptum will tear during autotomy.

The ability to autotomise appears to depend upon an unconstricted notochord. In *Sphenodon* and geckoes, which retain the primitive amphicoelous condition of the vertebral column, the notochord is not constricted. In *Lacerta* the notochord is constricted at the posterior end of the C by perichordal cartilage forming the knob which fits into the concavity at the anterior end of a procoelous vertebra. The walls of the concavity are also formed by perichordal cartilage. Both in *Sphenodon* and *Lacerta* the middle part of the C, where the notochord is unconstricted, is supported by chordal cartilage. In non-autotomous lizards and in other reptiles and mammals, the notochord becomes constricted throughout its length by perichordal chondrification.

Lepidosaurs, like other amniotes, have modified the anterior cervical vertebrae to produce an atlas-axis complex, which allows rotatory movements of the head. In *Sphenodon* an extra arch is present on each side between the neural arch of the atlas and the occipital region of the skull. This pro-atlas arch is present in many amniotes and, according to Barge (1918) is developed from the anterior sclerotomite of the embryonic segment which gives rise, from its posterior part, to the atlas arch. The pro-atlas can be regarded, then, as a separate DID element in series with the zygapophyseal portion of more caudal neural arches. The BD associated with the pro-atlas (i.e. that formed from the posterior sclerotomite of the preceding segment) is included in the occipital region of the skull. It grows up behind the last hypoglossal root, and is extended posteriorly to form the exoccipital portion of the occipital condyle.

Four hypoglossal foramina are present on each side in early embryos of *Sphenodon* (Howes and Swinnerton, 1901), which suggests that the last hypoglossal nerve root is XIId. The number of hypoglossal roots is subsequently reduced to two. In lizards such as *Lacerta*, the occipital cartilage is pierced by three hypoglossal roots, corresponding perhaps to XIIb, c and d.

Chelonians, crocodiles, birds

The cervical vertebrae of turtles possess well-marked ventral arches. In early embryonic stages, e.g. *Chrysemys* 3 mm, paired BVs form HCs below the neural arches. These fuse in the mid-line and become applied to the ventral part of the perichordal mesenchyme which gives rise to the back of the body of the vertebra in front. The bodies of the second cervical vertebra and of the odontoid are detached from one another and each is opisthocoelous. In each case the HCs contribute to the posterior concavity. The more caudal cervical vertebrae are procoelous, and the HCs fuse with the base of the perichordal cartilage which forms the posterior convexity. The anterior-most HCs fuse with the BDs of their own segment to form the ventral arch of the atlas. BVs are not formed in the trunk region and the single head of the rib, representing a tuberculum, articulates with the base of the neural arch. In later stages the ribs move forwards to an intervertebral position.

Three hypoglossal roots (? XIIb, c and d) are originally present in *Chrysemys* passing through separate foramina, but later they are reduced to two. If the last hypoglossal root is XIId, the exoccipital portion of the occipital condyle is formed by the BDs of the occipital arch. There is no pro-atlas. The atlas is formed from the BVs and BDs of the definitive segment of the second cervical nerve. The odontoid is formed from the C of this segment, with which are fused the BVs of the segment of the third cervical nerve. The tip of the odontoid is formed from the rudimentary C belonging to the segment of the first nerve.

The organisation of the vertebral column in Crocodilia resembles that of turtles, except that there is a pro-atlas, and ribs are present in the cervical region. There are also cranial ribs included in the basioccipital bone in the preoccipital region which fuse with the back of the otic capsule. The HCs are recognisable in early embryonic stages but later disappear or become incorporated in the backs of the preceding Cs, so that the capitula of the ribs come to articulate with the front of their own Cs. In the thoracic region the capitulum retreats from the C onto the front of the diaphysis. HCs are retained in the caudal region, where they form chevron bones which lie between adjacent Cs, but articulate mainly with the C in front.

In 10 mm crocodile embryos the cellular primordia of the myosepta run obliquely forwards towards the anterior border of the neural arches, thereby defining the definitive segments. The BD is beginning to chondrify, but the postzygapophyseal (DID) portion of the neural arch, lying behind the segmental arteries, is still mesenchymatous. The delayed development of the zygapophyseal articulation, which is a feature of tetrapods, reflects its shorter evolutionary history. A comparable stage of vertebral development is reached after $5\frac{3}{4}$ days in the chick.

The crocodilian pro-atlas has not yet differentiated by the 10 mm stage. The BDs of the occipital arch can be seen growing up behind the last hypoglossal root. This nerve supplies the fourth metotic myotome and can, therefore, be identified as XIId. The BDs of the occipital arch have not yet fused above XIId with the rest of the occipital cartilage derived from preoccipital arches which encloses the more anterior hypoglossal roots. The bases of the occipital arch form the exoccipital portion of the occipital condyles, as in other amniotes. There are, in all, four hypoglossal roots on each side at this stage, but XIIa is vestigial and later disappears. Four hypoglossal roots and four metotic myotomes behind the vagus nerve also occur in avian embryos (de Beer and Barrington, 1934; Jager, 1924).

Mammals

In mammals separate HCs persist into adult life in the tail as chevron bones, and in the lumbar region of Insectivora (hedgehog, mole) where they form median crescents. Mesenchymatous BVs interconnected by a membranous band beneath the notochord (the hypochordal bar or bow; Froriep, 1886) are found in embryonic stages. These soon disappear, except those which form the ventral arch of the atlas, and the capitula of the ribs articulate intervertebrally with adjacent Cs.

The occipital region of the skull in mammals was shown by Froriep (1886) to be formed from four vertebral elements (three preoccipital and an occipital). Cranial ribs in the preoccipital region form the paroccipital process of the basioccipital bone (Kernan, 1916). The exoccipital portion of the occipital condyles is formed from the BDs of the occipital arch as in reptiles. Four metotic myotomes develop, supplied by four hypoglossal roots XIIa, b, c and d. The first myotome and its ventral nerve root disappear at an early stage (Dawes, 1930). In very early embryos there may be a vestigial myotome in front of the four metotic myotomes, with no ventral nerve root. This myotome is probably in the segment of the vagus, and represents the hinder of two hypotic myotomes, below the otic capsule, belonging to segments of which the glossopharyngeal and vagus nerves are the dorsal roots. Such transitory segments have been described by Jager (1924) in the chick, by Matveiev (1925) in the sturgeon and by Goodrich (1918) and de Beer (1922) in elasmobranchs.

CONCLUSIONS

Vertebrae in relation to primitive and definitive segments

It appears that most of the confusions relating to the morphology of vertebrae have been due to a failure to distinguish between primitive embryonic segments, as defined by the position of the somatic arteries, and the definitive segments defined by the attachments of the myosepta. The function of the primitive segment is to provide for vascularisation of the structures of the somite: that of the definitive segment is to alternate bony elements of the vertebral column with somatic musculature. The relationships of the arcual elements of the vertebra to the boundaries of the primitive and definitive segments remain constant in different vertebrate groups. Either the position of the somatic arteries or the attachment of the myosepta can be used as criteria for identifying arcualia. In some vertebrates, e.g. modern Amphibia, the line of attachment of the myoseptum to the arcualia is very clear; in others it is in practice much more difficult to determine, on account of the development of secondary attachments on the side opposite to that of the primary attachment. The course of the somatic arteries remains very constant in relation to the dorsal arcualia so long as the arteries themselves are recognisable. If the segments become oblique, as in some primitive vertebrates, then the arteries slant with them. For this reason, the position relative to the somatic arteries is often the surest way of identifying the dorsal arcualia. Identification of ventral arcualia does not present comparable difficulties, owing to the relationship of the BVs to ventral ribs and haemal arches. Dorsal and ventral arcualia cannot always be identified from their relationship to one another, since they may, as in bony fish, be involved in the obliquity of the segments.

The fully resegmented vertebra is only found in amniotes. The neural arch is resegmented in tetrapods on account of the addition of postzygapophyseal elements to the back of the BDs. The formation of zygapophyses is associated with reduction of the original IDs. In amniotes the C is resegmented as a result first of perichordal chondrification in the territory of the anterior sclerotomite vacated by the vanished IDs, and secondly of perichordal chondrification, at a slightly later stage, in the territory of the posterior sclerotomite of the embryonic segment immediately in front, which is vacated by the reduced BVs. The reduction of the BVs is accompanied by elongation of the neck of the rib, which is functionally important in costal respiration.

The cranio-vertebral articulation and the segmentation of the head

In amniotes the exoccipital portions of the occipital condyles (or condyle) are formed from the BDs of the occipital arch, which is the neural arch of the defintive segment of the first cervical nerve. These BDs develop from the embryonic segment of the fourth hypoglossal root, which supplies the fourth metotic myotome. Their postzygapophyseal portion remains separate and forms the pro-atlas when that element is present. The basioccipital portions of the condyles are formed from the posterior ends of the parachordal cartilages, which are unsegmented and chondrify in the lateral part of the perichordal mesenchyme. Ventral arcualia do not make a significant contribution to the basiocciput, except possibly in the region of the cranial ribs, which connect with the perichordal region below the preoccipital arches and form the paroccipital process.

In vertebrates generally, the posterior termination of the occipital region of the skull is less arbitrary, in terms of head segments, than has been suggested by de Beer (1937). The variable element is the occipital arch (the basidorsals of the embryonic segment associated with XIId) which is included and forms the back of the true skull in amniotes, lungfish, *Cryptobranchus* and *Squalus*, and is excluded from the skull and forms the first free spinal arch in urodeles other than *Cryptobranchus*, in holostean and teleost fish, and in *Scyllium*. In a number of fish one or more anterior spinal (occipito-spinal) vertebrae may become secondarily included in the back of the adult skull, but are recognisably separate from it in larval stages.

In the lamprey, which has no occipital portion of the skull, the anterior neural arches of the spinal series, formed by BDs and IDs, correspond to the preoccipital and occipital arches of jawed vertebrates. To determine how many such arches are represented, it is necessary to consider the fate of the hypotic myotomes. There seems to be no sound basis for the commonly held view that the hypotic myotomes are retained in lampreys. Assuming these segments to be lost as in jawed vertebrates, the occipital arch is the one associated with the definitive segment of the fifth metotic myotome and its nerve. This is the third arch of the series, and there are two preoccipital arches.

Phylogeny

There has been a tendency during the evolutionary history of vertebrates for all arcual elements other than the BDs and their DID extensions to regress. BVs persist in amniotes only in the ventral arch of the atlas, in caudal chevrons, and in vestigial form as HCs. IDs are lost in all living tetrapods. They are incorporated into the C in holostean fish and form bases for the neural arch in teleosts. The basic similarity in vertebral composition in the two groups supports the view that teleosts evolved from a holostean stock. Modern Amphibia likewise possess features in common, particularly in the region of the cranio-vertebral articulation, which indicate that the three living Orders have a common origin.

REFERENCES

DE BEER, G. R. 1937. *The development of the vertebrate skull.* Clarendon Press. Oxford.
EMELIANOV, S. W. 1935. Die Morphologie der Fischrippen. *Zool. Jahrb. Abt. Anat.,* **60,** 133-262.
Fox, H. 1962. Study of the evolution of the amphibian and dipnoan mesonephros by an analysis of its relationship with the anterior spinal nerves. *Proc. zool. Soc., Lond.,* **138,** 225-256.
GADOW, H. 1933. *The evolution of the vertebral column.* Cambridge University Press.
GOODRICH, E. S. 1930. *Studies on the structure and development of vertebrates.* Macmillan. London.
PARSONS, T. S. and WILLIAMS, E. E. 1962. The relationships of the modern Amphibia: a re-examination. *Q. Rev. Biol.,* **38,** 26-53.
PRATT, C. W. M. 1946. The plane of fracture of the caudal vertebrae of certain lacertilians. *J. Anat. Lond.,* **80,** 184-188.
REMANE, A. 1936. Wirbelsäule und ihre Abkömmlinge. In *Handbuch der Vergleichenden Anatomie der Wirbeltiere,* ed. L. Bolk *et al.,* **4,** 1-206.
WILLIAMS, E. E. 1959. Gadow's arcualia and the development of tetrapod vertebrae. *Q. Rev. Biol.,* **34,** 1-32.

Other references (earlier than 1935) as cited in de Beer, G. R. (1937) and in Goodrich, E. S. (1930, chapters 1 and 5).

S. M. ANDREWS

The shoulder girdle of *'Eogyrinus'*

INTRODUCTION

In the course of his description of the labyrinthodont Amphibia from the Coal Measures of Northumberland, Watson (1926) interpreted a specimen in the Thomas Atthey Collection, in the Hancock Museum, Newcastle upon Tyne (DMSW No. 34, now registered as Hancock Museum No. G. 6.39), which showed some large dermal bones, as the shoulder girdle of the embolomere *Eogyrinus*. He did not illustrate the actual specimen, but gave only a reconstruction (*ibid.* text-fig. 25; Fig. 2 here); because his reconstruction displayed features intermediate between rhipidistian fish and early Amphibia it has received a prominent place in discussions of the origin of tetrapods, and has been reproduced in text-books ever since. The specimen itself is problematical, however, and the opinion has been expressed that it may be part of a fish (Romer, 1956). Postcranial remains of the British Carboniferous Rhipidistia have now been studied and described (Andrews and Westoll, 1970b), and new light on their structure has made possible the following re-evaluation of the specimen described by Watson.

The specimen is contained in two blocks of cannel coal from the Low Main Coal Shale, Newsham, Northumberland (see Plates IA, IIA). The bones are exposed on both sides of both blocks; they are large and somewhat crushed, and appear to be only slightly disarranged. The larger block contains two large plates of bone covered with dermal ornament of a type commonly found on labyrinthodont shoulder girdles, but also somewhat reminiscent of the shoulder girdle ornament of *Strepsodus* and *Rhizodus* (Barkas, 1873; see also Plate I here). These two bones are arranged in an apparently symmetrical manner, both overlying a smooth central bone. The larger block also contains some small, very thin fragments. The smaller block contains a second bone of the same type as the smooth central plate, and also two smaller, thin, dermally ornamented fragmentary bones. None of the bones is certainly determinable as belonging to any of the known Coal Measures vertebrates, although as just hinted, the ornament suggests that it is a shoulder girdle either of an amphibian or a fish.

HISTORY OF THE SPECIMEN

In 1868, Hancock and Atthey described these isolated bones, which were apparently from a "reptilian" shoulder girdle, interpreting them as the "sternal plates of *Pteroplax cornuta*", whose name they derived from the winged form of the smooth central plate. They figured parts of the specimen (1868, plate XIV, figs. 1 and 2, reproduced as Fig. 1 here), but their figures, as noted later by Watson, contain some inaccuracies. They took the two large ornamented plates in the main block to be symmetrical right and left lateral elements, and the smooth central plate to be median and ventral; these bones would nowadays be called the right and left clavicles and the interclavicle. The second block they therefore interpreted as a second specimen, since it contained a bone like the smooth central plate and both could not have come from one individual.

Watson's (1926, text-fig. 25) restoration of the specimen, reinterpreted as the shoulder girdle of *Eogyrinus*, is reproduced here as Fig. 2, while Fig. 3 shows how he seems to have interpreted the various bones. Like Hancock and Atthey, Watson took the two large, ornamented plates in the main block to be symmetrical right and left elements, but he recognised that the smooth central plate

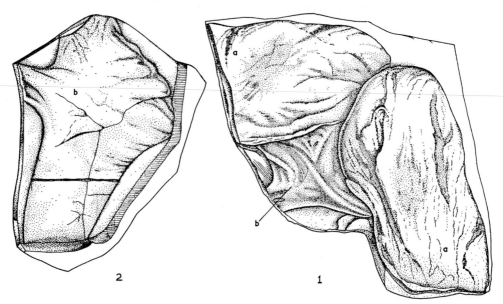

Fig. 1 The interpretation of Hancock and Atthey of specimen G. 6.39. According to these authors the two blocks belonged to two individuals of *Pteroplax cornuta*. Redrawn from Hancock and Atthey, 1868, Plate XIV, figs. 1 and 2. Abbreviations to all figures are found on p. 48.

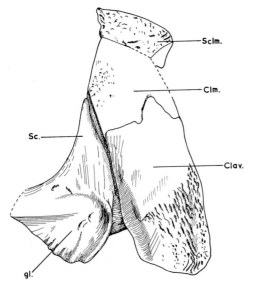

Fig. 2 Watson's interpretation of specimen G. 6.39. Bones from both blocks were reconstructed as the shoulder girdle of *Eogyrinus attheyi*. Redrawn from Watson, 1926, text-fig. 25.

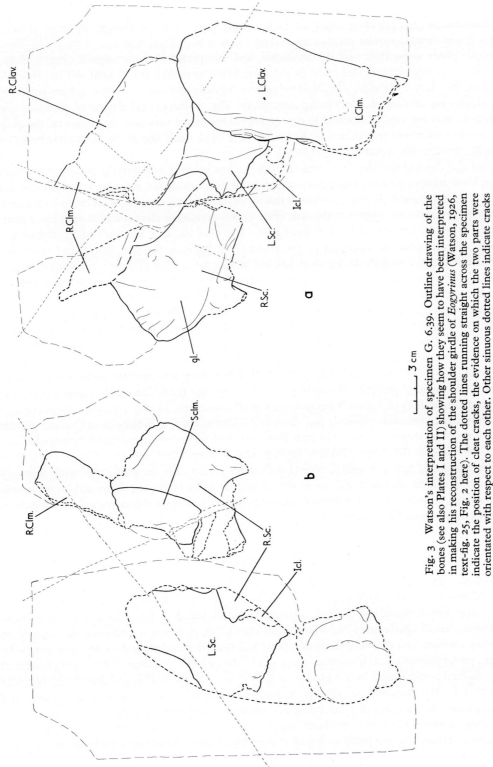

Fig. 3. Watson's interpretation of specimen G. 6.39. Outline drawing of the bones (see also Plates I and II) showing how they seem to have been interpreted in making his reconstruction of the shoulder girdle of *Eogyrinus* (Watson, 1926, text-fig. 25, Fig. 2 here). The dotted lines running straight across the specimen indicate the position of cleat cracks, the evidence on which the two parts were orientated with respect to each other. Other sinuous dotted lines indicate cracks or possibly sutures across the bones. (a) shows one side of the specimen, with most of the bones in external view, while (b) shows the reverse side.

was asymmetrical (and therefore could not be a median ventral plate as thought by these authors) and that it matched the similar plate in the second block in both shape and size. Watson took these two similar plates to be right and left antimeres, and interpreted them as scapulae (Fig. 3, *R.Sc.*, *L.Sc.*). The two large ornamented plates in the main block he interpreted as right and left cleithro-claviculars, the clavicles (*R.Clav.*, *L.Clav.*) being near together and almost in their natural positions, and both cleithra (*R.Clm.*, *L.Clm.*) being incomplete. The cleithrum and clavicle of each side are, according to Watson, rigidly attached to one another. One of the very thin fragments on the main block seems to have been interpreted as the interclavicle (*Icl.*), while one of the incomplete bones on the smaller block is his supracleithrum (*Sclm.*).

Watson also claimed that the "two slabs of shale . . . just touch at a spot in the middle of the right scapula, their relative position being obvious from the direction of the striae on this bone and the anterior edge of the right clavicle". Although this is *not* immediately obvious, it seems to be the case, since when the blocks are aligned in this way (Fig. 3) two prominent cleat cracks in each block come into line (shown by fine dotted lines in this figure). The elucidation of Watson's interpretation was not possible until after the discovery of the relationship between the two blocks, but his whole description fits very well with the bones as labelled in Fig. 3.

Watson's restoration had many primitive and fish-like features (Fig. 2) such as the lack of a coracoid plate, the small size of the scapula, the wide extent of the dermal bones and the dorsal attachment to the head through the supracleithrum, and in its shoulder girdle *Eogyrinus* thus appeared to be intermediate in structure between the fish and early amphibian types. It has been used in text-books and other works ever since, as a "missing link" between the fish and amphibian types of shoulder girdle (e.g. Romer, 1924; 1945; Goodrich, 1930; Gregory and Raven, 1941; Young, 1950; Gregory, 1951; Jollie, 1962—*Pteroplax* (sic); Yapp, 1966). However, its many peculiar features have caused problems, and as other primitive shoulder girdles have become known among tetrapods, doubt has been cast on the validity of Watson's interpretation of this specimen. Finally, when the shoulder girdle of the Permian embolomere *Archeria* was described (Romer, 1957) it became clear that the present specimen could no longer be associated with *Eogyrinus*, and Romer suggested that it belonged to one of the Carboniferous Rhipidistia (Romer, 1956, p. 164; 1957, p. 112).

Before examining this possibility, several difficulties which arise in connection with Watson's interpretation will be discussed and the results of further study of the specimen will be presented. First, however, a summary of the relevant features of the structure of the shoulder girdle in Rhipidistia will be given as a basis for comparison.

SOME FEATURES OF THE RHIPIDISTIAN SHOULDER GIRDLE

The shoulder girdle of *Eusthenopteron foordi* Whiteaves is well known, and exemplifies the basic pattern to which the shoulder girdles of most other Rhipidistia conform. It has been described by Jarvik (1944) and Andrews and Westoll (1970a), and is illustrated in Fig. 4 here. The largest bone is the paired dermal cleithrum (*Clm.*) which carries the small, primitive scapulocoracoid (*Sc.cor.*) on its internal surface. The glenoid (*gl.*) is elongated and slightly screw-shaped (as in early tetrapods) and faces posterolaterally; it is supported above and below by bony buttresses. The rest of the dermal girdle is formed by the paired clavicles (*Clav.*), firmly sutured to the cleithra and joined mid-ventrally by the interclavicle (*Icl.*), and dorsally by the paired anocleithra (*Aclm.*), supracleithra (*Sclm.*) and post-temporals (*Pt.*), a chain of small bones not commonly preserved, linking the cleithrum to the extrascapular series at the back of the head. Each of the bones in this chain overlaps the bone behind and below it, except the anocleithrum which is overlapped both by the supracleithrum dorsally and

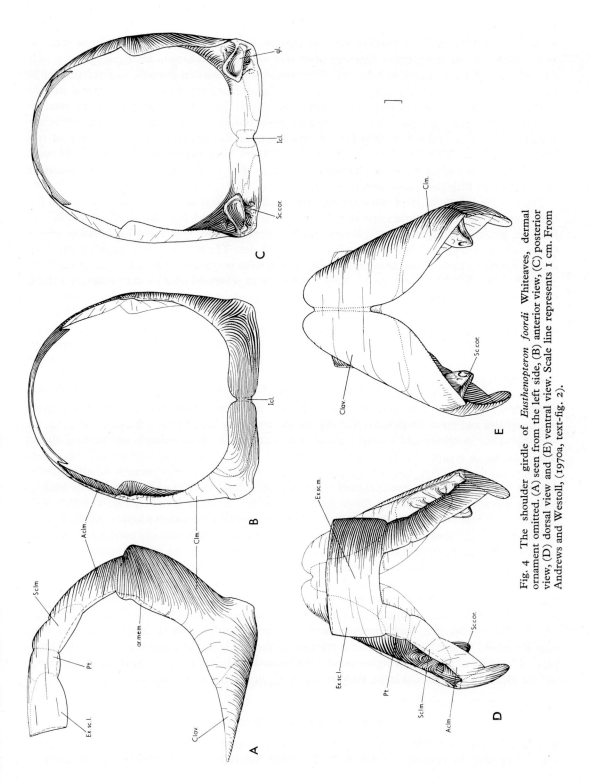

Fig. 4 The shoulder girdle of *Eusthenopteron foordi* Whiteaves, dermal ornament omitted. (A) seen from the left side, (B) anterior view, (C) posterior view, (D) dorsal view and (E) ventral view. Scale line represents 1 cm. From Andrews and Westoll, (1970a, text-fig. 2).

by the cleithrum ventrally; the suture with the cleithrum appears to have been very loose and was probably the site of considerable kinetism connected with the mechanism for opening the mouth (Andrews and Westoll, 1970a, text-fig. 29). The anocleithrum bears an inwardly directed process at its anteroventral corner, which may have been for a ligamentous attachment. The external surface of the dermal bones is covered by dermal ornament except for a narrow strip along the anterior border of the girdle (*ar.mem.*), from the post-temporal dorsally to part-way down the cleithrum ventrally. This was probably mainly for the attachment of the membrane lining the back of the branchial chamber. The dermal ornament is rather different from the ornament of the skull bones, the tubercles more often becoming confluent into ridges than on the head, and this is a common phenomenon in Rhipidistia and labyrinthodonts.

The shoulder girdle of the Carboniferous rhipidistians *Rhizodus* and *Strepsodus* is known so far only from the clavicle, cleithrum and scapulocoracoid, (Andrews and Westoll, 1970b). These bones, however, differ from the more normal type briefly described above (possessed by members of the Orders Osteolepidida and Holoptychiida) in many features. These two genera, together with the similar American form *Sauripterus taylori* Hall, which were previously included in the group here called Osteolepidida (Osteolepiformes), have therefore been separated (*ibid.*) and classified as a third Order, the Rhizodontida (the Family Rhizodontidae has been re-defined to include only these three forms). This group is characterised by large size, the largest, *Rhizodus hibberti* (Agassiz in Hibbert) probably reaching 6-7 m, in length, and the smallest known, *R. ornatus* Traquair, being 2 m. As far as it is known, the shoulder girdle seems to conform to the basic rhipidistian pattern in the arrangement of the component bones. An outline of the differences relevant to the present study, referring mainly to *R. hibberti*, follows.

The cleithrum of *Rhizodus* (Fig. 5) is a large bone firmly sutured to the clavicle (see Plate IC, which shows the similar overlap in *Strepsodus*), but the direction of overlap is the reverse of that in *Eusthenopteron* and other Rhipidistia. The clavicle carries a very long dorsal spine (*Clav. sp.*, Fig. 7) overlapping the cleithrum and having a complicated structure, twisting inwards and forming about three quarters of a turn of a loose spiral. The clavicle and cleithrum are very broadly expanded ventrally, but the cleithrum narrows as it passes dorsally, forming a posterior pectoral incision. Above this, the bone becomes very much thicker (more than doubling the thickness of the ventral blade), while posteriorly it bears a strong flange devoid of dermal ornament (*lm.Clm.*, Fig. 5) which was depressed well below the general body surface and was probably for muscle attachment. All the known overlap areas in the shoulder girdle in this group are characteristically strongly corrugated at right angles to the bone margins (Plates IC, IIB), possibly indicating that the fibrous connective tissue linking the bones was strongly developed in these large forms. The dermal ornament is of long ridges and grooves which anastomose and branch, but this is not uniformly developed. It is best seen on the ventral blade of the cleithrum (Plate IB) and in some other forms also near the pectoral incision (Plate IC), but it fades dorsally, and is lacking on the anterior margin (*ar. mem.*, Fig. 5) as in *Eusthenopteron*. This anterior smooth strip passes into a groove, dorsally, beneath a boss of ornamented bone. The scapulocoracoid (*Sc.cor.*, see Plate IIC) is attached to the inside of the cleithrum at the level of the pectoral incision, which is much higher above the ventrolateral angle of the body than in *Eusthenopteron*, and the glenoid fossa (*gl.*) is a large, circular concavity, in contrast to the last-named.

NEW STUDY OF THE SPECIMEN

Several difficulties arise from Watson's interpretation of specimen G. 6.39, and these must be dealt with before new observations are presented. The bones interpreted by Watson as scapulae are flat plates, of similar proportions to the acknowledged dermal bones, and although they do not seem to bear dermal ornament their internal structure (seen in section at the edges of the blocks) also resembles dermal bone. The supposed glenoid fossa is an extremely shallow, ill-defined area, lined with smooth, finished bone. If these bones were in reality scapulae, the glenoid would be expected to be an unfinished area, lined with cartilage in life, and the plate would not be expected to show a three-layered structure like dermal bone, and Watson's interpretation must therefore be rejected.

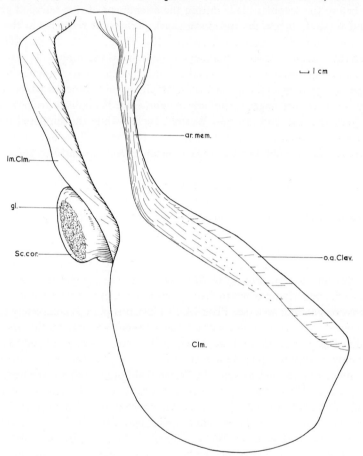

Fig. 5 The right cleithrum of *Rhizodus hibberti* (Agassiz in Hibbert). The dermal ornament is omitted and the bone is represented flattened into one plane. Re-drawn from Andrews and Westell (1970b).

Other problems concern the sutures and overlap relations of the cleithra and clavicles of Watson. A jagged crack between his cleithrum and clavicle of the left side (Fig. 3, *L.Clm., L. Clav.*) seems to have been taken as the suture and used in the restoration. On the other (Watson's right) side, where the evidence for the presence of two separate bones is better (two bony laminae may be seen in section at the edge of the main block), a suture is very difficult to make out. Possible sutures may be traced in several places, but not following the course indicated in the restoration. The great depth of the overlap area between cleithrum and clavicle and the broad dorsal extent of the clavicle are unusual when compared both with early tetrapods and crossopterygians.

The bones interpreted as supracleithra and the interclavicle are wafer thin, in contrast with the other bones which average 5 mm in thickness. The difficulty of overlap between bones of such different thicknesses is not mentioned by Watson, and nor is the possibility that the extremely thin bone interpreted as the interclavicle and the other thin fragments may be scales; they are rather similar to scales of *Strepsodus*.

The more recently discovered shoulder girdle of *Ichthyostega* seems to indicate that the coracoid plate developed before the scapular blade during the attainment of the tetrapod grade of evolution (see Andrews and Westoll, 1970a, for full discussion), perhaps in response to the need for larger, weight-bearing ventral muscles. '*Eogyrinus*', on the other hand, shows the scapula developing first. Watson identified the specimen as an amphibian on the basis of the supposed possession of an interclavicle, and of the dermal ornament. He related it more specifically to *Eogyrinus* on the basis of the ornament of ridges and grooves on the larger bones, which resembled the dermal ornament of the head of this form. This evidence no longer carries any weight, since Rhipidistia are now known to possess an interclavicle (Jarvik, 1944; Andrews and Westoll, 1970a) while the supposed interclavicle in the specimen is too poorly exposed to be identified, and too thin to be likely to be this bone; the dermal bone ornament somewhat resembles that of the cleithrum in *Rhizodus hibberti* and *Strepsodus*, as well as the head of *Eogyrinus*.

Fresh study of this problematical specimen has brought several new facts to light. The two large ornamented plates on the main block are not a pair of symmetrical antimeres, as thought by both Watson and Hancock and Atthey, since one of them (*L.Clav.*, *L.Clm.*, Fig. 3) has several features lacking in the other (*R.Clav.*, *R.Clm.*). On the external, ornamented side (Fig. 3a) of this oblong plate of bone, one of the long margins (the shorter, left margin) lacks the coarse ornament, and this smooth strip passes into a groove at one end in a manner reminiscent of the anterior edge of the dorsal lamina of the cleithrum (*ar.mem.* Fig. 5) in *Rhizodus*. The other large ornamented plate, although crushed down over the underlying bones to give the impression of a groove, has coarse ornament over the whole of the corresponding area (see Plate IA). In fact this plate is completely covered by coarse vermiculating grooves as far as its outer surface can be seen, whereas on the first plate the coarse ornament is confined to part of one end, as shown in Watson's restoration, and the rest of the bone has an ornament only of very fine pits. This first plate (*L.Clm.*, *L.Clav.*, Fig. 3) is very much thicker than the second (about 10 mm thick as noted by Watson) although both are crushed, and on the inner surface, the former has a thick crushed mass of endoskeletal bone with a circular unfinished concavity (*gl.*, see Fig. 6 and Plate IIA) resembling the scapulocoracoid and glenoid fossa of *Rhizodus* in size and shape (compare the similar scapulocoracoid of *Strepsodus*, Plate IIC). This is lacking on the other plate, and was not mentioned at all by Watson. Finally, the first plate is definitely undivided by sutures, whereas the other is composed of two bones seen overlapping in the thickness of the rock, although the suture between them is hard to locate.

POSSIBLE NEW INTERPRETATION

It can now be suggested that these bones represent the dorsolateral part of a large shoulder girdle of rhizodont (*sensu stricto*) type (see Figs 6 and 7), the dorsal lamina of the left cleithrum (*Clm.*) and the attached scapulocoracoid (*gl.*), the left supracleithrum (*Sclm.*) and post-temporal (*Pt.*), and both anocleithra (*Aclm.*) being present. The following evidence supports this idea.

The endoskeletal bone on the internal surface of the cleithrum (*gl.*) is suitably situated and of a suitable size to be a scapulocoracoid. The unfinished, circular glenoid fossa is also of the expected size and although crushed, faces more or less posteriorly. As already mentioned, the smooth anterior

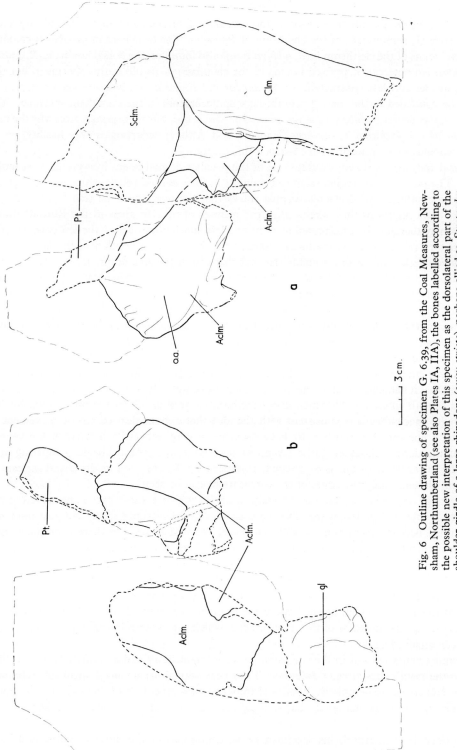

Fig. 6 Outline drawing of specimen G. 6.39, from the Coal Measures, Newsham, Northumberland (see also Plates IA, IIA), the bones labelled according to the possible new interpretation of this specimen as the dorsolateral part of the shoulder girdle of a large rhizodont (*sensu stricto*), perhaps allied to *Strepsodus* (see text). The two parts of the specimen are placed in their natural relationship. (*a*) shows the side of the specimen with most of the bones in external view; (*b*) shows the reverse side of the specimen.

border of the cleithrum, passing into a groove dorsally, and the ornament of anastomosing ridges and grooves are strongly reminiscent of the cleithrum of *Rhizodus*. The ornament of grooves is confined to the ventral lamina of the cleithrum in *R. hibberti* but is also found on the dorsal lamina in *R. ornatus* and *Strepsodus*. No posterior depressed lamina of the cleithrum is visible in this specimen, but this structure is smaller and less conspicuous in *Strepsodus* and *Rhizodus ornatus*, and such a depressed flange may be concealed in the matrix behind the scapulocoracoid in the Newsham specimen. The size of the bones is nearer *R. hibberti* than *R. ornatus* or *Strepsodus*, but there is no reason why the very large *Strepsodus* (= *Archichthys*) *sulcidens* (Hancock and Atthey) occurring at this locality, or an allied form, should not occasionally have reached this magnitude.

At its dorsal end, the cleithrum overlaps one of the peculiar dermal bones interpreted as scapulae by Watson. The bones in this region seem to be almost in their natural relationships to one another, and the separate antimere of Watson's scapula shows that the overlapped portion (*o.a.*) is deeply corrugated at right angles to the margin, after the manner of overlap areas in the Rhizodontidae. These bones may therefore be interpreted as right and left anocleithra (*Aclm.*) the left anocleithrum being in its natural relationship to the adjacent bones in the shoulder girdle both dorsally and ventrally. The corrugated overlap area is of a suitable size and shape to fit the area on the inner surface of the dorsal lamina of the cleithrum in *Rhizodus hibberti* (Plate IIB), or a large individual of *Strepsodus*, and the bone is overlapped ventrally by the cleithrum and dorsally by another bone which may be the supracleithrum, as in *Eusthenopteron*. There is a stout, forwardly and inwardly directed process on the anterior margin at the ventral end, which may be homologous with such a process of the anocleithrum in *Eusthenopteron* and *Osteolepis* (see Jarvik, 1944, text-fig. 8D; *p. Aclm.*). The anocleithrum is a very rarely preserved bone in other forms, and yet both right and left antimeres are present in this specimen: if this interpretation is correct, however, it shows that very little of this bone was visible externally and that it had no dermal ornament, being almost completely overlapped by adjacent bones, apart from a smooth anterior knob which may have been covered in life by the operculum. Such a structure of the anocleithrum is consistent with the idea that a great deal of kinetic movement in connection with opening the mouth must have taken place here, especially in view of the ventral expansion of the dermal shoulder girdle (Andrews and Westoll, 1970b). The long tapering area overlapped by the supracleithrum is roughened, as noticed by Watson, and its distal end appears to be corrugated, although only the edges of the corrugations can be seen.

The other large ornamented plate on the main block (of which a small part is preserved in the smaller block) is thus interpreted as the closely associated supracleithrum (*Sclm.*) and post-temporal (*Pt.*). The suture between these bones is difficult to see, as noted above, but there is no doubt that two bones are present (since they can be seen at the edge of the main block), and it seems that the more dorsal bone (*Pt.*) overlaps the more ventral (*Sclm.*) as in *Eusthenopteron*. The dorsal end of the post-temporal is obscured, so that the presence and nature of an overlap area for the extrascapulars could not be investigated. These two bones of the shoulder girdle have no smooth anterior strip for attachment of the membranes lining the back of the branchial chamber. There is, however, a rather narrow groove along the thickness of the anterior edge, which can be seen dorsally, and a membrane may have been attached to this.

The remaining bones present in the specimen are very fragmentary and extremely (1-2 mm) thin compared to the main plates already described. They bear dermal ornament of small tubercles and granulations, and appear to be imperfect scales of the *Strepsodus* type. One of these (Watson's supracleithrum: Fig. 3, *Sclm.*) is oblong in shape; its nature and position in the body are unknown at present.

It is thus possible to interpret this specimen as the lateral part of the shoulder girdle of a large

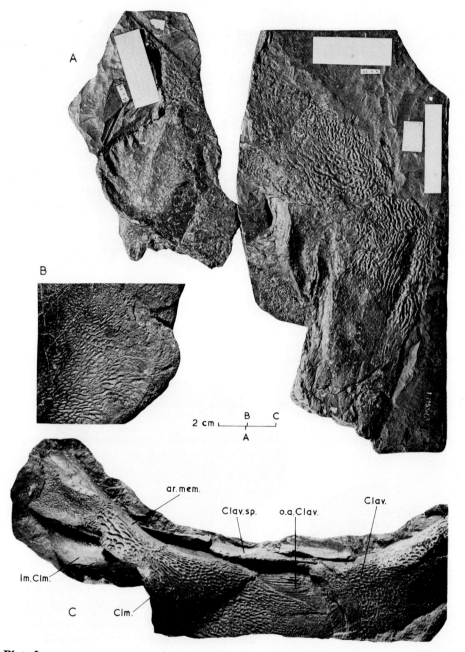

Plate I
A. Specimen G. 6.39 (Hancock Museum, Newcastle upon Tyne), from the Coal Measures, Low Main Seam, Newsham, Northumberland. The two parts of the specimen are placed in their natural relationship. ×c. 0·4.
B. *Rhizodus hibberti* (Agassiz in Hibbert). The dermal ornament of part of the ventral blade of the cleithrum. Specimen E. 4715 (Sedgwick Museum, Cambridge) from the Lower Carboniferous, Gilmerton Ironstone, Gilmerton near Edinburgh. ×c. 0·4.
C. *Strepsodus sauroides* (Binney). Latex cast of an incomplete right cleithrum and clavicle showing the dermal ornament. Specimen G. 4.43 (Hancock Museum, Newcastle upon Tyne), from the Coal Measures, Wemyss Parrot Coal, Pirnie Colliery, Fifeshire. ×c. 0·8.

Plate II
A. Specimen G. 6.39 (Hancock Museum, Newcastle upon Tyne), from the Coal Measures, Low Main Seam, Newsham, Northumberland. The same specimen as in Plate IA, seen from the reverse side. ×c. 0·4.
B. *Rhizodus hibberti* (Agassiz in Hibbert). The visceral surface of the dorsal lamina of the cleithrum—the area overlapped by the anocleithrum—showing characteristic corrugations. Specimen E. 4715 (Sedgwick Museum, Cambridge), from the Lower Carboniferous, Gilmerton Ironstone, Gilmerton near Edinburgh. ×c. 0·8.
C. *Strepsodus* sp. An incomplete scapulocoracoid. Specimen 1970.16.2 (Royal Scottish Museum, Edinburgh) probably from the Low Main Seam, Newsham, Northumberland. ×c. 1·2.

crossopterygian of rhizodont (*sensu stricto*) type. There are still difficulties in the interpretation, as mentioned above, but these may be due to present lack of knowledge of the dorsal dermal bones in the rhizodont shoulder girdle. It has not been possible to identify the specimen with certainty as far as genus and species because the only part available for direct comparison (the dorsal lamina of the cleithrum) shows several points of difference from all hitherto known rhizodont cleithra, but it may be allied to one of the larger species of *Strepsodus* (= *Archichthys*) occurring at Newsham. *Strepsodus* is at present known only from teeth, jaws and parts of the postcranial skeleton (Andrews and Westoll, 1970b) and is systematically very confused, although it is hoped that new material of the head at present being studied will help to remedy this situation.

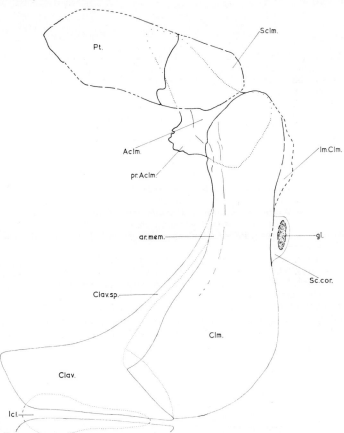

Fig. 7 Composite reconstruction made from specimen G. 6.39, interpreted as the dorsal dermal bones of a rhipidistian allied to *Strepsodus* (Coal Measures), and Royal Scottish Museum specimen 1963.16.16 and Sedgwick Museum specimen E. 4715, showing the cleithrum, clavicle and scapulocoracoid of *Rhizodus hibberti* (Lower Carboniferous); *R. hibberti* was used because it is nearest the Newsham specimen in size. The purpose of this diagram is to give a general idea of the possible form of a rhizodont shoulder girdle (see text), and it should not be taken as a serious attempt to reconstruct either the Newsham specimen or *Rhizodus*. The bones are represented flattened into one plane, and would appear foreshortened dorsally and ventrally in 'life'. Heavy lines are outlines from the Newsham specimen; fine lines from *Rhizodus hibberti*; dashed lines show structures not represented in the material; dotted lines show concealed structures.

As the shoulder girdle of *Strepsodus* and its allies has so far been known only from cleithra, clavicles and scapulocoracoids, it seemed worth while to make a composite reconstruction from the Newsham specimen (dorsal dermal bones) and the known rhizodont bones (Fig. 7) in order to give a general idea of what a complete rhizodont girdle may have looked like.

It may be said in conclusion that Romer's suggestion (1956, 1957) is probably correct, and that this specimen most likely belongs to a Carboniferous rhipidistian. In any event, it cannot be a labyrinthodont shoulder girdle, and can have nothing to do with the origin of the tetrapod shoulder girdle, since the group of Rhipidistia under discussion is unlikely to have given rise to any tetrapods (Andrews and Westoll, 1970b).

ACKNOWLEDGEMENTS

This work formed part of a thesis submitted for the Degree of Ph.D. in the University of Durham, and I should first like to thank Professor T. S. Westoll for his support and continued interest and encouragement. I am most grateful to the Staffs of the Royal Scottish Museum, Edinburgh, the Sedgwick Museum, Cambridge and particularly to Mr A. M. Tynan of the Hancock Museum, Newcastle upon Tyne, for permission to borrow and study fossils in their care. I am deeply indebted to Mr R. C. M. Thomson for the photography, and to Dr R. S. Miles for reading the manuscript. The Royal Society of Edinburgh has kindly granted permission to use the illustration here reproduced as Fig. 4.

REFERENCES

ANDREWS, S. M. and WESTOLL, T. S. 1970a. The postcranial skeleton of *Eusthenopteron foordi* Whiteaves. *Trans. R. Soc. Edinb.*, **68**, 207-329.

—— 1970b. The postcranial skeleton of rhipidistian fishes excluding *Eusthenopteron*. *Trans. R. Soc. Edinb.*, **69**, 391-489.

BARKAS, T. P. 1873. *Illustrated guide to the fish, amphibian, reptilian and supposed mammalian remains of the Northumberland Carboniferous strata.* W. M. Hutchings, London.

GOODRICH, E. S. 1930. *Studies on the structure and development of vertebrates.* Macmillan, London.

GREGORY, W. K. 1951. *Evolution emerging.* Macmillan, New York.

GREGORY, W. K. and RAVEN, H. C. 1941. Studies on the origin and early evolution of paired fins and limbs. *Ann. N.Y. Acad. Sci.*, **42**, 273-360.

HANCOCK, A. and ATTHEY, T. 1868. Notes on the remains of some reptiles and fishes from the shales of the Northumberland coal field. *Ann. Mag. nat. Hist.*, (ser. 4) **1**, 1-46.

JARVIK, E. 1944. On the exoskeletal shoulder-girdle of teleostomian fishes, with special reference to *Eusthenopteron foordi* Whiteaves. *K. svenska Vetensk-Akad. Handl.* (ser. 3), **21**, 1-32.

JOLLIE, M. 1962. *Chordate morphology.* Reinhold, New York.

ROMER, A. S. 1924. Pectoral limb musculature and shoulder-girdle structure in fish and tetrapods. *Anat. Rec.*, **27**, 119-143.

—— 1945. *Vertebrate paleontology.* (2nd Edition). University of Chicago Press, Chicago.

—— 1956. The early evolution of land vertebrates. *Proc. Am. phil. Soc.*, **100**, 157-167.

—— 1957. The appendicular skeleton of the Permian embolomerous amphibian *Archeria*. *Contr. Mus. Paleont. Univ. Mich.*, **13**, 103-159.

WATSON, D. M. S. 1926. The evolution and origin of the Amphibia. *Phil. Trans. R. Soc.*, B. **214**, 189-257.

YAPP, W. B. 1966. *Vertebrates, their structure and life.* Oxford University Press, Oxford.

YOUNG, J. Z. 1950. *The life of vertebrates.* Oxford University Press, Oxford.

ABBREVIATIONS

a.	The "lateral plates" of Hancock and Atthey's interpretation. Clavicles of modern terminology.	*L.Clav.*	Left clavicle.
		L.Clm.	Left cleithrum.
		lm.Clm.	Depressed posterior lamina of the cleithrum in *Rhizodus*.
Aclm.	Anocleithrum.		
ar.mem.	Area, probably mainly for the attachment of the membrane lining the posterior wall of the branchial cavity.	*L.Sc.*	Left scapula.
		o.a.	Overlap area for the dorsal part of the cleithrum.
b.	The "central plate" of Hancock and Atthey's interpretation. Interclavicle of modern terminology.	*o.a.Clav.*	Overlap area for the clavicle.
		pr.Aclm.	Anteroventral process of the anocleithrum.
		Pt.	Post-temporal.
Clav.	Clavicle.	*R.Clav.*	Right clavicle.
Clav.sp.	Clavicular spine.	*R.Clm.*	Right cleithrum.
Clm.	Cleithrum.	*R.Sc.*	Right scapula.
Ex.sc.l.	Lateral extrascapular plate.	*Sc.*	Scapula.
Ex.sc.m	Median extrascapular plate.	*Sc.cor.*	Scapulocoracoid.
gl.	Glenoid fossa.	*Sclm.*	Supracleithrum.
Icl.	Interclavicle.		

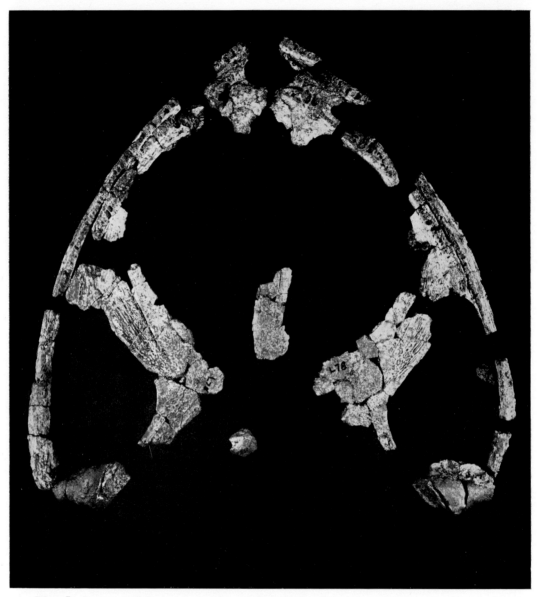

Plate I *Rewana quadricuneata* gen. et sp. nov. Ventral view of the palate of the type specimen ($\times \frac{2}{3}$).

A. A. HOWIE

On a Queensland labyrinthodont

INTRODUCTION

In 1859 T. H. Huxley described the first found Australian labyrinthodont as *Bothriceps australis*, a (probably) Permian brachyopid. The following 106 years produced three more brachyopids: another Permian species, *Trucheosaurus (Bothriceps) major* (Smith Woodward, 1909), the Lower Triassic *Blinasaurus (Platyceps) wilkinsoni* (Stephens, 1887) and an unidentified fragment from the Middle Trias (Watson, 1958), and five Triassic capitosaurs: the magnificent *Paracyclotosaurus davidi* (Watson, 1958), *Parotosaurus (Subcyclotosaurus) brookvalensis* (Watson, 1958), two scraps of '*Mastodonsaurus*' (Stephens, 1886, 1887), and a section of lower jaw of *Austropelor wadleyi* (Longman, 1941). All were isolated specimens and all came from New South Wales except *Austropelor* which was found near Brisbane, Queensland, as a reworked specimen in sediments of Jurassic age (Colbert, 1967).

During the 1960s, Cosgriff uncovered labyrinthodont faunas in the Trias of Tasmania and Western Australia. His two species of *Deltasaurus* from the Lower Triassic Blina Shale of the West Kimberley district, Western Australia, and the Lower Triassic Kockatea Shale of the Perth Basin proved to belong to a new superfamily, the Rhytidosteidae (Cosgriff, 1965) and were part of an extensive labyrinthodont fauna including *Blinasaurus* (Cosgriff, 1969), a capitosaurid, and two trematosaurids. The type of *Blinasaurus* (formerly *Platyceps wilkinsoni*, Stephens) is a larval labyrinthodont from New South Wales and the genus is also represented in the Tasmanian Lower Triassic fauna, which also includes *Deltasaurus*, a new genus and species of the Rhytidosteidae and a lydekkerinid. Cosgriff also discovered a primitive Lower Triassic parotosaur in the Wade fish collection at the Australian Museum, and a second specimen of the same species, as well as a tiny Upper Middle Triassic brachyopid, in the Mining Museum of the New South Wales Geological Survey.

Early in 1969, the Director of the Queensland Museum, Alan Bartholomai, showed me a collection of amphibian and reptilian remains which he had collected in the Rewan Formation of Southern Queensland during 1965, 1966 and 1968, and in September of 1969 a party consisting of Alex Ritchie and Kingsley Gregg of the Australian Museum, and myself, joined Bartholomai in the field and considerably added to the collection (Bartholomai and Howie, 1970).

Although remains of *Rewana quadricuneata* (described below) are not uncommon in the Rewan fauna, capitosaurid bones by far outnumber those of other species. There are also present a small aphanerammid interclavicle, a section of the snout from a much larger, apparently similar animal, and a length of mandible whose pustular ornament is similar to that of *Peltobatrachus pustulatus* (Panchen, 1959). Other components of the fauna include thecodont reptiles, short skulled eosuchians, dipnoan tooth plates and, preserved as scales in coprolites, actinopterygian fish. Small fresh water molluscs were also found in the coprolites.

All the material described below was collected from a single locality in the lower beds of the upper part of the Rewan Formation of the Mimosa Group: Queensland Museum field locality number L78, approximately 45 miles South-West of Rolleston, and 7.5 miles South of Rewan homestead. Some doubt still exists as to the position of the Permo-Triassic boundary in Eastern Australia; a recent review of the area by Balme (1969) concludes, on the basis of palynological and palaeobotanical evidence, that the Rewan Formation of the Bowen Basin may range in age from Late Permian to lower Middle Triassic. The fauna under consideration from the lower beds of the upper part of the Rewan is thus probably of Lower Triassic age, a conclusion supported by the Lower Triassic age of Cosgriff's similar labyrinthodont assemblage from the Blina Shale.

The red beds of the Rewan Formation are found in a large area of the Bowen Basin and the adjacent Springsure-Corfield Shelf and are typically composed of red-brown silty mudstones with occasional thin green layers and, especially in the lower part of the Rewan, lenses of lithic sandstone. The beds were almost certainly laid down by a meandering stream system.

The Rewan mudstones weather readily to produce a thick soil cover and so are rarely seen, the chief exposures being in stream banks and landslides in areas where the Rewan Formation is still overlaid by the more durable Clematis Sandstone. While all exposures looked essentially similar only one produced bone in quantity, although scraps of bone were found in three others.

The most productive locality, L78, is a steep-sided basin some 200 yards across, apparently initially formed by a landslide. Heavy rain results in a rubble of gravel and bone scraps being deposited in the centre of the basin so that only occasionally is a specimen found weathering out *in situ*. In general the larger fossils occurred in the mudstone while most of the small labyrinthodonts and all the small reptiles were contained in reddish-purple ironstone nodules. The sandstone lenses yielded a few isolated bones.

SYSTEMATIC DIAGNOSIS

Class Amphibia
Subclass Labyrinthodontia
Order Temnospondyli

Superfamily uncertain

Rewana. gen. nov.

Type species *Rewana quadricuneata* gen. et sp. nov.[1]

Generic diagnosis: A temnospondylous amphibian with skull as broad as long, shallow posteriorly, lateral skull margins curved but snout pointed anteriorly. Sensory canals present, ornament of the usual labyrinthodont type. Orbits small and rounded, situated near the lateral skull margins well forward in the anterior half of the skull. Lacrimal and jugal form a part of the orbital margin. External nares probably ovate, parallel to skull margins, very far forward. Palate with large interpterygoid and subtemporal vacuities, anterior palatal vacuity present and undivided, additional median vacuity bordered by the vomers posterior to the anterior palatal vacuity. Internal nostrils probably ovate, at least partly separated from the lateral tooth row by a denticulate flange of the vomers. Pterygoids ornamented, do not reach palatines lateral to the interpterygoid vacuities, and medially form a moderately elongate suture with the body of the parasphenoid. Cultriform process of the parasphenoid wide, flat, and dorsoventrally thin. Exoccipitals exposed on the palate. Vomers, palatines, ectopterygoids,

---

[1] The generic name is taken from the type locality while the specific name refers to the four wedges found in each vertebral centrum.

pterygoids, and parasphenoid with a dense covering of denticles; palatal tusk pairs on vomers, palatines and ectopterygoids.

Vertebrae of neorhachitomous type except that both neural arches and intercentra are divided into left and right halves; haemal arches ossified. Some ribs with heavily ridged proximal ends; sacral rib of the type found in primarily aquatic labyrinthodonts.

Appendicular skeleton moderately well ossified. Femur with strong adductor ridge. Fibula without characteristic distal groove on the median surface. Dorsal blade of the ilium narrow and nearly vertical.

Rewana quadricuneata gen. et sp. nov.
Holotype. An incomplete skull and postcranial skeleton: the Queensland Museum number F 6471.
Referred Specimen. A fragmentary skull associated with vertebrae and ribs: the Australian Museum, number AMF 54126.
Type locality. The Queensland Museum field locality L78 approximately 45 miles South-West of Rolleston, Southern Queensland.
Horizon. Lower beds of the upper part of the Rewan Formation of the Mimosa Group: Lower Trias.
Diagnosis. As for the Genus.

DESCRIPTION

The type specimen of *Rewana quadricuneata* was recovered from some two square yards of decomposing red mudstone on the floor of the basin at L78. Unfortunately the skull had weathered dorsally leaving a large part of the palate with fragments of the skull roof attached, and much of the postcranial skeleton. Left and right scapulocoracoids, humeri, radii, ulnae, ilia, ischia, femora, tibiae and fibulae were recovered as well as several smaller limb bones, neural arches from both pre- and post-sacral regions, haemal arches, intercentra, and pleurocentra, and several kinds of rib including the left and right sacral ribs. No positive trace of the dermal pectoral girdle was found.

Although most of the material was fragmented, the bone surface was well preserved and most of the breaks appear to be recent.

Skull

Fig. 1 shows the palate of *Rewana quadricuneata* with slight distortion in the snout region corrected. Individual bones differ little in shape from those found in most short faced labyrinthodonts from the Lower Trias.

The pointed snout found in *Rewana* is reminiscent of the short faced trematosaurs, but the lateral curvature of the maxillae, jugals and quadratojugals is natural, so that the skull has a true parabolic outline, distinguishing it from the straight-sided trematosaur skulls. Both the subtemporal and interpterygoid vacuities are well developed and alongside the latter the pterygoids do not extend anteriorly as far as the palatines, a character often associated with the 'advanced' Triassic metoposaurs, trematosaurs and brachyopids. An anterior palatal vacuity is present indicating a flattened skull (Romer, 1947) while posterior to this and level with the leading edge of the interpterygoid vacuities are indications of at least one, and possibly two additional smaller vacuities enclosed by the vomers. The more anterior of these is tiny and probably housed a blood vessel, but the more posterior one is over half the size of the anterior palatal vacuity. The curved section of 'true edge' on the right vomer which indicates the presence of this vomerine vacuity could simply be the edge of the vomer where it sutures with the cultriform process of the parasphenoid, but if so it is unlike other sutures.

Typical labyrinthodont teeth are present on the premaxillae and maxillae and tusk pairs are carried by the vomers and palatines with a smaller pair on each ectopterygoid. Bounding the inner edge of the nostril and apparently extending across the palate behind the anterior palatal vacuity is a regular row of smaller teeth, while an irregular row of teeth of similar size is found on a raised strip of ectopterygoid and palatine beside the maxillary teeth. Denticles of varying density and size cover the parasphenoid and pterygoid and are particularly dense on the raised strip of ectopterygoid and palatine, and on a similar strip of vomer which runs from the anterior palatal vacuity posteriorly towards the palatine, lateral to the internal nostril. Along part of its lateral border the maxilla is separated from the nostril by this strip of vomer. There is sufficient room between the nostril and the premaxillary-maxillary tooth row for a process from the palatine to suture with the vomerine process (above) and thus exclude the maxilla completely from the nostril, a condition found in some metoposaurs and

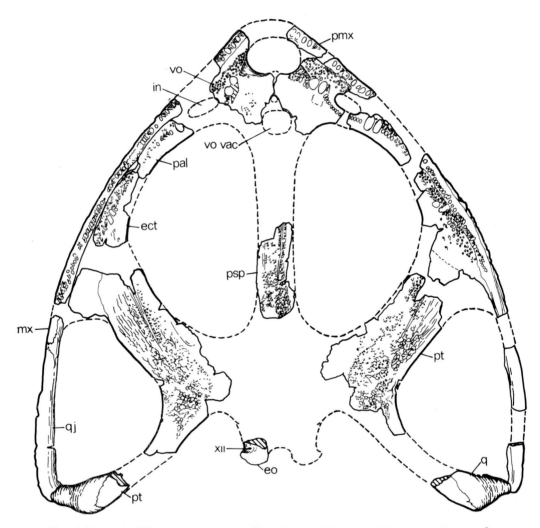

Fig. 1 *Rewana quadricuneata* gen. et sp. nov. Ventral view of the palate of the type specimen ($\times \frac{2}{3}$). Abbreviations to Figs. 1-6 will be found on p. 64.

brachyopids but rarely elsewhere. No evidence can be found of denticles on the maxillary bone as noted by Cosgriff (1965) in *Deltasaurus*.

Reticulate sculpture of a much finer construction than that found on the dorsal skull roof occupies the body and palatal ramus of the pterygoids, mostly posterolateral to any denticulate areas. The width of the surviving central areas of the pterygoids indicates that there must have been a moderately long pterygoid-parasphenoid suture. There is no evidence for a distal downturning of the quadrate ramus of the pterygoid to form the deep cheek and ∩-shaped palate characteristic of brachyopids.

No hint of an enlarged paraquadrate foramen could be found on either quadratojugal.

The screw-shaped quadrate condyles are better ossified than they are in many other Triassic genera but no extension of their articulating surfaces onto the adjacent pterygoids is present.

Some exposure of the exoccipital on the palate was present in *Rewana*, probably an equivalent amount to that seen in *Wetlugasaurus magnus* (Watson, 1962). A small foramen for the XIIth nerve passes through the exoccipital just anterior to the condyle. Comparison with the palatal structure in contemporary neorhachitomes suggested the reconstruction of the position of the exoccipitals and the width of the parasphenoid anterior to them (Fig. 1): the condyles may have been positioned more posteriorly, nearer the level of the quadrates.

All that remains of the skull roof is a rim of bone which fortunately contains parts of the orbits on the lacrimal and jugal, and part of the left external nostril on the premaxilla. It can be seen (Fig. 2) that the preorbital face is exceptionally short so that the external nares approach the tip of the snout and the orbits lie in the anterior half of the skull, further forward than in any amphibian other than *Trimerorachis* (as restored by Williston, 1914), *Erpetosaurus*, and some of the nectridians.

Just above its suture with the maxilla the left jugal carries a section of the infraorbital sensory groove passing close beneath the lateral margin of the orbit, which is preserved anteriorly on the jugal. No positive trace can be found of a median extension across the jugal of the infraorbital canal.

On the right side a long strip of the maxilla appears to suture anterodorsally with a piece of lacrimal bone which bears a fraction of the anterior border of the orbit. The lacrimal carries a well defined length of the infraorbital sensory canal. I cannot tell whether this maxillary-lacrimal suture is real, or merely a deceptive break in the maxilla, but if the latter were true the maxilla would take part in the orbital margin, a condition rarely found in labyrinthodonts. If the suture is real it is clear that the bone divided off from the maxilla is not the premaxilla (as it is in the Rhytidosteidae) as the premaxilla is never known to carry a sensory canal which passes laterally to the orbit.

The orbits as reconstructed are particularly small but little flexibility is possible in positioning the left jugal relative to the right lacrimal.

Although no tabular horns were associated with the type specimen three distinct tabular shapes were present in the collection. The most common horn shape is capitosaurid, and the next most common can probably be assigned to *Rewana* as there are more of this genus in the collection than anything except capitosaurids. The third tabular horn is much reduced and could have belonged to a brachyopid. The (?) rewanid tabular horn is shorter than a typical *Parotosaurus* horn and is a little pointed distally as shown in Fig. 2: the otic notch as restored is purely hypothetical.

No trace of the lower jaw remained with the type specimen, but of the two dissociated kinds of lower jaw found, one with a moderately elongate retroarticular process must have belonged to the parotosaur (Bartholomai and Howie, 1970, plate 1), while a deeper jaw with an elongate tapered retroarticular process probably belonged to *Rewana*.

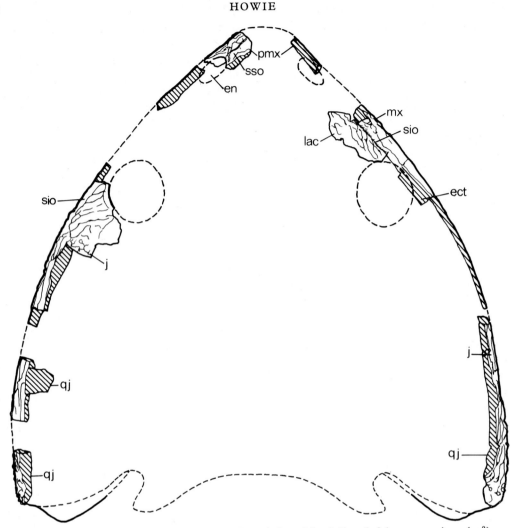

Fig. 2 *Rewana quadricuneata* gen. et sp. nov. Dorsal view of the skull roof of the type specimen ($\times \frac{2}{3}$).

Vertebrae

Since von Meyer (1856) published drawings of several *Archegosaurus* vertebral columns in which the neural and haemal arches were split into right and left halves many writers have noted a similar division in the rhachitomous neural arch, at least in part of the vertebral column. Few, however, have reported division of the intercentrum or its supposed derivative, the haemal arch, despite the fact that this condition must have been sought after as evidence supporting Gadow's (1933) theory of vertebral development. In 1935 Case, describing a collection of skeletons of *Trimerorachis*, noted that "the intercentra are so thin and fragile that they are almost invariably broken across in the median line into right and left halves" but he used this apparent break to cast doubt upon the bipartite condition of the anterior cervicals reported in earlier papers. Steen (1937) found that, in *Acanthostoma vorax*, "a pleura- or intercentrum consists of a right and left hemicylinder, unfused in mid-dorsal or mid-ventral line, right and left halves together completely enclosing the notochord." This evidence for the occurrence of divided intercentra was challenged in 1947 by Romer whose opinion was accepted by most palaeontologists, although Watson (1956) supported Steen's findings. Parrington also had doubts about Romer's view, these doubts being supported by his description (1948) of a series of six lydekkerinid intercentra which were split in the midline. In his opinion the division was not due to *post mortem* crushing.

All components of the vertebrae of *Rewana quadricuneata* are paired. Thus, each vertebra consists of left and right neural arch elements, left and right pleurocentra, and left and right intercentral halves. Unfortunately, in no case were the elements of a vertebra found in articulation but there is no doubt as to their identity.

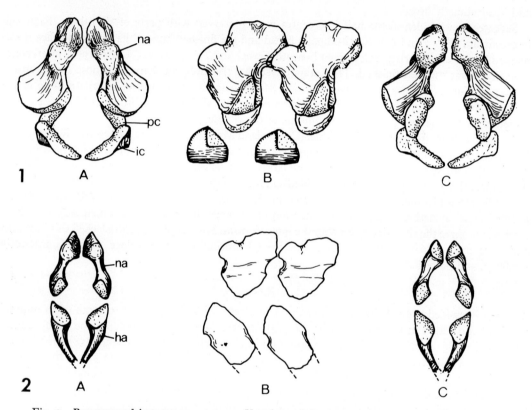

Fig. 3 *Rewana quadricuneata gen. et sp. nov.* Vertebrae of the type specimen. 1: presacral vertebra, 2: caudal vertebra. **A**, anterior view; **B**, left lateral view of two vertebrae; **C**, posterior view ($\times 1\frac{1}{3}$).

Thirty-eight fairly complete neural arch halves from pre- and postsacral regions were found, with sections from about twenty others (Fig. 3). It is not possible to distinguish axis, atlas or sacral vertebrae. The presacral neural arches are poorly ossified so that the neural spine is absent above the level of the postzygapophyses, and the transverse processes are short. No area of finished bone separates the latter from the ventral, cartilage finished facets for articulation with the pleurocentra. Some of the larger arches have a well defined ridge dorsolaterally on what represents the neural spine, indicating the position of the segmental boundary (Panchen, 1967). Several arches have a small canal which runs parallel to the midline above the neural canal and turns a little ventrally to exit from the canal area between the postzygapophysis and the articular area for the pleurocentrum. In the more posterior vertebrae the transverse processes and the postzygapophyses decrease and finally disappear, the neural spine is further reduced and the overall size of the element decreases to half the size of the anterior neural arches.

Seventeen haemal arch halves remain (Fig. 3). Above the haemal canal these are pointed dorsally, this area forming a crescentic, dorsomedially directed, cartilage-finished facet for articulation with the neural arch. Ventromedially the haemal canal is marked by a longitudinal ridge which is prominent anteriorly. The largest haemal arch bends sharply posteriorly below this ridge and its ventral shaft narrows away from the dorsal areas in contrast to the other arches where the ventral shafts are straight and longitudinally broad.

Seventeen easily identifiable half intercentra were preserved with parts of six others. Each half intercentrum (Figs 3 and 4) is longitudinally broad and the finished bone of its ventral surface is unmarked by the characteristic ridges seen beneath *Eryops* and *Parotosaurus* (Howie, 1970) intercentra. Dorsally it was cartilage finished. The parapophyses are triangular in shape like those found on the more posterior intercentra of *Parotosaurus* and *Mastodonsaurus* (von Huene, 1922). In none of the halves is there any sign that the median division is a break: all intercentra in which this area is preserved taper to a natural point near the middle of the column.

An intercentrum from another individual (Queensland Museum no. F6472) shows that *Rewana* must have reached at least twice the size of the type specimen while still retaining divided intercentra (Fig. 4: 1 A-D). This size would compare with the larger Triassic capitosaurs and metoposaurs and shows that, unless *Rewana* was neotenous, the divided condition of the vertebrae was not an immature feature.

The fourteen pleurocentra preserved varied in size. Larger ones showed a rounded lateral face of finished bone, sometimes with a small intrusion of cartilage-finished surface which may indicate a tendency to share the rib articulation with the intercentrum. Smaller pleurocentra lacked this intrusion and were more dorsoventrally elongate. All the pleurocentra were cartilage finished, anteriorly medially, and posteriorly.

Ribs

The poorly ossified state of the vertebral column contrasts with the apparent strength of an unusual proximal end found in some of the ribs. Several rib types are present including the usual temnospondyl rib (Fig. 4: 2F) found in a dorsal position in *Paracyclotosaurus* (Watson, 1958), and *Parotosaurus pronus* (Howie, 1970). These have a double rib head (reduced almost to single in *Rewana*), an uncinate process, and a 'knee' bend in the shaft. A rib with its distal end expanded in a similar manner to the one thought by Watson to support the pectoral girdle in *Paracyclotosaurus* is present in the referred specimen. The rib shown in Fig. 4: 2A is complete and could have been a cervical. Left and right sacral ribs of a similar type to those seen in *Benthosuchus sushkini* (Bystrow and Efremov, 1940) and *Parotosaurus pronus* (Howie, 1970), and thought by Howie to be indicative of the flattened pelvis of labyrinthodonts which are primarily aquatic, are present in *Rewana* (Fig. 4: 2G).

However, the remaining rib type is exceptional (Fig. 4: 2B-D). Each rib head is strongly built and has distinct capitular and tubercular areas. At a varying distance lateral to the rib head the dorsal border of the rib turns abruptly downwards forming an uncinate process at the beginning of a narrowed shaft. The expanded proximal end so formed has a complex system of ridges which run almost diagonally across its anterior surface, from the capitulum to the uncinate process. Associated with this ridge system in all specimens is a foramen which varies in position. In the rib illustrated in Fig. 4: 2C it enters the ventral border and exits from the mid-anterior surface, while in another rib it enters the posterior surface beneath the capitulum and exits in a similar position to that seen in Fig. 4: 2C. In both these cases a groove runs from the exit of the foramen between two of the bone ridges to a position immediately distal to the uncinate process. This groove alone is present in three of the ribs and leads in these cases to a clearly defined notch beside the uncinate process (Fig. 4: 2B-D). An equally prominent rib foramen was described by Panchen (1959) in *Peltobatrachus* but in this case the foramen is near the distal end of the rib which is expanded to support the dermal armour. The most likely position for these ribs in the body is immediately posterior to the sacrum. The rib shown in Fig. 4: 2E is apparently intermediate in structure between 'normal' dorsal ribs (Fig. 4: 2F) and these probable postsacral ribs.

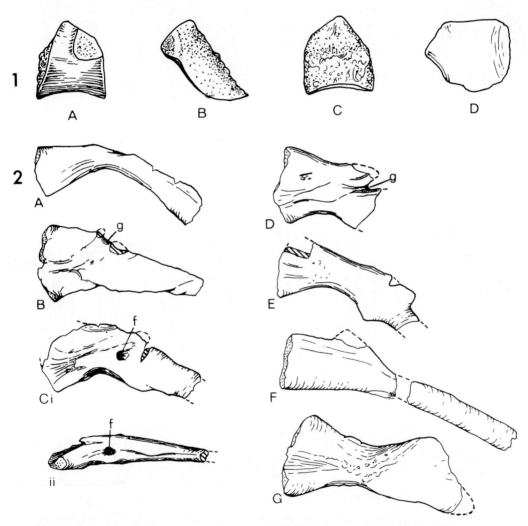

Fig. 4 *Rewana quadricuneata* gen. et sp. nov. 1: left half intercentrum (Queensland Museum no. F6472). **A**, lateral view; **B**, posterior view; **C**, median view; **D**, ventral view. 2: ribs of the type specimen. **A**, ? cervical; **B-D**, ? postsacrals; **E**, intermediate between **D** and **F**; **F**, dorsal; **G**, sacral. All ribs are drawn as if from the left hand side of the animal and all are shown in anterodorsal view except Cii which is an anteroventral view of Ci. ($\times 1\frac{1}{3}$).

Appendicular skeleton (Fig. 5)

Overall the appendicular skeleton is intermediate in development between that found in the well ossified terrestrial Permian eryopoids and the presumably more aquatic Middle Triassic capitosaurs. The skeleton will be compared with that seen in *Parotosaurus pronus* (Howie, 1970) as no fully illustrated skeletal description of the less specialised rhinesuchids exists. Thus the scapulocoracoid and humerus resemble those of *P. pronus* but are a little more ossified. The radius is more flattened posteriorly. Rather different, however, is the ulna: it is shorter and broader and the 'incisura radialis ulnae' is more prominent.

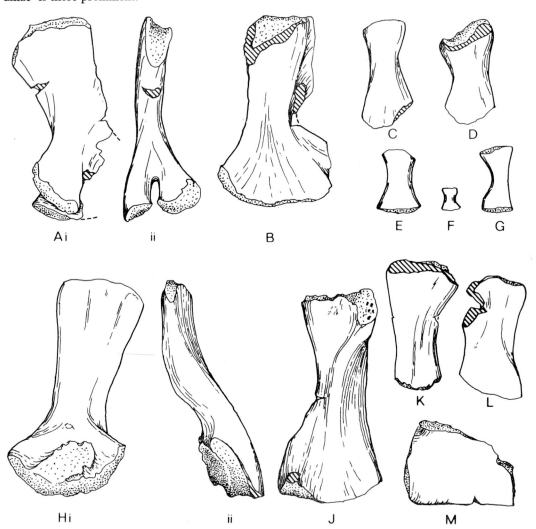

Fig. 5 *Rewana quadricuneata* gen. et sp. nov. Appendicular skeleton of the type specimen. **A**, right scapulocoracoid (drawn as a left) i, median view, ii, posterior view; **B**, right humerus in distal dorsal view; **C**, right radius in dorsal view; **D**, left ulna in dorsal view; **E**, ? phalanx; **F**, ? terminal phalanx; **G**, ? phalanx; **H**, right ilium (drawn as a left) i, lateral view, ii, posterior view; **J**, left femur in ventral view; **K**, left tibia in dorsal view; **L**, left fibula in dorsal view; **M**, left ilium in dorsal view ($\times 1\frac{1}{3}$).

In the pelvic girdle, the dorsal blade of the ilium is slim, leans posteriorly less than it does in *P. pronus*, and bears an additional knob of bone above the acetabulum.

Surprisingly the proximal and distal ends of the femur are a little less ossified than they are in *P. pronus*, and the fourth trochanter is barely present. Conversely the ventral adductor crest is strongly developed. The fibula is stout and lacks a characteristic posterodistal groove as found in *Benthosuchus* (Bystrow and Efremov, 1940) and *Buettneria* (Case, 1932) as well as in *P. pronus*.

Three kinds of smaller limb bones, thin (?) metapodials, shorter (?) phalanges, and a single much smaller (?) terminal phalanx were found.

TAXONOMIC POSITION OF *Rewana quadricuneata*

Among the more advanced rhachitomes characterised by an immovable basal articulation in the palate, a double occipital condyle and wide interpterygoid vacuities, only the metoposaurs, the brachyopids, and some of the trematosaurs are short faced. The anterior position of the eyes in these groups results in a correspondingly elongate postorbital region which is found also in the long snouted trematosaurs, the long snout in these animals being considered here as secondary.

Rewana is clearly excluded from the brachyopids by its shallow cheek region, unarched palate, the small size of its orbits and the probable presence of an otic notch and well developed tabular horns. It resembles them, however, in its broad flat parasphenoid and the presence of double tusk pits on the vomers, palatines and ectopterygoids.

Rewana has the broad parasphenoid found in metoposaurs and the shape and proportions of the subtemporal and interpterygoid vacuities, and the length of the pterygoid-parasphenoid suture, are common to both. However, in *Rewana* the skull is narrower antorbitally and broader postorbitally, its lateral borders are curved rather than straight as they are in metoposaurs, and its external nostrils are oval rather than rounded.

Its convex skull margins exclude *Rewana* from the conventional concept of the Trematosauridae. In addition, its pterygoid-parasphenoid suture is shorter than is usually found in that group and the cultriform process of its parasphenoid is much wider. The Rhytidosteidae, proposed by Cosgriff (1965) and included by Romer (1966) in the Superfamily Trematosauroidea, also have triangular skulls, and the peculiar character of the ornament "dominated by radiating ridges which bear prominent nodes at points of junction and bifurcation" (Cosgriff, 1965) is not apparent in *Rewana*. *Mahavisaurus* and possibly *Lyrosaurus* (Lehman, 1966) should probably be added to the Rhytidosteidae. *Rewana* differs from *Deltasaurus*, the Australian member of the Rhytidosteidae, in the shape and more anterior position of the orbits, the inclusion of the ectopterygoid in the margin of the interpterygoid vacuity, the smaller parasphenoid-pterygoid contact resulting in larger interpterygoid vacuities, and the position of the internal nostrils nearer to the skull margins. If the suture dividing the right maxilla from the lacrimal is real, then the presence of a lacrimal also excludes *Rewana* from the Rhytidosteidae, allying it more with the metoposaur-brachyopid line where the lacrimal enters the orbit, than with the trematosaur line where this rarely occurs. However, *Rewana* resembles both the trematosaurs and rhytidosteids in the smallness and position of its orbits, well forward near the lateral margins of the skull, and in its pointed snout. No other group shares these features. A character apparently exclusive to *Rewana* and *Deltasaurus* is the presence of denticles between the internal nostril and the maxilla.

The metoposaurs and brachyopids on one hand and the trematosaurs on the other must have arisen from an early line of rhachitomes and Romer (1964) suggested an origin among the less specialised rhinesuchids, but he qualified this in 1968, remarking that the long snouted trematosaurs may alternatively have sprung from the eryopoid or pre-eryopoid level. *Rewana* shares characters diagnostic of both the metoposaur-brachyopid and the trematosaur lines. But it could only be placed in a purely horizontal classification in the Rhinesuchoidea and should undoubtedly be separated ultimately from this group. It differs markedly from the rhinesuchids in the size and position of the orbits, the width and dorsoventral depression of the skull, the pointed snout, and the inclusion of ectopterygoid in the margin of the interpterygoid vacuity and the lacrimal in the orbit.

Until the dorsal skull roof, especially in the temporal and tabular regions, and the occiput are known *Rewana* has been left in an uncertain taxonomic position.

The unusual combination of characters found in *Rewana* does point to an alliance between the metoposaurs, brachyopids and trematosaurs, as suggested by Säve-Söderbergh (1936) but doubted by Romer (1957). Conversely, the marked separation of the rhinesuchids from *Rewana* indicates that the former three groups diverged from the rhachitomes before the rhinesuchid level.

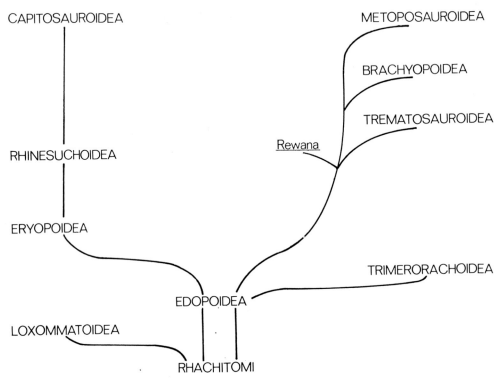

Fig. 6 Basic scheme for the division of the temnospondyls into long faced (left hand side) and short faced (right hand side) lines.

It is apparent when considering relationships within the temnospondyls that the distinction between long and short faced (or posterior- and anterior-eyed) types, recognised taxonomically in the later groups, was also present in the earlier eryopoids and edopoids. The temnospondyl line may have been divided almost from its origin into short-faced and long-faced lines. The scheme proposed in Fig. 6 leans more towards Romer's tree pattern than Watson's bush (Romer, 1968). The division may have started at a pre-edopoid level, but if not, a group possibly ancestral to both lines is the Dendrerpetontidae. Langston (1953) places this group in an ancestral position to his two divisions of the Edopidae. Chase (1965) noted that a trend already well established in the dendrerpetontids, characterised by withdrawal of the lacrimal from the orbit, barred them from ancestry of a later edopoid group, the trimerorachids; however, Carroll's more recent (1967) reconstruction of *Dendrerpeton* shows the lacrimal entering the anterolateral border of the orbit. Carroll considers that *Dendrerpeton* is the only edopoid sufficiently early and generalised to be ancestral to eryopids. It is possible that a labyrinthodont such as *Dendrerpeton* gave rise on one hand to long faced edopoids, and through these to long faced eryopoids (or all eryopoids) and thence to the rhinesuchid-benthosuchid-capitosaurid line culminating in *Cyclotosaurus*, and on the other hand to short faced edopoids leading to a group at the eryopoid level (early dissorophids ?) and thence to a stem from which the trematosaurs, brachyopids and metoposaurs branched. Genera such as *Eobrachyops* and *Dvinosaurus* would have been early deviants from this line, while *Rewana* belonged to a lateral branch not far removed from the main stem which culminated in the Upper Triassic metoposaurs. It may be premature to propose this division before adequate skull roof material of *Rewana* is available, and a detailed analysis of the changes necessary in modern concepts of temnospondyl phylogeny must wait until that time.

ACKNOWLEDGEMENTS

I would like to thank Alan Bartholomai particularly for drawing my attention to the fauna, for showing me the type locality, and for lending me material collected by the Queensland Museum. Dr Frank Talbot allowed me to join an Australian Museum field party in Queensland, and without the help of Alex Ritchie and Kingsley Gregg of his staff much of the field work would have been impossible. Ralph Jensen of the Bureau of Mineral Resources was instrumental in showing us new exposures in the Rewan and Mr R. McKinley kindly allowed us to work on 'Rewan' property. This work was supported by a Sydney University Postdoctoral Research Fellowship.

REFERENCES

BALME, B. E. 1969. The Permo-Triassic boundary in Australia. *Geological Society of Australia*. Special Publication No. 2., 99-112.
BARTHOLOMAI, A. and HOWIE, A. A. 1970. Vertebrate fauna from the Lower Trias of Australia. *Nature*, Lond., **225**. p. 1063 only.
BYSTROW, A. P. and EFREMOV, J. A. 1940. *Benthosuchus sushkini* Efr.—A labyrinthodont from the Eotriassic of Shargenga River. *Trúdy paleozool. Inst.*, **10**, 1-152.
CARROLL, R. L. 1967. Labyrinthodonts from the Joggins Formation. *J. Paleont.*, **41**, 111-142.
CASE, E. C. 1932. A collection of stegocephalians from Scurry County, Texas. *Contr. Mus. Paleont. Univ. Mich.*, **4**, 1-156.
—— 1935. Description of a collection of associated skeletons of *Trimerorachis*. *Contr. Mus. Paleont. Univ. Mich.*, **4**, 227-274.
CHASE, J. N. 1965. *Neldasaurus wrightae*, a new rhachitomous labyrinthodont, from the Texas Lower Permian. *Bull. Mus. comp. Zool. Harv.*, **133**, 155-225.
COLBERT, E. H. 1967. A new interpretation of *Austropelor*, a supposed Jurassic labyrinthodont amphibian from Queensland. *Mem. Qd. Mus.*, **15**, 35-41.
COSGRIFF, J. W. 1965. A new genus of Temnospondyli from the Triassic of Western Australia. *J. Proc. R. Soc. West. Aust.*, **48**, 65-90.
—— 1969. *Blinasaurus*, a brachyopid genus from Western Australia and New South Wales. *J. Proc. R. Soc. West. Aust.*, **52**, 65-88.
GADOW, H. 1933. *The evolution of the vertebral column*. Cambridge University Press.
HOWIE, A. A. 1970. A new capitosaurid labyrinthodont from East Africa. *Palaeontology*, **13**, 210-253.
HUENE, F. VON 1922. Beitrage zur Kenntnis dur Organisation einiger Stegocephalen der Schwabischen Trias. *Acta zool.*, Stockh. **3**, 395-460.
HUXLEY, T. H. 1859. On some amphibian and reptilian remains from South Africa and Australia. *Proc. geol. Soc.*, **15**, 642-658.
LANGSTON, W. JR. 1953. Permian amphibians from New Mexico. *Univ. Calif. Publs. geol Sci.*, **29**, 349-416.
LEHMAN, J. P. 1966. Nouveaux stegocephales de Madagascar. *Annls Paleont.*, **52**, 117-139.

LONGMAN, H. 1941. A Queensland fossil amphibian (*Austropelor*). *Mem. Qd Mus.*, **12**, 29-32.

MEYER, H. VON 1856. Reptilien aus der Steinkohlen-Formation in Deutschland. *Palaeontographica*, **6**, 59-219.

PANCHEN, A. L. 1959. A new armoured amphibian from the Upper Permian of East Africa. *Phil. Trans. R. Soc.*, B. **242**, 207-281.

—— 1967. The homologies of the labyrinthodont centrum. *Evolution*, **21**, 24-33.

PARRINGTON, F. R. 1948. Labyrinthodonts from South Africa. *Proc. zool. Soc. Lond.*, **118**, 426-445.

ROMER, A. S. 1947. Review of the Labyrinthodontia. *Bull. Mus. comp. Zool. Harv.*, **99**, 1-352.

—— 1964. Problems in early amphibian history. *J. Anim. Morph. Physiol.*, **11**, 1-20.

—— 1966. *Vertebrate Paleontology*. 3rd edition. Chicago University Press. Chicago.

—— 1968. *Notes and Comments on Vertebrate Paleontology*. Chicago University Press. Chicago.

SÄVE-SÖDERBERGH, G. 1936. On the morphology of Triassic stegocephalians from Spitzbergen, and the interpretation of the endocranium in the Labyrinthodontia. *K. svenska VetenskAkad. Handl.*, **16**, 1-181.

SMITH WOODWARD, A. 1909. On a new labyrinthodont from oil shale at Airly, New South Wales. *Rec. geol. Surv. N.S.W.*, **8**, 317-319.

STEEN, M. C. 1937. On *Acanthostoma vorax* Credner. *Proc. zool. Soc. Lond.*, B. **107**, 491-500.

STEPHENS, W. J. 1886. Note on a labyrinthodont fossil from Cockatoo Island, Port Jackson. *Proc. Linn. Soc. N.S.W.*, (2), **1**, 931-940.

—— 1887. On some additional labyrinthodont fossils from the Hawkesberry Sandstones of New South Wales. *Proc. Linn. Soc. N.S.W.*, (2) **1**, 1175-1192.

WATSON, D. M. S. 1956. The brachyopid labyrinthodonts. *Bull. Br. Mus. nat. Hist.*, **2**, 317-392.

—— 1958. A new labyrinthodont (*Paracyclotosaurus*) from the Upper Trias of New South Wales. *Bull. Br. Mus. nat. Hist.*, **3**, 233-264.

—— 1962. The evolution of the Labyrinthodontia. *Phil. Trans. R. Soc.*, B. **245**, 219-265.

WILLISTON, S. W. 1914. *Trimerorachis*, a Permian temnospondyl amphibian. *J. Geol.*, **23**, 246-255.

ABBREVIATIONS USED IN FIGS. 1-6

ect	ectopterygoid	*pal*	palatine
en	external nostril	*pc*	pleurocentrum
eo	exoccipital	*pmx*	premaxilla
f	foramen	*psp*	parasphenoid
g	groove	*pt*	pterygoid
ha	haemal arch	*q*	quadrate
ic	intercentrum	*qj*	quadratojugal
in	internal nostril	*sio*	infraorbital sulcus
j	jugal	*sso*	supraorbital sulcus
lac	lacrimal	*vo*	vomer
mx	maxilla	*vo vac*	vomerine vacuity
na	neural arch	*XII*	foramen for XIIth nerve

A. L. PANCHEN

The interrelationships of the earliest tetrapods

INTRODUCTION

The problem of the origin of reptiles is one that has preoccupied vertebrate palaeontologists for many years. In the early part of this century a powerful stimulus to discussion of the subject was administered by Watson's (1917) classification of pre-Jurassic tetrapods, followed by his discussion of the anatomy of the Lower Permian tetrapod *Seymouria* (Watson, 1919a), and his classic work on the origin and evolution of the labyrinthodont Amphibia (Watson, 1919b, 1926).

For many decades *Seymouria* occupied the position, in the speculations of evolutionists, of an almost perfect intermediate between primitive ('cotylosaur') reptiles and Palaeozoic labyrinthodont Amphibia. Broili who had named *Seymouria* described it as a reptile, Williston, in describing further material, agreed, and in Watson's classification it was representative of one of the three groups of cotylosaurs. Gradually, however, the emphasis shifted to the amphibian features of *Seymouria*. It was first pronounced an amphibian by Broom (1922) and Sushkin (1925) agreed, basing his conclusion largely on the evident labyrinthodont features of the allied *Kotlassia*.

In 1939 White gave a definitive account of *Seymouria* and, while concluding that it was a reptile, noted possible traces of the lateral line system on the skull roof. Later Špinar (1952) was able to describe growth stages in small related forms, from Czechoslovakia, which bore external gills when young and were thus undoubted Amphibia. Thus *Seymouria* and its allies are now generally accepted as amphibian and are placed within the labyrinthodonts as allies of the Carboniferous and early Permian anthracosaurs (Romer, 1947, 1966; Panchen, 1970). The relationship of *Seymouria* to reptiles is, however, in dispute.

Nobody disputes the reptiliomorph nature of the postcranial skeleton of *Seymouria* but the structure of the skull presents a number of anomalies.

In his 1917 classification Watson's three groups of cotylosaurs were the Seymouriamorpha, the Diadectomorpha and the Captorhinomorpha. Of these it is now generally agreed that only the Captorhinomorpha should be regarded as ancestral forms for post-Triassic and thus living reptiles (e.g. Romer, 1968). However, if *Seymouria* is at all closely related to the two remaining groups of cotylosaurs, it is with the Diadectomorpha that its affinities lie. For this reason there have been a number of attempts to suggest an alternative derivation for the captorhinomorphs. Westoll in 1942 suggested their relationship to, and possible descent from, the small non-labyrinthodont microsaur amphibians. The Microsauria were then relatively unknown and there was much confusion both ways between microsaurs and small early captorhinomorphs. This was largely resolved by Romer (1950), who rejected the microsaur ancestry of reptiles, as did Watson. Westoll's idea was, however, adopted by Olson (1947), who proposed a diphyletic grouping of reptiles, with seymouriamorphs, diadectomorphs and Chelonia as one ramus and with captorhinomorphs and all other reptiles forming the other, with presumed microsaur ancestry. Olson (1965, 1966) has subsequently rejected this arrangement but reaffirmed the close relationship of seymouriamorphs and at least *Diadectes* amongst the diadectomorphs.

Watson, while not accepting the microsaur origin of captorhinomorphs, nevertheless came to accept that derivation from the seymouriamorphs was unsatisfactory and thus suggested the origin of captorhinomorphs directly from the Carboniferous anthracosaurs (Watson, 1954). This, however, was more due to the assumption that all reptiles could ultimately be traced back to the embolomerous anthracosaurs, rather than to any exclusive resemblances between anthracosaurs and captorhinomorphs.

Thus *Seymouria* and its allies have become rather removed from the mainstream of reptile phylogeny and there are more specific reasons for dismissing them from consideration. Firstly the anatomy of the skull is specialised, notably in the structure of the ear region, in which the fenestra ovalis is situated at the end of a tubular extension of the otic capsule, floored by the parasphenoid. It is this character that has been used to ally *Seymouria* and *Diadectes*. The latter is a very aberrant tetrapod, even within the 'diadectomorphs', and both Olson and Romer (1964b) have suggested that it should be regarded as a seymouriamorph amphibian. The remaining 'diadectomorphs' are then separated as the suborder Procolophonia.

The other objection to *Seymouria* and its immediate allies is one of time. Most speculation on reptile origins has until recently used as its raw materials the rich tetrapod fauna from the Lower Permian of Texas and neighbouring states, of which *Seymouria* is a member. The earliest reptiles, however, are known from early in the Pennsylvanian or Upper Carboniferous (Carroll, 1964; Baird and Carroll, 1967) and it is to the Carboniferous that one must look for potential reptile ancestors. At best *Seymouria* could only be a very late relict form of the ancestral group.

Recent work, particularly that of Carroll, has immensely increased our knowledge of all those groups represented in the Carboniferous which should be considered in relation to reptile ancestry, as well as of the earliest known reptiles. As a result of restudy of very difficult material from the well-known Upper Carboniferous site at Nýřany, Czechoslovakia, Carroll (1969, 1970c) has proposed new candidates for relict reptilian ancestry in the genus *Gephyrostegus* and in a related form from the roughly contemporary shales of Linton, Ohio. *Gephyrostegus* is an anthracosaur in the broad sense in which Romer uses the term Anthracosauria, i.e. to include the seymouriamorphs, the anthracosaurs *s.s.* and their allies. It was, judging from its anatomy, a terrestrial form with a generally "reptiliomorph" post-cranial skeleton and a skull which lacked the specialisations of the ear region seen in *Seymouria*.

Thus both in anatomy and in time *Gephyrostegus* is a much better relict reptile ancestor than *Seymouria* and Carroll's views on reptile ancestry will probably gain well-deserved acceptance. There are, however, certain disquieting features in a theory of reptilian descent from an animal anatomically similar to *Gephyrostegus*.

The very history of speculation on the subject reviewed above suggests that a more generalised relation of *Seymouria* could not be expected to have any necessary skull resemblance to captorhinomorphs, as the whole orthodox story of reptilian ancestry has been built on the seymouriamorph-diadectomorph resemblance. It does not follow, however, that the microsaurs have any stronger claim. Several recent reviews (Gregory, 1965; Baird, 1965; Carroll and Baird, 1968) have reinforced Romer's rejection of microsaur ancestry of the reptiles, although the opposite view has been taken by Vaughn (1962).

My aim is to look at the anatomical problem afresh and as far as possible without prejudice. There is no alternative reptile ancestor and because of this I have used the present title. I want thus to suggest that the problem of the ancestry of reptiles is a part of the problem of the interrelationship of all early tetrapods and the mode of their origin from the fish level. I shall also try to base my conclusions as far as possible on Carboniferous tetrapods. The reptiles originated in the Upper Mississippian (late Lower or early Upper Carboniferous) at the very latest. From this period to the top of the Carboniferous covers a time span of some thirty or forty million years. Speculation about the origin of reptiles based on animals beyond that time span should, therefore, be automatically suspect.

THE SKULL IN EARLY 'REPTILIOMORPHS'

In a recent account of a number of important Carboniferous Amphibia, including *Gephyrostegus*, Brough and Brough (1967) suggested an unorthodox classification in which a number of early amphibian groups were placed together as Eoreptilia. These included anthracosaurs, seymouriamorphs (*s.l.* including *Gephyrostegus*) and the various groups of lepospondyl Amphibia. These animals were described as 'reptiliomorph', and while the grouping violates the unity of the labyrinthodonts by separating the anthracosaurs, gephyrostegids and seymouriamorphs (Batrachosauria) from the temnospondyls the concept is useful to characterise the batrachosaurs and microsaurs. In this sense a reptiliomorph postcranial skeleton agrees with that of true reptiles in having the pleurocentrum as the principal element in the vertebral centrum and in commonly being adapted for terrestrial locomotion.

There is now little doubt that the single centrum of microsaurs is a pleurocentrum (Carroll, 1968) a conclusion reinforced by Carroll's (1969d) recent description of what is essentially an embolomerous microsaur. The nature of the centrum in the other lepospondylous Amphibia (the adelogyrinids, the Nectridea and the Aïstopoda) is, however, still unresolved and their skulls are so aberrant that they have never been proposed for even relict reptile ancestry.

The skull in Carboniferous batrachosaurs may be characterised by that of *Gephyrostegus* and the British Coal Measure anthracosaur *Palaeoherpeton* ("*Palaeogyrinus*") (Panchen, 1964, 1970). Both are certainly labyrinthodont and, characteristically for batrachosaurs, have a well developed skull table in which the median paired parietals and postparietals are flanked on each side, from front to back, by intertemporal, supratemporal and tabular bones. The latter, again characteristically, has a suture with the parietal. Also, unlike all described temnospondyl labyrinthodonts, the supratemporal is kinetically joined to the squamosal of the cheek region and there is apparently some mobility between intertemporal and postorbital. The kinetic skull table is certainly a remnant of the condition found in rhipidistian fish but it can be assigned a function in these Amphibia (Watson, 1926; Panchen, 1964, 1970).

The condition of the middle ear region is highly characteristic of most labyrinthodonts. The tympanum was held in an otic 'notch', a rounded excavation bordered antero-ventrally by the squamosal and dorsally by the tabular and supratemporal. In lateral view the posterior border of the cheek extends postero-ventrally down to the jaw articulation.

The occiput of labyrinthodonts is also diagnostic. Powerful buttresses of the tabulars extend ventromedially in occipital view to contact the opisthotics. These in turn contact powerful pillars of the exoccipital on each side of the foramen magnum. The series of three bones, tabular—opisthotic—exoccipital, forms the principal brace of the back of the skull against vertical compression forces. That part of the opisthotic involved is known as the paroccipital process.

In all temnospondyls, including the Carboniferous loxommatids (Watson, 1926), as well as in seymouriamorphs and possibly *Gephyrostegus*, a posttemporal fossa is present on each side. It extends forward below the skull roof as an elongate pocket above the otic capsule. The fossae are absent in *Palaeoherpeton* but this is certainly a secondary condition confined to true anthracosaurs. Anthracosaurs are primitive on the other hand in retaining a supraoccipital ossification, missing in most temnospondyls and apparently in seymouriamorphs. As with the fossa, the presence of a supraoccipital ossification in *Gephyrostegus* is unconfirmed.

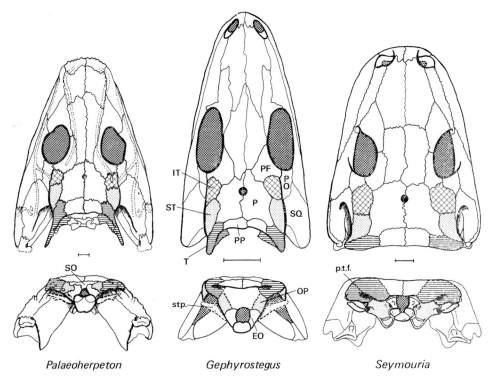

Fig. 1 Skull roof and occiput in batrachosaur labyrinthodonts. Scale lines in all figures represent 1 cm.
  EO, exoccipital; IT, intertemporal; OP, opisthotic (paroccipital process); P, parietal; PF, postfrontal; PO, postorbital; PP, postparietal; SO, supraoccipital; SQ, squamosal; ST, supratemporal; T, tabular; p.t.f., posttemporal fossa; stp., stapes. *Gephyrostegus* after Carroll, *Seymouria* after White.

Primitive Carboniferous reptiles are represented by the earliest undisputed reptile *Hylonomus* from Joggins, Nova Scotia (Carroll, 1964) and a later but still primitive form *Paleothyris* from Florence, Nova Scotia (Carroll, 1969a). Both are captorhinomorphs of the family Romeriidae and both antedate *Gephyrostegus* (late Westphalian D): *Hylonomus* is of Westphalian B time and roughly contemporary with the holotype of *Palaeoherpeton*. *Paleothyris* is Westphalian C, or early D.

Unfortunately the back of the skull table and occiput are not sufficiently preserved for accurate reconstruction in *Hylonomus*, but they have been described in *Paleothyris* and the early Permian (Moran formation) romeriid *Protorothyris* (Watson, 1954).

The skull table in romeriids is very reduced in comparison with batrachosaur labyrinthodonts. The intertemporal is unknown in true reptiles: its former territory in romeriids and most other primitive reptiles is occupied by a lappet of the parietal bone. Also, as in all early reptiles, the reduction of the back of the skull table has resulted in the supratemporal lying lateral to the tabular and forming the corner of the table. The tabular is small and the supratemporal itself reduced to a slender splint. Supratemporal, tabular and postparietal have a mainly occipital exposure in many early reptiles due to the embayment of the skull table for occipital musculature. In advanced captorhinomorphs such as the Permian *Captorhinus* the tabular has gone and the supratemporal is minute.

There is no sign of an otic notch in captorhinomorphs, but it is probable that most retained the loose junction between skull table and cheek region. This is certainly the case in the romeriids, and even in the advanced *Captorhinus* the sutural junction is still a weak one (Fox and Bowman, 1966).

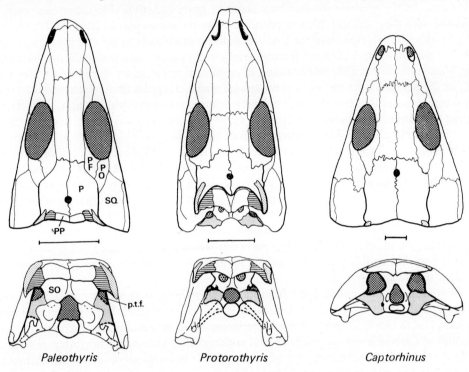

Fig. 2 Skull roof and occiput in captorhinomorph reptiles. Abbreviations and conventions as in Fig. 1. Scale lines represent 1 cm. *Paleothyris* after Carroll, *Protorothyris* after Watson, *Captorhinus* after Watson, Fox and Bowman.

The occiput of romeriids is totally different from that of labyrinthodonts. In primitive reptiles the main compression member on each side still incorporates the paroccipital process but this now extends more or less directly sideways bracing the squamosal of the cheek region and the occipital arch. The tabular, if present, is relatively superficial. The opisthotic is not known in *Hylonomus* or *Paleothyris* but is seen in other captorhinomorphs. Dorsally there is a large supraoccipital between the postparietals and the exoccipitals, these three bones forming a compression member against dorso-ventral compression. In many reptile groups there is a tendency for all the occipital ossifications of the braincase, including the opisthotic, to form a solid occipital plate.

It is also characteristic of some early reptiles to have a strongly ossified quadrate on each side to brace the skull against vertical compression in the plane of the jaw articulation. The feature of a vertical quadrate was assigned great importance by Watson in a cumulative series of papers in which he discussed the origin of reptiles (Watson, 1914a, 1914b, 1917, 1919a, 1951, 1954). He derived the reptilian condition from the labyrinthodont one in which the quadrate, even if well ossified, lies obliquely in the skull relative to the transverse plane in primitive forms.

Watson's views on the derivation of the 'cotylosaurs' from the labyrinthodont condition are well-known. In primitive labyrinthodonts the quadrate is oblique because of the dorsally placed otic notch, which extends anteriorly to both the back of the skull table and the very posteriorly placed quadrate condyles. According to Watson the condition seen in diadectomorphs was achieved by swinging forward the quadrate about its upper end thus enlarging the otic notch, while the captorhinomorph condition arose by the quadrate rotating back about its lower end, thus eliminating the notch.

Correlated with these changes Watson postulated that the stapes moved from the labyrinthodont position, in which its axis runs from the fenestra ovalis dorso-laterally to the otic notch, to a reptilian position in which the stapes is directed ventro-laterally in the general direction of the quadrate condyle.

Thus Watson postulated a diphyletic origin of reptiles, as noted in the introduction. He also concluded that the captorhinomorph line, and the related synapsid reptiles from which mammals evolved, lost the tympanum and thus that the mammalian tympanum is a neomorph. It is now, however, generally agreed that this is improbable (e.g. Hotton, 1959) and there is good reason to doubt the origin of the captorhinomorph skull from a labyrinthodont condition.

In 1927 Sushkin pointed out that the relatively massive ventrally directed stapes of early reptiles resembled the hyomandibular of rhipidistian fishes from which it was ultimately derived more than either resembled the usually slender dorsally directed rod in labyrinthodonts. The significance of this fact lay largely unappreciated until it was taken up by Parrington (1958), although Westoll (1943b) had suggested an intermediate position of the stapes as the primitive one. Gregory (1965) has also emphasised this and other divergent features between labyrinthodonts and captorhinomorphs.

Much of the ground covered so far in this essay was reviewed by Parrington in his important paper and it is appropriate at this point to acknowledge the stimulus of Dr. Parrington's work for the speculations that follow.

The labyrinthodont otic notch, apart from its dorsal position, is also characterised by extending forward to a point corresponding to the rhipidistian spiracular opening. It thus separates the squamosal and tabular, and the supratemporal may also be incorporated into its border. Parrington noted that most groups of Carboniferous Amphibia as well as the possibly Devonian ichthyostegids do not have the fully developed labyrinthodont type of otic notch. It is incipient or absent in ichthyostegids, adelogyrinids, microsaurs and early nectrideans. Thus among early tetrapods only the labyrinthodonts have a fully developed dorsal notch and in living amniotes the middle ear cavity develops as a relatively ventral diverticulum of the spiracular gill pouch.

Also relevant is the position of the chorda tympani (internal mandibular VII) nerve. In frogs, representing the extant Lissamphibia, the chorda tympani is post-tympanic in position: the middle ear is relatively degenerate in the urodeles and coecilians. In amniotes on the other hand the nerve is pre-tympanic. The nerve in Rhipidistia appears to have run in a tunnel along most of the length of the hyomandibular. Westoll (1943b) pointed out that from this neutral position a dorsal rotation of the distal end of the stapes (hyomandibular) with 'escape' of the nerve ventrally would lead to the frog condition while some degree of ventral rotation would lead to the amniote one.

Parrington suggested that the frog condition is paralleled by the labyrinthodont one and there is now stronger evidence of a labyrinthodont/frog relationship than there was when he wrote (Bolt, 1969). Such an interpretation, however, also involves the implication that the labyrinthodont and captorhinomorph stocks diverged either at a prototetrapod level as Parrington argued or, conceivably, at a rhipidistian one.

Hotton (1960) also restored the orientation of the chorda tympani in anthracosaurs as a froglike post-tympanic one but regarded this as a primitive, pre-amniote condition. He derived the reptile condition from it by a change in stapes-tympanum contact.

Parrington's discussion, however, was interpreted by Hotton and others as a rejection of anthracosaur or seymouriamorph affinity for the captorhinomorphs and possibly as being in favour of a microsaur origin of reptiles. This was not Parrington's intention and in 1959 he demonstrated that the high position of the otic notch and thus tympanum in labyrinthodonts could be regarded as due to differential growth of the cheek region in ontogeny. This is correlated with the lengthening of the quadrate ramus of the pterygoid, which results in a backward migration of the quadrate condyles.

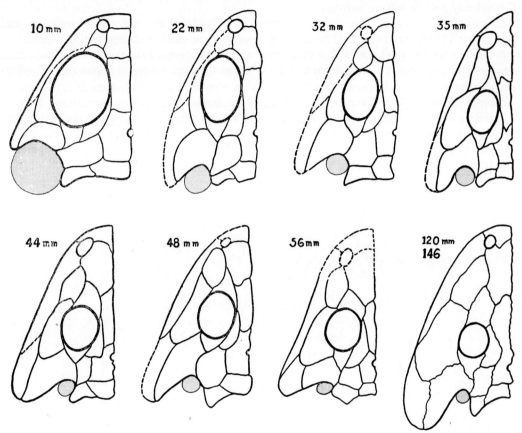

Fig. 3 Allometric growth in labyrinthodonts. Series after Romer with tympana restored.

He cited Romer's (1939) 'growth' series of skulls culminating in the mature labyrinthodont *Onchiodon*. While certainly not a true ontogenetic series (A. R. Milner—personal communication), it equally certainly demonstrates the allometric changes that would have occurred in the growth of a primitive labyrinthodont skull (Watson, 1963; Panchen, 1970). In very small or young labyrinthodonts the tympanum was relatively enormous, spanning the whole distance from skull table to quadrate. Thus the back of the cheek down to the level of the quadrate condyle was pretympanic. Parrington used the Carboniferous temnospondyl *Dendrerpeton* to illustrate a similar arrangement in a small labyrinthodont and the condition, as he notes, is equally true of the small seymouriamorph *Discosauriscus* described by Špinar (1952).

However, the small labyrinthodont differs in its overall skull morphology from an early captorhinomorph as already described. In *Dendrerpeton*, as restored by Steen (1934), the paroccipital process is more or less horizontal and thus analogous with that of captorhinomorphs, but it still contacts the tabular. In *Dendrerpeton*, and even more strikingly in the first member of Romer's series and in *Discosauriscus*, while the whole squamosal and even the quadrates are pretympanic the back of the skull table is not. Thus, if the captorhinomorph condition is derived from that in a small labyrinthodont it was not by any swing of the quadrate, as Watson suggested, but by reduction of the bones of the back of the skull table. Therefore, the tympanum of captorhinomorphs was not just post-squamosal but completely post-cranial, as in frogs.

The skull of microsaurs on superficial inspection appears to be of the captorhinomorph type and before Romer's (1950) review several so-called microsaurs were cited as demonstrating captorhinomorph features for the good reason that they were captorhinomorphs. Like the latter, all described microsaurs lack any trace of an otic notch and the skull has the same truncated appearance. Furthermore, the skulls of Carboniferous microsaurs such as *Asaphestera* from Joggins (Carroll, 1963), *Tuditanus* from Linton (Carroll and Baird, 1968) and *Microbrachis* from Nýřany (Brough and Brough, 1967) retain a kinetic junction between skull table and cheek region. The last genus also has a skull table in which the former territory of the intertemporal is occupied by a lappet of the parietal as in captorhinomorphs. This character is also present in *Hyloplesion* from the same site. A further character unites *Microbrachis* with the captorhinomorphs, the very reduced post-parietals.

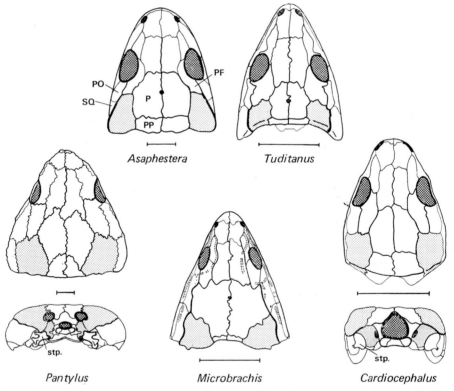

Fig. 4 Skull roof and occiput in microsaur Amphibia. Abbreviations and conventions as in Fig. 1. Scale lines represent 1 cm. *Asaphestera* after Carroll, *Tuditanus* after Carroll and Baird, *Pantylus* after Romer, *Microbrachis* after Steen and the Broughs, *Cardiocephalus* after Gregory, Peabody and Price.

In the gymnarthrid microsaurs of the Permian, including *Pantylus*, the kinetism has apparently sealed but in *Pantylus* at least this may be correlated with the durophagous dentition. Neither gymnarthrids nor the Tuditanidae, represented by *Tuditanus* and *Asaphestera*, have a parietal lappet: the region of the missing intertemporal is occupied by a backward extension of the postfrontal.

Most diagnostic of all microsaurs and divergent from captorhinomorphs is the presence of a large single bone on each side of the skull table occupying the territory of supratemporal and tabular. This is normally interpreted as the supratemporal, but where the occipital surface is known it is seen to have a considerable descending lamina in posterior view.

Little is known of the occiputs of microsaurs with the exception of the gymnarthrids. Romer (1969b) has given a very detailed account of the skull of *Pantylus*: the 'supratemporal' has exactly the occipital relationships of the labyrinthodont tabular, contacting the paroccipital process of the opisthotic and forming the dorso-lateral border of the posttemporal fossa. A supraoccipital is present in *Pantylus* and *Asaphestera* as in anthracosaurs and reptiles, but has apparently been lost in the Permian gymnarthrid *Cardiocephalus* (Gregory, Peabody and Price, 1956). In *Microbrachis* Brough and Brough reconstruct the occiput with remarkably small opisthotics bearing laterally directed paroccipital processes.

The condition of the stapes in microsaurs is anomalous and variable. In *Pantylus* Romer describes a complex element with a double footplate and a lateral branch diverging at less than half the length of the shaft. The main ramus again divides at the distal end to give a projection standing out anteriorly. Despite the complexities, however, the general orientation as restored is the fish–reptile one. (Romer also notes a similar stapes preserved in *Ostodolepis*, to be described by Dr Carroll, and in Dr Olson's gymnarthrid from Oklahoma.) However, other microsaurs are described with a much simpler stapes consisting of a large footplate with a variably reduced stem. In *Cardiocephalus* (Gregory, Peabody and Price, 1956) the stem is a mere tubercle: in *Tuditanus* it is virtually absent, but in *Microbrachis* there is a short stout shaft (Carroll and Baird, 1968).

In spite of all the variation it is improbable that any microsaur had a labyrinthodont orientation of the stapes. It is, therefore, reasonable to assume that what I have called the fish–reptile orientation of the stapes is both primitive and the normal one for early tetrapods and that in this respect the labyrinthodonts are aberrant. They, or at least the anthracosaurs and gephyrostegids as well as some temnospondyls, are also remarkable in retaining the three bones of the temporal series on each side of the skull table. All labyrinthodonts have the tabular-paroccipital brace in the occiput.

The characters of skull table, tympanum position and architecture of the occiput are critical in discussion of the interrelationships of early tetrapods. To these characters must be added the fate of the rhipidistian kinetic skull table, and those characters used by other authors to characterise early reptiles. An attempt will now be made to assess their significance.

ORIGIN OF THE TETRAPOD SKULL TABLE AND OCCIPUT

The origin of the critical features of the skull table and occiput can easily be traced back to the rhipidistian fish level. The most thoroughly described of all Rhipidistia is *Eusthenopteron foordi*, thanks to the detailed anatomical investigations of Jarvik. In considering the homologies of the dermal roofing elements with those of tetrapods, however, the correct terminology for the median paired skull bones and those of the temporal series is that demonstrated by Westoll (1938, 1943a). The truth of this has been confirmed by Parrington (1967) and Jarvik's nomenclature must be amended accordingly.

If the position of the dermal bones of the skull table is considered in relation to the braincase the origin of the labyrinthodont pattern of table and occiput immediately becomes clear. Each tabular bone ('supratemporal' of Jarvik) overlies one of the fossae bridgei of the otic region of the braincase.

The lateral wall of the fossa bridgei is formed by a thick dorso-lateral extension of the otic capsule which contacts the lateral part of the underside of the tabular throughout its length. This process of the otic capsule becomes the paroccipital process in tetrapods and the fossa bridgei is agreed to be the homologue of the posttemporal fossa.

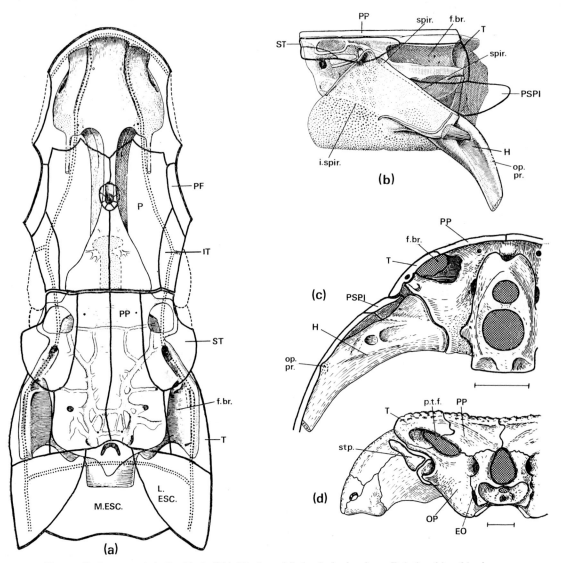

Fig. 5 Braincase and skull table in Rhipidistia and Labyrinthodontia. *a*, Relationship of braincase and median roofing bones in *Eusthenopteron*, lateral line canals—dotted lines; overlap surfaces of dermal bones—broken lines. *b*, *Eusthenopteron*, restoration of spiracular cavity and otic region in lateral view. *c*, Occipital view of *Eusthenopteron*, extrascapular bones removed, dermal bones of cheek in section. *d*, Occipital view of the seymouriamorph labyrinthodont *Kotlassia*.
H, hyomandibular; L.ESC, lateral extrascapular; M.ESC, median extrascapular; PSPl, prespiracular plate; f.br., fossa bridgei; i.spir., internal spiracular opening; spir., external spiracular opening; op.pr., opercular process. Other abbreviations as in Fig. 1. Scale lines represent 1 cm. *a-c*, modified after Jarvik; *d* after Bystrow.

The position of the otic notch of labyrinthodonts corresponds to the slit between the skull table and cheek region in *Eusthenopteron*, extending forward from the front of the prespiracular plate to the spiracular fenestra. This is bordered by the whole length of the tabular above and just reaches the supratemporal ('intertemporal'). According to Jarvik (1954) there was an open slit extending back as a continuation of the spiracular fenestra along this exact course.

It is generally accepted that the homologue of the tympanic process ('extrastapes') of tetrapods, which inserts into the tympanum, is the opercular process of Rhipidistia (Eaton, 1939; Romer, 1941; Westoll, 1943b). In *Eusthenopteron* and other rhipidistians this is situated much further distally on the hyomandibular than the spiracular slit but is probably in a similar morphological position to the extrastapes of captorhinomorphs and possibly microsaurs (as in Parrington's prototetrapod).

It seems unnecessary to postulate that the labyrinthodont tympanum or tympanic process was not the homologue of that of other early tetrapods. If the prototetrapod was a very small animal the tympanum would be relatively very large. It would thus span the whole distance from the spiracle to the quadrate and easily come into the appropriate relationship with the opercular process of the hyomandibular. Thomson (1966) has also noted this necessity for a large tympanum at this early tetrapod stage, but did not draw the conclusion that the most likely way in which this could be achieved would be for the prototetrapod to be a very small animal.

It is also important to note that in a small animal the hyomandibular or incipient stapes would be both absolutely and relatively smaller. This is because for closely related animals the relative size of the braincase to the whole skull is greater the smaller the animal: thus the relative size of the hyomandibular is diminished in the smaller animal by the closer approach of its otic articulation to the quadrate. An exact parallel is provided by Watson's (1931) similar argument concerning the size of skull and stapes in the ancestry of mammals. In both cases the small stapes would be a relatively more efficient conductor of sound waves.

It is the most important contention of this essay that the *phyletic lines leading to labyrinthodonts on the one hand and the captorhinomorphs on the other diverged from this small prototetrapod level.*

The principal characteristic of the labyrinthodont line was an increase in overall body size producing, with an increase in skull size, the characteristic features of primitive labyrinthodonts. Thus, as the skull enlarged, two factors would give the labyrinthodont configuration of the otic region. Firstly differential growth would result in the backward movement of the quadrates relative to the skull table and a relative reduction in the size of the tympanum. Secondly the tympanum would come to occupy a position as near as possible to the fenestra ovalis in the braincase to minimise the length and mass of the stapes: thus the distinctive dorsal otic notch and dorso-laterally directed stapes of labyrinthodonts would result.

Because the otic notch on each side was deeply incised into the posterior border of the skull roof a horizontal transverse compression member in the occipital plane could not evolve. The powerful paroccipital brace was therefore retained dorso-laterally. For this reason the tabular bone is literally the cornerstone of the labyrinthodont skull.

The reptile line on the other hand was not characterised by a size increase from the prototetrapod level. As a result the stapes retained essentially the configuration of the fish hyomandibular; there was no backward migration of the quadrates and the large tympanum occupied the whole area behind the essentially vertical posterior border of the cheek region.

With a post-squamosal tympanum and a vertical quadrate, together with a short ventro-laterally directed stapes, a horizontal brace was possible. Thus in the reptile line the paroccipital processes are horizontal, having rotated downwards, just as described by Watson, to achieve contact with the squamosal or, in heavily ossified skulls, with the dorsal pillar of the quadrate.

The orientation of the stapes is primitive in early reptiles, but like the occiput the form of the skull table is far removed from the more primitive labyrinthodont condition. In captorhinomorphs the intertemporal has gone to be replaced by a lappet of the parietal, and the back of the skull table, with loss of the importance of the tabular brace, has been reduced to the transverse level of the cheek region.

The reduction of the supratemporal and tabular is very far advanced in even the earliest known captorhinomorphs. The tabular is minute and has lost all contact with the paroccipital process and even the supratemporal is not structurally important. Significantly, however, earlier stages in the loss of the tabular brace are represented by other primitive reptiles. In these the tabular generally has a purely occipital exposure but retains the same morphological relationship to the neurocranial bones of the occiput as in the primitive pattern exemplified by anthracosaurs. With descent of the end of the paroccipital process the lateral end of the tabular is dragged down lateral to the posttemporal fossa. Finally this lateral part of the tabular is lost. A (very non-phyletic) series of pelycosaurs demonstrating how this probably happened is given by *Eothyris* (Watson, 1954), *Edaphosaurus* and *Ophiacodon*

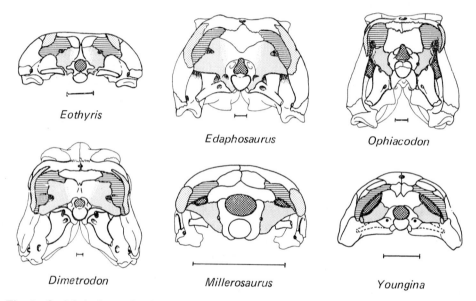

Fig. 6 Occipital views of pelycosaurs and "sauropsids". Conventions as in Fig. 1. Scale lines represent 1 cm. *Eothyris*, *Millerosaurus* and *Youngina* after Watson, others after Romer and Price.

(Romer and Price, 1940). The advanced sphenacodont *Dimetrodon* still retains the tabular-paraoccipital contact but the paroccipital processes are ventro-laterally directed, probably due to the great depth of the skull. Stages of the process can also be seen in 'sauropsids', e.g. in the possible diapsid relative *Millerosaurus*, and the eosuchian *Youngina* (Watson, 1957). If the nearest common ancestor of these and the pelycosaurs was found it would probably be described as a captorhinomorph but one more primitive and probably earlier than any yet known.

The position of the 'diadectomorphs' and other anomalous anapsids will be considered later.

The microsaurs have also achieved a levelling off of the back of the skull roof by reduction of the skull table, but in a different way. As with reptiles the intertemporals must have been lost at an early stage, conceivably before the reptile and microsaur lines diverged, but instead of shortening the table by moving the postparietals and tabulars forward mesial to the supratemporals, either the supratemporal or the tabular has been lost.

If the former was the case and the large table bone retained is the tabular, then the occiput of *Pantylus* is surprisingly primitive. Apart from the diagnostic occipital condyle and the loss of the kinetism the configuration would be close to that of the primitive tetrapod, except of course for the relative levels of the back of the table and the cheek. On this interpretation the shortening of the skull table would have been, as it were, by removal of a section in front of the tabular.

If, on the other hand it is the supratemporal which is retained one would expect the structure of the occiput seen in *Cardiocephalus*, in which loss of the tabular would also seem to have eliminated the posttemporal fossa and the characteristic tabular-opisthotic contact. The 'supratemporal' and opisthotic are shown as adjacent in the restoration by Gregory *et al.*, but do not appear to form a paroccipital process.

If the disputed bone is the supratemporal it would also imply, on the other hand, that in *Pantylus* it has taken over the function of the tabular and that the posttemporal fossae have extended forward from their primitive position. Interpretation of the bone as the tabular therefore seems rather more likely.

THE FATE OF THE RHIPIDISTIAN KINETISM

The kinetism of the braincase of rhipidistians is one of their most characteristic features. The oticooccipital region of the braincase and the overlying skull table articulate with the ethmoid region and the rest of the skull roof. The function of the kinetism has been most effectively discussed by Thomson (1967). In the evolution of the tetrapod skull this kinetism is eliminated to a variable degree. The events contributing to the consolidation of the skull in tetrapods are as follows:

1. Consolidation of the braincase by suture or fusion of the bones comprising the two regions;
2. Extension of the parasphenoid back under the otico-occipital region;
3. Locking of the transverse hinge at the front of the skull table by backward migration of the parietal-postparietal junction relative to the intertemporal-supratemporal one and the conversion of these mobile junctions into immobile transverse sutures.
4. Locking of the longitudinal junction between skull table and cheek region on each side by relative lateral movement between different parts of that junction, and the establishment of sutures.
5. Loss of mobility in the basicranial articulation between the braincase and palatoquadrate.

The sequence in which these events occur and the extent to which they occur is critical in understanding the interrelationships of the various groups of early tetrapods.

The consolidation of the braincase has probably occurred at some stage in the ancestry of all known tetrapods, but in many lines there is a secondary reduction of ossification in the braincase which is again divided into two regions. In this case, however, in contrast to the rhipidistian condition, the basisphenoid, bearing the basipterygoid process for articulation with the palatoquadrate, is incorporated into the posterior region of the braincase.

The next two features are also concerned with elimination of the relative movement between the two braincase regions. As Thomson has pointed out the scope of this relative movement diminishes with a relative increase of snout length in rhipidistians and some of these fish already had an akinetic neurocranium. It is therefore likely that the mobility within the braincase was lost at the inception of tetrapod evolution.

It is a mark of the primitiveness of *Ichthyostega* that there is still a line of division visible across the braincase in palatal view and that the parasphenoid is restricted to in front of that line (Jarvik, 1952). If the interpretation of the skull table and temporal region is correct this seems to have been compensated for by a remarkably precocious consolidation of the skull roof. Not only has the transverse hinge been locked but the longitudinal kinetic lines have been lost also. This has taken place by loss of the intertemporal and expansion inward of the postorbital to occupy at least part of its territory. Also the squamosal is sutured to supratemporal and tabular.

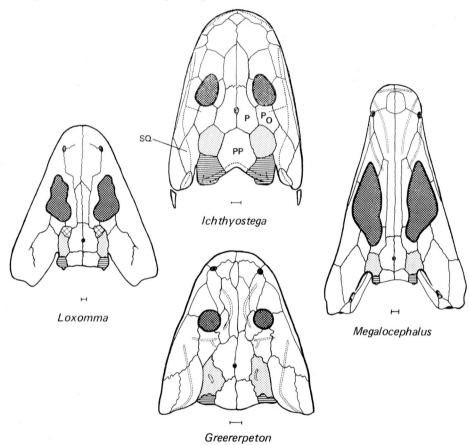

Fig. 7 Skull roof in primitive labyrinthodonts. Conventions and abbreviations as in Fig. 1. Scale lines represent 1 cm. *Ichthyostega* after Jarvik, *Greererpeton* after Romer, loxommatids after Watson.

This very early skull consolidation is surprising. It is doubly surprising to find that the earliest known temnospondyl labyrinthodont, apart from *Otocratia* (which is probably an ichthyostegid) and the loxommatids, has also lost the intertemporal. This is *Greererpeton* from the late Viséan of West Virginia (Romer, 1969a). The pattern of dermal bones behind the orbits is virtually identical to that of *Ichthyostega* but with reduced tabulars and postparietals. There may, however, have been some mobility between squamosal and supratemporal.

Significantly both *Ichthyostega* and *Greererpeton* lack a true otic notch. They probably diverged from the labyrinthodont stock before that diagnostic feature was established. This is also true of the colosteids from Linton to which, as Romer notes, *Greererpeton* could well be directly ancestral.

All remaining temnospondyls have also consolidated the skull roof and the pattern of bones in the temporal region suggests that even in those cases in which the intertemporal has been lost, consolidation preceded that loss. Apparently when the intertemporal is lost the postorbital moves into its territory as well as the postfrontal and supratemporal. The meeting of the latter two bones is in contrast to the condition in ichthyostegids and colosteids and is probably correlated with the relative shortening of the skull table which had already occurred in the more advanced forms. The process of loss of the intertemporal can be seen in the Carboniferous loxommatids (Watson, 1926, 1929).

Once the skull was consolidated in temnospondyls the usefulness of the basal articulation diminished and all the evolutionary trends which characterise the Permian and Triassic forms (Watson, 1919b, 1951) were set in train.

The early evolution of the batrachosaurs, microsaurs and reptiles, unlike that of the temnospondyls, is characterised by a retention of the lateral kinetism and with it the mobility of the basal articulation.

The batrachosaur line is further characterised by retention of the intertemporal. The probable reasons for this are the late retention both of the lateral kinetism and of a long skull table. The kinetism is retained in all known true anthracosaurs and in *Gephyrostegus* and when lost in seymouriamorphs there is remarkably little disturbance of the skull table pattern.

It is, therefore, unlikely that an early batrachosaur could be ancestral to the first reptiles. In the captorhinomorphs there was presumably a functional reason for the retention of the kinetism as there was in the anthracosaurs (Panchen, 1964, 1970). With the loss of the paroccipital brace to the table, characteristic of a small animal, it was therefore inevitable with uniform bone growth (Parrington, 1956) that the intertemporal should be succeeded by the parietal lappet. No encroachment of the postorbital through the kinetic line was possible.

The pelycosaurs are interesting in this respect. Sealing of the lateral kinetism seems to have taken place, possibly in direct correlation with the opening of the temporal vacuity, before reduction of the back of the table had gone very far. Nevertheless their derivation from the same stock as the captorhinomorphs, albeit with an early divergence, is emphasised by the presence of a parietal lappet.

Turning to the microsaurs, *Microbrachis* and to a lesser extent *Hyloplesion* stand out in strong contrast to all the others. The skull table is exceptionally long in *Microbrachis* and with the reduced post-parietals this must represent a secondary condition. With the retention of the kinetism one might, but for the huge 'supratemporal', imagine that it had undergone a similar evolution to that resulting in the Captorhinomorpha. However, *Microbrachis* bears traces of the lateral line system during its development and must therefore be an amphibian.

Romer in 1941 demonstrated a steady reduction in the posterior part of the skull table in the series rhipidistian-primitive amphibian-primitive reptile. One of the reasons given for this was the diminution in size of the acoustico-lateralis region of the brain with decreasing importance of the lateral line system. Possibly the retention of the 'supratemporal' in *Microbrachis* is correlated with the retention of the lateral lines.

In the other well known microsaurs, comprising the families Tuditanidae and Gymnarthridae the proportions of the skull and position of the orbits is much more reptilian. The huge 'supratemporal' in a relatively short table ensures that the territory of the intertemporal is occupied by junction of the 'supratemporal' and postfrontal. In the kinetic Tuditanidae the kinetism ensures that the postorbital is excluded from the skull table. In the probably descendant gymnarthrids the most important bone migration in the sealing of the kinetism is the lateral expansion of the 'supratemporal' so that its ventro-lateral border on the cheek continues the line of that of the postorbital.

OTHER DIAGNOSTIC CHARACTERS

One of the most striking features of any comparison between the major groups of early tetrapods is the degree to which it is complicated by parallel or convergent evolution of the postcranial skeleton. The parallels between seymouriamorphs and reptiles are well known: those between microsaurs and reptiles are no less striking and have been commented on recently by Brough and Brough (1967), Carroll and Baird (1968) and Carroll (1968). Carroll and Baird quote the apt analogy of the parallel evolution found in mammal-like reptiles.

Thus in early tetrapods the anthracosaurs, gephyrostegids, seymouriamorphs, very occasionally temnospondyls (e.g. *Doleserpeton* Bolt, 1969), plagiosaurs (Panchen, 1959), diadectids, captorhinomorphs, pelycosaurs and microsaurs all have vertebrae in which the pleurocentrum is dominant. It is also possible that in the other lepospondyls the single holospondylous centrum is the homologue of the pleurocentrum. This is known to be the case in the microsaurs, as intercentra with haemal arches are known in the tail region of *Pantylus* (Carroll, 1968). There is also parallelism in the development of dome-shaped neural arches, which occur in *Pantylus*, *Seymouria* and *Diadectes* amongst others. The lack of trunk intercentra in microsaurs can, therefore, only be ascribed to precocious development of a reptilian feature, as was noted by Brough and Brough.

Another striking feature is the development of a reptilian astragalus and calcaneum in the microsaurs *Tuditanus*, *Pantylus* and *Ricnodon* (Carroll and Baird, 1968). In the case of the batrachosaurs a morphological series in the development of the reptilian condition can be seen with a primitive condition represented by the anthracosaur *Archeria* (Romer, 1957) and an intermediate stage in *Gephyrostegus*; *Seymouria*, however, is primitive.

Again the reptilian phalangeal formula of 2, 3, 4, 5, 4 in the pes is found not only in primitive reptiles but in *Tuditanus* and, with the addition of an extra phalanx in the fifth digit, in *Archeria* and possibly *Gephyrostegus*. The manus of microsaurs is not known to have more than four digits. Four are known in *Tuditanus* and *Pantylus* with a count of 2, 3, 4, 3, and three in *Microbrachis* and *Hyloplesion*. As Carroll and Baird note, the discovery of a microsaur with five digits would be no surprise. In the batrachosaurs five digits are known in the manus of *Archeria* and *Gephyrostegus*.

Thus the only possible conclusion is that in the postcranial skeleton of tetrapods the 'reptiliomorph' pattern is normal and conceivably primitive for tetrapods. In the case of the vertebrae a gastrocentrous vertebra with a tendency to reduction or even loss of the intercentrum is normal and in this view only the labyrinthodont types are aberrant. Within the labyrinthodonts the embolomerous vertebra is easily derived from the type seen in *Gephyrostegus* and only the rhachitomous vertebra of temnospondyls is widely divergent from the normal type. Now that it is difficult to accept *Pholidogaster* as demonstrating the derivation of the embolomerous vertebra from the rhachitomous pattern as Romer (1964a) postulated (Panchen, 1970), the temnospondyls stand alone, as they do in the reduced pattern of the pes.

On all the criteria noted, therefore, both the batrachosaurs (particularly *Gephyrostegus*) and the microsaurs (particularly *Tuditanus* and *Pantylus*) are products of evolutionary lines closely paralleling that of reptiles, just as Brough and Brough concluded, but both probably diverged from the line leading to the reptiles at a very early stage of tetrapod evolution. Reptiliomorph femora, awaiting description by Dr Baird, are now known from the Tournaisian at the bottom of the Lower Carboniferous and may conceivably predate the ichthyostegids (Westoll, 1940).

Thus it is necessary to ask if there are any other criteria bearing on the relationship of either microsaurs or batrachosaurs to reptiles. In an excellent discussion on microsaur affinities taking into account all previous work, Carroll and Baird (1968) refine significant differences of microsaurs from reptiles down to five:

1. The presence of one large bone (supratemporal or tabular) in the temporal series.
2. The absence of a transverse flange of the pterygoid.
3. The structure and relationship of the occipital condyle and the atlas-axis complex.
4. The absence of trunk intercentra.
5. The structure of the scales.

Of these, numbers 1 and 3 are important diagnostic features and are the only characters which debar them from a position of relict reptilian ancestry. Number 2 is a primitive feature. The transverse flange of the pterygoid, often bearing small teeth or denticles, is very characteristic of early reptiles and may come to occupy the same position in deciding reptile status as the squamosal-dentary joint has as an indicator of mammalian status. It is almost certainly related to the differentiation of the anterior pterygoideus muscle (Fox, 1964). Number 4, as discussed above, is indicative of a more advanced 'reptiliomorph' condition in microsaurs and is thus no bar to relationship and number 5, although microsaur scales are distinctive, does not seem a weighty difference.

From the present discussion it is now possible to draw up a similar list differentiating batrachosaurs, or, more specifically, *Gephyrostegus* from reptiles. Carroll (1970) has produced a list of characters differentiating reptiles from seymouriamorphs (*s.l.*) all of which he regarded as features in which the reptiles have advanced from the batrachosaur condition. In some cases the contrasting character is a primitive tetrapod one but others, if the present thesis is accepted, debar the batrachosaurs from even relict reptilian ancestry. Carroll's characters for reptiles are:

(1) Loss of intertemporal bone.
(2) Reduction in size of supratemporal, tabular and postparietal.
(3) Loss of palatine fangs and development of transverse flange on pterygoid.
(4) Loss of labyrinthine infolding of enamel.
(5) Absence of otic notch.
(6) Development of specialised atlas-axis complex.
(7) Reduction in relative size of trunk and caudal intercentra.
(8) Development of astragalus.
(9) Wheat-shaped ventral scales; absence of ossified dorsal scales.

Of these, numbers (3), (4), (6), (7) and (8) are simple characters whose opposites represent a primitive condition. It should be noted that the microsaurs have reached the reptilian condition in part of (3) and in (4), (7) and (8). Microsaurs have also lost the intertemporal (1), but characters (1) and (2) represent in the batrachosaur condition features in which *Gephyrostegus* and its allies show no trend whatsoever towards a reptilian condition. This persistent primitiveness, together with the retention of the tabular-paroccipital brace, is almost as strong an indicator of lack of affinity as is the characteristic otic notch and stapes of labyrinthodonts (5). Finally number (9), as with the microsaurs, represents a diagnostic but not vitally important distinction.

A little more needs to be said on the axis-atlas complex and occipital condyle. Those of the microsaurs are certainly specialised. The first cervical vertebra of at least *Pantylus*, as described by Carroll and Baird, appears to be a fused atlas and axis. Also the occipital condyle, in *Lysorophus* and *Palaeomolgophis* as well as typical microsaurs, is concave and forms the socket of a ball-and-socket joint with the 'atlas'. In typical microsaurs the condyle is also widened with a roughly equal contribution from basioccipital and each exoccipital in its make up.

The whole arrangement is in very strong contrast to the reptilian condition with a convex spherical condyle and separate proatlas, atlas and axis. It is also certain that neither condition could give rise to the other and probable that, bearing the Viséan *Palaeomolgophis* in mind, the microsaur condition diverged at an early Mississippian stage. If, however, one looks at the condition of *Gephyrostegus* and its allies the picture is less clear.

The condyle in *Palaeoherpeton* and other anthracosaurs is concave like that of microsaurs but is circular in outline. The axis-atlas complex is unknown, but, as the intercentra of anthracosaurs fit as the ball in the socket of the pleurocentrum, it is not inconceivable that the atlas intercentrum has a similar relationship to the condyle. Turning to *Gephyrostegus* the form of the condyle itself is not certainly known and was evidently rather poorly ossified. The atlas-axis complex is primitive but could easily give rise to the reptilian condition.

REPTILIOMORPHS INCERTAE SEDIS

So far the discussion has been confined to forms whose taxonomic position is generally agreed whatever the disagreement as to the interrelationships of the major taxa to which they belong. There are, however, several primitive tetrapods whose reptilian status is in doubt. As they might appear to weaken the present argument it would be disingenuous to ignore them. They are the limnoscelids, the diadectids and *Solenodonsaurus*.

Limnoscelis from low in the Permian of New Mexico is the best known limnoscelid (Romer, 1946) but three rather similar Lower Permian genera have also been described (Romer, 1952; Lewis and Vaughn, 1965; Langston, 1965). Perhaps more important, Carroll (1967) has attributed to the Limnoscelidae a contemporary of *Paleothyris* from the Middle Pennyslvanian (Westphalian C or D) of Florence, Nova Scotia and a specimen of Westphalian A age also [from Nova Scotia (Baird and Carroll, 1967). The former is attributed almost entirely on postcranial remains, but the latter, *Romeriscus*, includes a skull.

Briefly, *Limnoscelis* is a massive apparently primitive reptile conventionally placed in the captorhinomorphs but not by any means like any other captorhinomorph. The skull is reptilian in having a transverse flange to the pterygoid, a truncate occipital surface with supratemporals in part flanking the tabulars, no intertemporal, no palatine tusks and no otic notch.

The occiput, as most workers have agreed, is primitive and morphologically intermediate between that of labyrinthodonts and reptiles. The teeth are also labyrinthodont. In the reptiliomorph postcranial skeleton the astragalus is not developed.

On all these features therefore *Limnoscelis* is an ideal very primitive reptile as much on the present theory as on rival ones. However, there are two other important features of the skull which may be variously interpreted: there is a kinetic junction between the supratemporal and squamosal, and the intertemporal is replaced, not by a parietal lappet, but by a process of the postorbital.

In his account of *Limnoscelis* Romer postulated that the kinetic line was formed by the closure of a labyrinthodont type of otic notch, but this was on the assumption that the otic notch of anthracosaurs was a 'fissure' effectively formed by widening of the back of the kinetic junction. This is not the case and there is no need to regard the loose junction of supratemporal and squamosal in *Limnoscelis* as other than a remnant of the original kinetism, just as in captorhinomorphs (Panchen, 1964). Romer's account presupposes the correctness of Watson's theory of the origin of the captorhinomorph skull, but without postulating the loss of the tympanum.

The condition of the postorbital in *Limnoscelis* is more puzzling and is in striking contrast to the captorhinomorph condition. One must suppose that there was an early sealing of the intertemporal-postorbital junction, possibly due to the large size of the ancestral skull, followed by loss of the intertemporal and its replacement by the postorbital.

Romeriscus from Westphalian A is attributed to the Limnoscelidae on the presence of a captorhinomorph configuration of the atlas-axis, the expansion of the rib-heads (a limnoscelid feature) and the restored form of the very incomplete skull. Baird and Carroll restore a primitive skull table in which there is some paralleling of the tabular by the supratemporal. However, a postorbital is restored with considerable doubt in the limnoscelid position and, more controversially, an otic notch extending right forward to the postorbital, but the latter "feature cannot, however, be established conclusively."

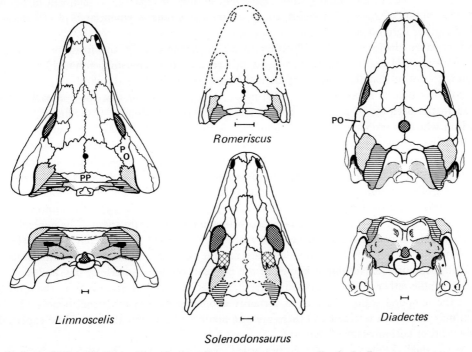

Fig. 8 Skull roof and occiput in 'reptiliomorphs' *incertae sedis*. Scale lines represent 1 cm. *Limnoscelis* after Romer; *Romeriscus* data after Baird and Carroll; *Solenodonsaurus* type restored as a batrachosaur labyrinthodont, data after Carroll; *Diadectes* after Olson and Watson.

It would certainly not be surprising if a limnoscelid were to be found as early or earlier than *Romeriscus*, but I do not think the evidence of the affinities of *Romeriscus* itself are very strong. Nothing is known of the palate to check its reptilian status and the postorbital is very doubtful. The axis-atlas is of reptilian type, but so is that of *Gephyrostegus*, and the remaining vertebrae are primitive with well-developed intercentra. The condition of the ribs is paralleled by *Seymouria*. Thus, if an otic notch really is present interpretation as a gephyrostegid would not be improbable, and the evidence that limnoscelids ever had an otic notch is thus very slight.

It is difficult to add anything to the many reviews and discussions of the status of *Diadectes*. The most recent are by Watson (1954, 1957), Parrington (1958, 1962), Olson (1965, 1966), Romer (1964b) and Carroll (1969a), The diadectids are a uniform highly aberrant group extending from the Stephanian to the Clear Fork, Lower Permian. *Diadectes* is by far the best known.

It lacks a transverse flange of the pterygoid but this is likely to be correlated with the probably herbivorous diet. Recent practice (Romer, 1964b, 1966) has been to regard it as a seymouriamorph. However, as pointed out by Carroll, the otic notch of *Diadectes* is essentially supported by the massive columnar quadrate and in this respect is analogous to the secondary otic notch found in early diapsids and in the millerettids (Parrington, 1958). In his account of *Limnoscelis* Romer showed the hypothetical derivation of the otic notch region of *Diadectes* from either an anthracosaur condition or from that in *Limnoscelis*. Both are still possible but the morphological gap in both cases is so great that not much more than that can be said. The detailed comparison by Olson between *Seymouria* and *Diadectes* slightly enhances the probability of anthracosaur-seymouriamorph descent, but as Carroll points out the specialisations of the otic region of *Diadectes* can only be regarded as incipient in *Seymouria* at best. Description of *Tseajaia* (Vaughn, 1964), claimed as a relict seymouriamorph-diadectid link, may shed further light on the problem.

The remaining 'diadectomorphs' (Procolophonia) do not appear in the fossil record until the Middle Permian and are outside the scope of this review. It is amusing, although probably not significant, to notice that the pattern of dermal bones of the back of the skull of procolophonids closely parallels that of microsaurs (e.g. Parrington, 1962, fig. 2) and that the comparison can be extended to the occiputs of prolocolophonids and *Pantylus* (e.g. Romer, 1956, fig. 47). A similar comparison which is probably not more significant is that between the skull roof of the microsaur *Hyloplesion* (Baird, 1965, fig. 5E) and the primitive pelycosaur *Oedaleops* (Langston, 1965, fig. 5).

The holotype of *Solenodonsaurus janenschi* from Nýřany comprises the skull and anterior skeleton (and their counterparts) of a reptiliomorph of rather less than two thirds of the size of *Limnoscelis*, or twice the size of *Gephyrostegus*. It is slightly larger than *Seymouria*. Size has been emphasised because the proportions of the skull are rather surprising. As restored by Carroll (1970) the impression is of a much larger specimen with small widely-spaced orbits and an elongate snout. The general outline, particularly in the unrestored skull, suggests very strongly the skull of *Anthracosaurus russelli* (Panchen, 1970) but the teeth, possibly labyrinthodont, are far too numerous to allow for membership of the family Anthracosauridae. Nothing is known of the palate or occiput. Anteriorly the skull is of gephyrostegid or reptile rather than anthracosaur pattern, in that the lacrimal extends from the orbit to the naris. In the temporal region there is no certainty about the pattern of dermal bones. The lateral kinetism may have been retained and there are otic notches of the type that would be expected in a gephyrostegid or anthracosaur of that size.

The postcranial skeleton is, as far as preserved, wholly reptilian. The vertebrae have massive pleurocentra with fused neural arches, which are not, however, swollen in the *Seymouria* manner, and the humerus is of reptilian type.

On its own the holotype of *Solenodonsaurus janenschi* would be *incertae sedis,* the combination of labyrinthodont and reptile characters perhaps suggesting a gephyrostegid whose relatively large size was correlated with the unusually reptilian vertebrae. However, Carroll (1969c and 1970, but not 1969b) has associated with it a second specimen attributed to the same species by Pearson (1924). As in the type only the preorbital part of the skull is at all well preserved. The right parietal is, however, in place and suggests a romeriid orientation of the supratemporal: however, tabulars, postparietals and supratemporals are all missing, and an intertemporal might or might not have been present in the intact skull. The fragmentary remains of the palate are restored in romeriid fashion by Carroll with a pterygoid flange. An oblique contour to the back of the cheek and thus a possible otic notch is restored on the basis of reconstructed jaw and reconstructed post-parietal length of the skull.

The vertebrae are either poorly ossified or poorly preserved although one large pleurocentrum is apparently preserved. This condition Carroll attributes to immaturity: Pearson's specimen, as restored, is considerably less than half the size of the type.

I cannot see that the anatomical information available in either specimen or the resemblance between them is sufficiently strong for attribution to the same genus. Pearson's specimen might cautiously be referred to the captorhinomorphs: the restored slope of the cheek is very similar to that also restored by Carroll in the notchless *Paleothyris*. As Carroll notes, the only possible close relation of the holotype is *Tseajaia*, which is, however, a considerably larger animal with swollen neural arches.

SUMMARY OF CONCLUSIONS

1. The rhipidistian-tetrapod transition was characterised by size reduction, so that the prototetrapod was a relatively small animal in which the neomorph tympanum extended from the skull table to the quadrate, and the stapes (hyomandibular) was relatively small.
2. The labyrinthodonts diverged from all other tetrapods in a relatively sudden increase in size from the prototetrapod level. This was correlated with an antero-dorsal retreat of the (relatively) small tympanum to the position of the fish spiracle, in order to retain a light effective stapes in a large skull. As a necessary correlate of the labyrinthodont otic notch the tabular-paroccipital brace is retained in all labyrinthodonts including the Batrachosauria (anthracosaurs, gephyrostegids and seymouriamorphs).
3. In all other primitive tetrapods (including early reptiles and microsaurs) the tympanum, if present, retained its position behind the cheek and eventually was post-cranial in position. Reduction of the posterior skull table to a variable degree resulted in a truncated occiput and freed the paroccipital processes to rotate ventrolaterally. Large forms (*Limnoscelis*, some pelycosaurs) must have already evolved irreversibly in this direction before their ancestors enlarged.
4. In the labyrinthodont line the temnospondyls are characterised by early sealing of the lateral kinetic junction on each side of the skull table, in the case of *Ichthyostega* before all traces of the transverse kinetic articulation had gone. The ichthyostegids, *Greererpeton* and the colosteids may never have developed a fully labyrinthodont type of otic notch.
5. The batrachosaur pattern of the skull table, with intertemporal, supratemporal and tabular retained and well developed, is explained by retention of the lateral kinetic junction. In the latter they parallel most captorhinomorphs and Carboniferous microsaurs.
6. In early reptiles the kinetism is retained after the reduction of the temporal series with the reduction of the tabular-paroccipital brace. Sealing of the lateral junction in pelycosaurs, possibly correlated with the establishment of the temporal vacuities, took place at a relatively primitive stage.
7. The pattern of the microsaur skull table is conditioned by initial retention of the kinetism and retention and expansion of the 'supratemporal' (in fact probably the tabular). Reduction of the skull table brings the 'supratemporal' in contact with postfrontal and postorbital on loss of the intertemporal and 'true' supratemporal. In *Microbrachis et al.*, however, a secondary expansion of the skull table, correlated with a well-developed lateralis system, gives a parietal lappet paralleling captorhinomorphs.
8. Only the temnospondyl labyrinthodonts amongst early tetrapods do not show 'reptiliomorph' features in the postcranial skeleton. Microsaurs, batrachosaurs and early reptiles also all show parallel trends correlated with terrestrial locomotion. Apart from the skull characters discussed only the atlas-axis complex significantly distinguishes microsaurs from the other two, and likewise only the pterygoid flange distinguishes reptiles.
9. The evidence that a transverse flange was ever developed in a group with labyrinthodont ancestors, i.e. ancestors with a labyrinthodont otic notch (*Solenodonsaurus* or limnoscelids), is far too slight to refute the conclusion that labyrinthodonts are excluded, by the anatomy and evolution of their skull table and occiput, from being the group ancestral to reptiles.

ACKNOWLEDGEMENTS

The immediate stimulus for the development of the ideas embodied in this review was a vigorous correspondence with Dr C. B. Cox on the relationships of early tetrapods: it gives me pleasure to acknowledge this and the help of my wife in typing the manuscript.

REFERENCES

BAIRD, D. 1965. Paleozoic lepospondyl amphibians. *Am. Zool.*, **5**, 287-294.

BAIRD, D. and CARROLL, R. L. 1967. *Romeriscus*, the oldest known reptile. *Science, N.Y.*, **157**, 56-59.

BOLT, J. R. 1969. Lissamphibian origins: possible protolissamphibian from the Lower Permian of Oklahoma. *Science, N.Y.*, **166**, 888-891.

BROOM, R. 1922. On the persistence of the mesopterygoid in certain reptilian skulls. *Proc. zool. Soc. Lond.*, (for **1922**), 455-460.

BROUGH, M. C. and BROUGH, J. 1967. Studies on early tetrapods. I. The lower Carboniferous microsaurs. II. *Microbrachis*, the type microsaur. III. The genus *Gephyrostegus*. *Phil. Trans. R. Soc.*, (B) **252**, 107-165.

BYSTROW, A. P. 1944. *Kotlassia prima* Amalitzky. *Bull. geol. Soc. Am.*, **55**, 379-416.

CARROLL, R. L. 1963. A microsaur from the Pennsylvanian of Joggins, Nova Scotia. *Nat. Hist. Pap. natn. Mus. Can.*, No. 22, 1-13.

—— 1964. The earliest reptiles. *J. Linn. Soc. (Zool.)*, **45**, 61-83.

—— 1967. A limnoscelid reptile from the Middle Pennsylvanian. *J. Paleont.*, **41**, 1256-1261.

—— 1968. The postcranial skeleton of the Permian microsaur *Pantylus*. *Can. J. Zool.*, **46**, 1175-1192.

—— 1969a. A Middle Pennsylvanian captorhinomorph, and the interrelationships of primitive reptiles. *J. Paleont.*, **43**, 151-170.

—— 1969b. The origin of reptiles. In *The biology of the Reptilia*, edited by Gans, C., Bellairs, A. d'A., Parsons, T. S. 1-44. London: Academic Press.

—— 1969c. Problems of the origin of reptiles. *Biol. Rev.*, **44**, 393-432.

—— 1969d. A new family of Carboniferous amphibians. *Palaeontology*, **12**, 537-548.

—— 1970. The ancestry of reptiles. *Phil. Trans. R. Soc.*, (B) **257**, 267-308.

CARROLL, R. L. and BAIRD, D. 1968. The Carboniferous amphibian *Tuditanus* [*Eosauravus*] and the distinction between microsaurs and reptiles. *Amer. Mus. Novit.*, No. 2337, 1-50.

EATON, T. H. 1939. The crossopterygian hyomandibular and the tetrapod stapes. *J. Wash. Acad. Sci.*, **29**, 109-117.

FOX, R. C. 1964. The adductor muscles of the jaw in some primitive reptiles. *Publ. Mus. nat. Hist. Univ. Kans.*, **12**, 657-680.

FOX, R. C. and BOWMAN, M. C. 1966. Osteology and relationships of *Captorhinus aguti* (Cope) (Reptilia: Captorhinomorpha). *Paleont. Contr. Univ. Kans.*, Art. **11**, 1-79.

GREGORY, J. T. 1965. Microsaurs and the origin of captorhinomorph reptiles. *Am. Zool.*, **5**, 277-286.

GREGORY, J. T., PEABODY, F. E. and PRICE, L. I. 1956. Revision of the Gymnarthridae, American Permian microsaurs. *Bull. Peabody Mus.*, **10**, 1-77.

HOTTON, N. 1959. The pelycosaur tympanum and early evolution of the middle ear. *Evolution*, **13**, 99-121.

—— 1960. The chorda tympani and middle ear as guides to origin and divergence of reptiles. *Evolution*, **14**, 194-211.

JARVIK, E. 1952. On the fish-like tail in the ichthyostegid stegocephalians. *Meddr. Grønland*, **114**, No. 12, 1-90.

—— 1954. On the visceral skeleton in *Eusthenopteron* with a discussion of the parasphenoid and palatoquadrate in fishes. *K. svenska Vetensk Akad. Handl.*, (4) **5**, No. 1, 1-104.

LANGSTON, W. 1965. *Oedaleops campi* (Reptilia: Pelycosauria) new genus and species from the Lower Permian of New Mexico, and the family Eothyrididae. *Bull. Texas mem. Mus.*, No. 9, 1-47.

LEWIS, G. E. and VAUGHN, P. P. 1965. Early Permian vertebrates from the Cutler Formation of the Placerville area Colorado. *U.S. geol. Surv. pro. Paper*, 503—C, 1-46.

OLSON, E. C. 1947. The family Diadectidae and its bearing on the classification of reptiles. *Fieldiana, Geol.*, **11**, 1-53.

—— 1965. Relationships of *Seymouria*, *Diadectes*, and Chelonia. *Am. Zool.*, **5**, 295-307.

—— 1966. Relationships of *Diadectes*. *Fieldiana, Geol.*, **14**, 199-227.

PANCHEN, A. L. 1959. A new armoured amphibian from the Upper Permian of East Africa. *Phil. Trans. R. Soc.*, (B) **242**, 207-281.

—— 1964. The cranial anatomy of two Coal Measure anthracosaurs. *Phil. Trans. R. Soc.*, (B) **247**, 593-637.

—— 1970. *Teil 5a Anthracosauria. Handbuch der Paläoherpetologie*. Stuttgart: Fischer.

PARRINGTON, F. R. 1956. The patterns of dermal bones in primitive vertebrates. *Proc. zool. Soc. Lond.*, **127**, 389-411.

—— 1958. The problem of the classification of reptiles. *J. Linn. Soc. (Zool.)*, **44**, 99-115.

PARRINGTON, F. R. 1959. A note on the labyrinthodont middle ear. *Ann. Mag. nat. Hist.*, (13) **2**, 24-28.
—— 1962. Les relations des cotylosaures diadectomorphes. *Colloq. int. Cent. nat. Rech. sci.*, No. 104 (*Problemes actuels de Paléontologie*), 175-185.
—— 1967. The identification of the dermal bones of the head. *J. Linn. Soc. (Zool.)*, **47**, 231-239.
PEARSON, H. S. 1924. *Solenodonsaurus* Broili, a seymouriamorph reptile. *Ann. Mag. nat. Hist.*, (9) **14**, 338-343.
ROMER, A. S. 1939. Notes on branchiosaurs. *Am. J. Sci.*, **237**, 748-761.
—— 1941. Notes on the crossopterygian hyomandibular and braincase. *J. Morph.*, **69**, 141-160.
—— 1946. The primitive reptile *Limnoscelis* restudied. *Am. J. Sci.*, **244**, 149-188.
—— 1947. Review of the Labyrinthodontia. *Bull. Mus. comp. Zool. Harv.*, **99**, 1-368.
—— 1950. The nature and relationships of the Palaeozoic microsaurs. *Am. J. Sci.*, **248**, 628-654.
—— 1952. Fossil vertebrates of the Tri-State area. Art. 2. Late Pennsylvanian and early Permian vertebrates of the Pittsburgh-West Virginia region. *Ann. Carneg. Mus.*, **33**, 47-112.
—— 1956. *Osteology of the reptiles*. Chicago University Press. Chicago.
—— 1957. The appendicular skeleton of the Permian embolomerous amphibian *Archeria*. *Contr. Mus. Geol. Univ. Mich.*, **13**, 103-159.
—— 1964a. The skeleton of the Lower Carboniferous labyrinthodont *Pholidogaster pisciformis*. *Bull. Mus. comp. Zool. Harv.*, **131**, 129-159.
—— 1964b. *Diadectes* an amphibian? *Copeia*, (for **1964**) 718-719.
—— 1966. *Vertebrate paleontology*. 3rd Edn. Chicago University Press. Chicago.
—— 1968. *Notes and comments on vertebrate paleontology*. Chicago University Press. Chicago.
—— 1969a. A temnospondylous labyrinthodont from the Lower Carboniferous. *Kirtlandia* No. 6, 1-20.
—— 1969b. The cranial anatomy of the Permian amphibian *Pantylus*. *Breviora*, No. 314, 1-37.
ROMER, A. S. and PRICE, L. I. 1940. Review of the Pelycosauria. *Spec. Pap. geol. Soc. Am.*, **28**, 1-538.
ŠPINAR, Z. V. 1952. Revise nêkterých moravských Discosauriscidů (Labyrinthodontia). *Rozpr, ústred. Úst. geol.*, **15**, 1-160.
STEEN, M. C. 1934. The amphibian fauna from the South Joggins, Nova Scotia. *Proc. zool. Soc. Lond.*, (for **1934**), 465-504.
SUSHKIN, P. P. 1925. On the representatives of the Seymouriamorpha, supposed primitive reptiles, from the Upper Permian of Russia, and on their phylogenetic relations. *Occ. Pap. Boston Soc. nat. Hist.*, **5**, 179-181.
—— 1927. On the modifications of the mandibular and hyoid arches and their relations to the brain case in early Tetrapoda. *Palaont. Z.*, **8**, 263-321.
THOMSON, K. S. 1966. The evolution of the tetrapod middle ear in the rhipidistian—amphibian transition. *Am. Zool.*, **6**, 379-397.
—— 1967. Mechanisms of intracranial kinetics in fossil rhipidistian fishes (Crossopterygii) and their relatives. *J. Linn. Soc. (Zool.)*, **46**, 223-253.
VAUGHN, P. P. 1962. The Paleozoic microsaurs as close relatives of reptiles, again. *Am. Middl. Nat.*, **67**, 79-84.
—— 1964. Vertebrates from the Organ Rock Shale of the Cutler Group, Permian of Monument Valley and vicinity, Utah and Arizona. *J. Paleont.*, **38**, 567-583.
WATSON, D. M. S. 1914a. On the skull of a pareiasaurian reptile, and on the relationship of that type. *Proc. zool. Soc. Lond.*, (for **1914**), 155-180.
—— 1914b. *Procolophon trigoniceps*, a cotylosaurian reptile from South Africa. *Proc. zool. Soc. Lond.*, (for **1914**), 735-747.
—— 1917. A sketch classification of the pre-Jurassic tetrapod vertebrates. *Proc. zool. Soc. Lond.*, (for **1917**), 167-186.
—— 1919a. On *Seymouria*, the most primitive known reptile. *Proc. zool. Soc. Lond.*, (for **1918**), 267-301.
—— 1919b. The structure, evolution and origin of the Amphibia.—The "Orders" Rhachitomi and Stereospondyli. *Phil. Trans. R. Soc.*, (B) **209**, 1-73.
—— 1926. Croonian Lecture.—The evolution and origin of the Amphibia. *Phil. Trans. R. Soc.*, (B) **214**, 189-257.
—— 1929. The Carboniferous Amphibia of Scotland. *Palaeont. hung.*, **1**, 219-252.
—— 1931. On the skeleton of a bauriamorph reptile. *Proc. zool. Soc. Lond.*, (for **1931**), 1163-1205.
—— 1951. *Paleontology and modern biology*. Yale University Press. New Haven.
—— 1954. On *Bolosaurus* and the origin and classification of reptiles. *Bull. Mus. comp. Zool. Harv.*, **111**, 297-450.
—— 1957. On *Millerosaurus* and the early history of the sauropsid reptiles. *Phil. Trans. R. Soc.*, (B) **240**, 325-400.
—— 1963. On growth stages in branchiosaurs. *Palaeontology*, **6**, 540-553.
WESTOLL, T. S. 1938. Ancestry of the tetrapods. *Nature, Lond.*, **141**, 127-128.
——1940. (In discussion on the boundary between the Old Red Sandstone and the Carboniferous). *Advanc. Sci., Lond.*, **1**, 258.
—— 1942. Ancestry of captorhinomorph reptiles. *Nature, Lond.*, **149**, 667-668.
—— 1943a. The origin of the tetrapods. *Biol. Rev.*, **18**, 78-98.
—— 1943b. The hyomandibular of *Eusthenopteron* and the tetrapod middle ear. *Proc. R. Soc.*, (B) **131**, 393-414.
WHITE, T. E. 1939. Osteology of *Seymouria baylorensis* Broili. *Bull. Mus. comp. Zool. Harv.*, **85**, 325-409.

D

A. R. I. CRUICKSHANK

The Proterosuchian thecodonts

INTRODUCTION

The Proterosuchia are known from a number of descriptions of skulls, some of which also contain accounts of some post-cranial bones. The best description of a skull is that of Broili and Schröder (1934), but the majority of the descriptions of post-cranial bones have come from China (Young, 1936, 1958, 1963). Whereas there is little doubt as to the generic, and probably specific, similarity of the South African and Chinese specimens, no single complete skeleton is as yet known. Going a long way to fill this gap is specimen NM C. 3016, collected and given a preliminary description by Hoffman in 1965 (Table 1).

All previous accounts of the skull (Broom, 1903, 1932, 1946; Haughton, 1924; Broili and Schröder, 1934; Brink, 1955) have emphasised the archosaurian relationship of *Proterosuchus*. The two main characters which were used by these authors were the antorbital fenestra and a supposed laterosphenoid in the braincase. The former is undoubtedly possessed by *Proterosuchus*, but the latter is certainly not. In addition, close examination of the temporal region shows that small supratemporals still existed in this genus, and that overall the skull is very primitive in its general build. This taken together with the conservative nature of the post-cranial skeleton, leads to the conclusion that the Proterosuchia are little evolved beyond the level of the Eosuchia. They are, however, also clearly antecedent to the later archosaurs and thus fill the gap between the Eosuchia and these archosaurs.

The earliest archosaur yet described (*Archosaurus rossicus*) comes from zone IV of Vladimir, U.S.S.R. (Tatarinov, 1961; Hughes, 1963), and its similarity to *Proterosuchus* was noted by both writers. However, while zone IV of Vladimir is equated with the Upper Permian, all specimens of the genus *Proterosuchus* have come from the Lower Triassic *Lystrosaurus* zone of South Africa, and its equivalent in China. An exception to this should be noted in Specimen BPI 4220 (see Table 1), which is figured as a proterosuchian sixth cervical vertebra (Fig. 4a).

The purpose of this paper is to give as full an account of the osteology of *Proterosuchus* as is possible. While some incidental remarks can be made at this stage on functional matters, a full analysis of this aspect of the genus *Proterosuchus* must remain for another occasion. Thus, the several points raised by Walker (1968) on the details of the temporal region of the skull can be clarified and the general proportions of the body and limbs can be detailed.

Taxonomic History (Fig. 1). The history of the sub-order Proterosuchia is confused. Broom (1903) described the specimen on which the sub-order is based (Romer, 1966, p. 368) as *Proterosuchus fergusi*. In the sub-order Romer (1966) includes two families, the Chasmatosauridae and the

Erythrosuchidae. *Proterosuchus* is included in the Chasmatosauridae as a questionable junior synonym of *Chasmatosaurus*, a genus described by Haughton in 1924. *Chasmatosaurus* is therefore of later date than *Proterosuchus*.

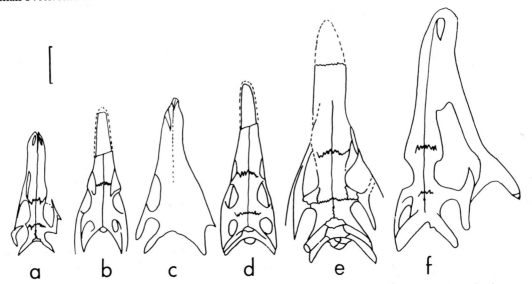

Fig. 1 "Growth series" of *Proterosuchus vanhoepeni* (Haughton). Specimens with their parietals aligned.
a. RC 59. Type of *Elaphrosuchus rubidgei* Broom.
b. BPI 4016.
c. SAM K. 140.
d. NM C. 3016. Type of *Chasmatosaurus alexanderi* Hoffman.
e. NM C. 500.
f. BPI 3993.
The type of *Proterosuchus fergusi* Broom was probably bigger than 'd' but smaller than 'e'.
The type of *Chasmatosaurus vanhoepeni* Haughton is slightly larger than 'f'.
Specimen 'c' is dorso-ventrally squashed, and specimen 'f' is laterally distorted.
Scale-mark is 5 cm.

Abbreviations to Figs 1-10 will be found on p. 119.

The type of *Proterosuchus fergusi* had been mislaid for a number of years, but recently came to light again in the South African Museum, Cape Town (SAM 591). The figures given by Broom (1903) are an accurate representation of the specimen, and show it to be the middle two-thirds of a skull of a small thecodont, which must have been about 25 cm overall in skull length when complete. A horizontal split along the upper surface of the pterygoids reveals the structure of the palate from the transverse flanges of the pterygoids forward to a point just anterior to the internal narial openings.

Unfortunately, the type of *Chasmatosaurus vanhoepeni* (Haughton, 1924) was not available for study until too late for inclusion here. However from an examination of the figures given by Haughton it is apparent that all subsequently described specimens of *Chasmatosaurus* agree in all essentials with the type and belong to the species *vanhoepeni*.

Haughton could see only a single row of palatal teeth and thought that the ectopterygoid was different from that of *Proterosuchus fergusi* and because of these points ascribed his specimen to a different genus. However, as all the other specimens of *Chasmatosaurus* in which the palates are preserved have palatal teeth of the same size and distribution as the type of *Proterosuchus*, and as Broom could only see the upper surface of the ectopterygoids in his specimen, the presumed differences between the palates of *Proterosuchus* and *Chasmatosaurus* are due entirely to the differing states of preservation.

Therefore, as the distinguishing characters of *Proterosuchus* are also seen in *Chasmatosaurus*, the latter becomes a junior synonym of the former. As the type of *P. fergusi* is such a poor specimen when compared with the other proterosuchians, it is impossible to refer these to the species *fergusi*. Therefore, the genus *Proterosuchus* will be assumed here to contain two South African species, viz. *P. fergusi* and *P. vanhoepeni*. It is unlikely that any other specimen will ever be assigned to *P. fergusi* in the future.

The generic identity of *Elaphrosuchus* (Broom, 1946) must also be discussed in this context. This genus was established for a small thecodont in the Rubidge Collection (RC 59) and distinguished from both *Proterosuchus* and *Chasmatosaurus* on the following grounds (Broom, 1946, p. 346). "It certainly cannot be the young of *Proterosuchus fergusi*, as the maxillae are very differently shaped in the two forms. Nor can it be the young of such an animal as has been described by Broili and Schröder (1934). The lower temporal fossa is very differently shaped in the two types. Further, the premaxilla on the newly found skull is so very different from that of the German specimen that the two appear to belong to different, though allied, genera. Then there is clearly a postfrontal bone in the German specimen, but not in Mr Rubidge's, and the interparietal is very different in the two types. Further, there is no lateral opening in the jaw of either of the specimens described by Haughton or by Broili and Schröder, while there is in this specimen. I am therefore definitely of opinion that this new specimen must be placed in a new genus, and propose to call it *Elaphrosuchus rubidgei*."

Taking Broom's points one by one, firstly no comparisons can be made between the maxillae of *Proterosuchus fergusi* and *E. rubidgei*, because there is nothing worthwhile left of the maxillae of the former. On similar grounds a comparison between their lower temporal fossae is invalid. Furthermore, there has been a certain amount of damage to the posterior region of the type of *Elaphrosuchus rubidgei*, which makes a comparison with the same region in any other skull difficult. The premaxilla of Broili and Schröder's specimen is not so different from that of *Elaphrosuchus*, if distortion in the latter is corrected. It is obvious from an examination of the type of *Elaphrosuchus* that the premaxillae have been dorsally displaced, to the extent that the rearmost tooth socket in the right premaxilla, for instance, lies about 11 mm above the level of the maxillary tooth row. This fact is not inadequately figured in the original description (Broom, 1946, fig. 2a). That the postfrontal bones in the type of *Elaphrosuchus* are small and damaged there is no doubt, but they are readily seen on careful examination. The presence of small postfrontals is a constant feature in all proterosuchians with this part of the skull preserved, and thus this point of difference also falls away. A careful examination of the interparietals also reveals no real difference between those of *Elaphrosuchus* and the other proterosuchians. Finally, the lateral opening in the lower jaw of *Elaphrosuchus* identified by Broom is the result of erosion of the posterior part of the dentary. This is confirmed when the matrix in this region is examined and the marks of the missing bone are identified on the underlying rock surface. However, in several other specimens of *Proterosuchus* there is a genuine gap between the dentary and post-dentary bones due to the movement of the dentary relative to the other bones. Thus the presumed mandibular fenestra appears to be concave posteriorly, but convex anteriorly (e.g. Hoffman, 1965, fig. 2a). Also, in SAM K. 140 this gap is seen clearly on one side while the bones of the other jaw fit closely together.

Therefore all the distinguishing characters used by Broom to separate *Elaphrosuchus* from the previously described proterosuchians are invalid, and thus *Elaphrosuchus* is a junior synonym of *Proterosuchus*. The type of *Elaphrosuchus rubidgei* represents a juvenile *Proterosuchus vanhoepeni* (Fig. 1).

The identity of Hoffman's (1965) specimen will now be discussed. This comprises the most perfectly preserved skeleton of a proterosuchian yet found (NM C. 3016). Hoffman ascribed it to a new species, *Chasmatosaurus alexanderi*, distinguishing it from *C. vanhoepeni* on the grounds that the preorbital region and the remarkable curvature of the snout differ from these regions in the type of *C. vanhoepeni*. It has already been shown in the discussion on *Elaphrosuchus* that the premaxillae are subject to displacement, and the snout of NM C. 3016 is in addition badly eroded, a combination of these effects causing the apparent difference. On the question of the preorbital region, Hoffman (1965, p. 38) states: "The prefrontals and lacrimals in *C. alexanderi* and in the specimens described by Brink and Broili and Schröder differ remarkably from *C. vanhoepeni* as described and figured by Haughton, Broom and von Huene. In the latter forms the lacrimal is shown to lie immediately and completely below the prefrontal. In *C. alexanderi* the lacrimal lies in front of the prefrontal. It articulates with the nasal dorsally and joins the prefrontal posteriorly and the maxillary anteriorly, so that the nasal is excluded from the margin of the preorbital fossa. Both prefrontal and lacrimal send down lateral processes, but only the lacrimal process meets the maxilla to form the division between the orbital and preorbital fossae."

However, as the bones forming the boundaries to the various openings in the skull are capable of great movement relative to one another and in some cases have even fallen away completely from the skull, e.g. the lacrimals in the type of *Elaphrosuchus rubidgei*, and as in some cases the relationship of the lacrimals on each side of the same skull is not the same, e.g. NM C.500, the lacrimal is a very doubtful taxonomic character. It is concluded that *Chasmatosaurus alexanderi* is yet another junior synonym for *Proterosuchus vanhoepeni*.

The question of the identity of the Russian and Chinese proterosuchians is best left until all the material can be assembled for direct comparisons. There is no doubt of the close relationship of *Archosaurus* to *Proterosuchus* and the Chinese specimens represent species very close to *Proterosuchus vanhoepeni*.

MATERIAL AND METHODS

(Fig. 1, Tables 1, 2, 3 and 4)

Nine South African specimens were used as a basis for this study (see Table 1 and Fig. 1). Preparation had been taken to varying degrees of completion in all the specimens over the years, but in all except the types of *Proterosuchus fergusi* and *Elaphrosuchus rubidgei* considerable improvement was effected in the state of preparation. Particular attention had to be paid to NM C. 3016 in order to clean the post-cranial skeleton which had not been touched at the time of publication by Hoffman (1965). The temporal region of specimen BPI 3993 was prepared in detail to help elucidate the relationships of the braincase and the bones of the temporal region.

Whereas the majority of the preparation was done manually, BPI 4016 was immersed in 5% acetic acid sufficiently long to etch the bone outlines. Further preparation in acid proved impossible because of the nature of the bone. The braincase of *Euparkeria capensis* (SAM 7696) was removed completely from the matrix by a combination of acetic acid and needle preparation. The broken parasphenoid rostrum was held in its correct relative position by a piece of wire cemented on with 'Glyptal' (Fig. 3d).

Table 1. List of Material.

Number	Identity	Locality, Horizon and Remarks
SAM 591	Type *Proterosuchus fergusi* Broom	Wheatlands, Tarkastad, C.P. Fragment of skull. "Procolophon" zone (Broom, 1903), Low *Lystrosaurus* zone (Kitching, pers. comm.).
SAM 7696	*Euparkeria capensis* Broom	Aliwal North, C.P. Braincase. *Cynognathus* zone. (Ewer, 1965).
SAM K. 140	*Chasmatosaurus vanhoepeni* Haughton	Visgat, Conway, Middelburg, C.P. Crushed skull and post-cranial remains. Low *Lystrosaurus* zone. (Hughes, 1963).
RC 59	Type *Elaphrosuchus rubidgei* Broom	* Barendskraal, Middelburg, C.P. Skull and eight caudal vertebrae. Low *Lystrosaurus* zone. (Broom, 1946).
NM C. 500	*Chasmatosaurus vanhoepeni* Haughton	† "Kruisvlei, 1095, East of Winburg", O.F.S. Untraceable locality. ? *Lystrosaurus* zone. Major portion of large skull. (Brink, 1955).
NM C. 3016	Type *Chasmatosaurus alexanderi* Hoffman	Zeekoeigat, Venterstad, C.P. Skull and post-cranial skeleton. Low *Lystrosaurus* zone. (Hoffman, 1965).
BPI 3993	*Chasmatosaurus vanhoepeni* Haughton	Nooitgedacht, 68, Bethulie, O.F.S. Large skull, partially disarticulated. Low *Lystrosaurus* zone.
BPI 4016	*Chasmatosaurus vanhoepeni* Haughton	Nooitgedacht, 68, Bethulie, O.F.S. Small skull. Low *Lystrosaurus* zone.
BPI 4220	? *Proterosuchus* sp.	‡ Gegund, Harrismith, O.F.S. ? Uppermost *Cistecephalus* zone. Cervical vertebra.

* Brink (1955) gives the locality of this specimen as being 'Brandskraal', which is incorrect (Kitching, pers. comm.).

† Known *Lystrosaurus* zone deposits occur in the district of Winburg, and it is thus assumed that this specimen also came from this horizon.

‡ Kitching (in press) has re-divided the Beaufort on up-to-date information and this horizon will be renamed.

Abbreviations

SAM—South African Museum of Natural History, Cape Town.
RC —S.H. Rubidge Collection, Wellwood, Graaff Reinet, Cape Province.
NM —Nasionale Museum, Bloemfontein, Orange Free State.
BPI —Bernard Price Institute for Palaeontological Research, University of the Witwatersrand, Johannesburg.

Notes on the reconstruction of Proterosuchus. (Tables 2, 3 and 4). The majority of the reconstructions are based on specimen NM C. 3016 and the bones are illustrated as though they were all of the left side. The hind limb, is however, based on SAM K. 140 and reversed left-for-right, the tarsus being from NM C. 3016. It was assumed here that because the skulls of these two specimens are so very nearly the same size that their post-cranial bones would have been much about the same also. Some details of the skull are also from other specimens, the snout being based on BPI 3993 and the quadratojugal on BPI 4016. The anterior part of the palate is based on fig. 1 of Young (1936).

The skull is reconstructed with the quadrate in the upright position and the dentary in natural articulation with the post-dentary bones. This gives a slightly different appearance to the skull when compared with other reconstructions e.g. Broili and Schröder (1934).

It will be noted from the reconstruction given by Broili and Schröder that they show the quadrate making a large angle with the skull roof, and that in this position the teeth at the tip of the dentary meet the maxillary teeth at a point about level with the rear limit of the premaxilla. Thus the premaxillary teeth are entirely free from contact with those of the dentary. However, if the quadrate is placed so that it makes more or less a right angle with the skull roof, then the symphyseal teeth on the lower jaw only just fail to meet with those on the tip of the premaxillar. It is thought that this is a more likely state of affairs than that figured by Broili and Schröder. To emphasise this point, specimen NM C. 3016 was preserved with the quadrate and jaw as reconstructed.

Tooth counts on the premaxilla, maxilla and dentary of all the specimens vary to a greater or lesser degree, apparently according to the size (? age) of the specimen. The skull here is figured with a tooth count thought to be typical of a specimen the size of NM C. 3016.

The humerus of NM C. 3016 was incomplete and had to be reconstructed from figures given by Young (1963). The epipodials of the forelimb and the fore-foot are from SAM K. 140. The latter is incomplete.

The vertebral column of NM C. 3016 is complete distally as far as the 16th caudal vertebra, though Young (1936) figures 27 caudals in his reconstruction of the tail region.

Covering the entire body portion of NM C. 3016 were a series of blue-grey markings, which in some regions were concentrated into thick masses. It is thought that these represent some form of poorly ossified scute. The state of preservation was such that little could be made of them, but an analysis of these dermal remains has now been completed (Thornley, 1970).

THE OSTEOLOGY OF *Proterosuchus vanhoepeni* (HAUGHTON)

The skull and lower jaw

(Figs. 2 and 3, Table 2)

The skull is at first sight typically archosaurian in pattern, with the characteristic downturned snout of the genus and gently elongated shape. It has, however, a number of surprisingly primitive features, which will be described below.

Both temporal arches are complete, with an L-shaped quadratojugal present at the bottom rear corner of the lower arch. There is no 'knee-bend' formed by the squamosal and quadratojugal at the back of the lower temporal fenestra so typical of some of the pseudosuchians. There is an antorbital fenestra. The bones of the skull roof have a prominent midline suture, and supratemporals are present.

The lower jaw is slender with a well developed retroarticular process turned up behind the articulation with the quadrate, but has no mandibular fenestra in the accepted sense of the word.

The skull is metakinetic, with well developed basipterygoid articulations and the braincase inset in the skull roof. There is no laterosphenoid. The epipterygoid is well developed and contacts the parietal through a slender process. The opisthotic has ventral processes similar to those in the braincase of *Captorhinus* and a range of diapsid reptiles up to and including *Massospondylus* (SAM K. 1314), an Upper Triassic dinosaur of the Red Beds of South Africa. These ventral processes are not as highly developed as in the primitive lizards of the English Trias (Robinson, 1962 and 1967).

In all the specimens long hyoids are seen, which presumably are associated with the feeding or breathing activities of the animal.

The cranial roof (Fig. 2a).

The cranial roof comprises the nasals, frontals, parietals, prefrontals, postfrontals, postorbitals, interparietal and possibly the premaxillae. The last named, however, might be considered a separate, independent unit. All these bones are to a greater or lesser degree fused together, and form the main structural unit of the skull. It should be noted that in many cases the skulls of these reptiles are found only with the foregoing bones in association, and that the remainder have fallen away (Brink, 1955, fig. 2; Tatarinov, 1961, fig. 2; Young, 1964, figs 1-6). Thus although Ewer (1965) assumes that the skull of *Euparkeria* had the bones adapted for reducing the amount of relative movement possible, it would seem that in the proterosuchians the bones other than those of the skull roof were capable of considerable movement relative to the roof bones. As *Euparkeria* and *Proterosuchus* are closely comparable in many of the skull characters, it might be assumed that *Euparkeria* also had a kinetic skull

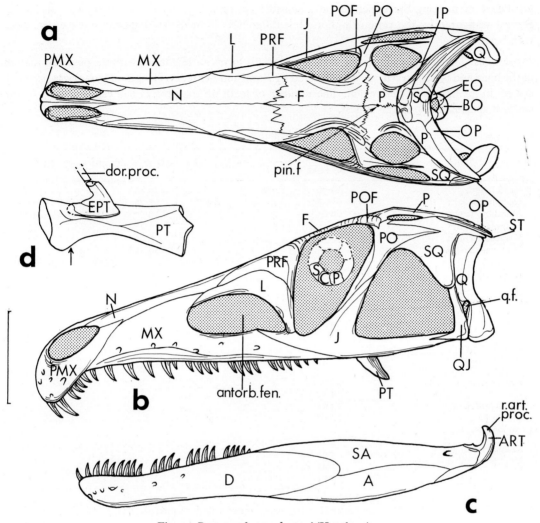

Fig. 2 *Proterosuchus vanhoepeni* (Haughton)

   a. Skull, dorsal view.
   b. Skull, side view.
   c. Lower jaw, outer view.
   d. Left epipterygoid and part of pterygoid. BPI 3993.

Arrow marks approximate position of basipterygoid articulation.

Scale-mark is 5 cm.

It becomes even more likely that *Proterosuchus* had a kinetic skull when various reconstructions of the skull are examined (Hoffman, 1965, figs 1 and 2). The great slope of the quadrate which is commonly figured (Broili and Schröder, 1934) demonstrates but one extreme position that the skull could adopt. The opposite extreme is shown in NM C. 3016 (Hoffman, 1965) and Fig. 2b in this paper, where the quadrate is almost vertical in position. In the latter case the lower jaw has at least a possibility of meeting the upper jaw in a sensible position. The bones of the cranial roof are in general as figured by the various authors, but differences in interpretation exist and they are dealt with below.

Premaxillae. The dentigerous border of each premaxilla forms a straight line with the posterior margin of the bone and is not curved, as reconstructed by Broili and Schröder. Haughton (1924) was perfectly correct in his figures of this region. The number of teeth on the premaxillae varies from 6 in the smallest specimen to 9 in the largest.

Table 2

	SAM 591	SAM K.140	RC 59	NM C. 500	NM C. 3016	BPI 3993	BPI 4016
† Skull length		19.70	14.82		20.85*	27.50	16.19*
Occipital height				8.81*	8.76		4.98
Mandible length		26.65			24.70	34.95	21.20
Scapula height					9.28		
Coracoid length					8.40		
Humerus length					10.30*		
Radius length			8.20				
Ulna length			8.83				
Metacarpal lengths I			1.70				
II			3.37				
III			3.48				
IV			3.66				
Ilium height					5.30		
Sacral width					7.18		
Midline length of sacrum					9.80		
Femur length			14.93/14.52				
Tibia length			13.00				
Fibula length			12.56*	18.3*			
Metatarsal lengths I			2.33				
II			4.42				
III			5.56				
IV			6.47				
V			3.51*		2.87		

* Estimated value
† Measured from tip of snout to anterior limit of interparietal.

Nasals. The nasals are long sub-rectangular bones which do not contribute to the margins of the antorbital fenestrae. There was a possible hinge between them and the frontals, marked by the coarse interdigitations of the two sets of bones, and in some specimens by a bending of the skull bones along the line of the naso-frontal suture. On their ventral margins there are grooves for the reception of the dorsal rims of the maxillae.

Frontals. These are shorter rectangles than the nasals, with firm sutures with the prefrontals, postfrontals and parietals. They contribute to the dorsal rims of the orbits and are indented strongly by them.

Prefrontals. These bones lie on the dorsal and lateral surfaces of the skull, forming most of the anterodorsal margin of the orbit. The laterally facing portion is facetted for the reception of the lacrimal. The prefrontal does not reach the maxilla ventrally but the relationship of the prefrontal to the jugal and maxilla differs from that of *Euparkeria* (Ewer, 1965, fig. 2).

Postfrontals. These are small triangular bones occupying positions on the posterior dorsal corners of the orbit. They are firmly fixed to the roof of the skull.

Parietals. A small parietal foramen is sometimes visible on the roof of the skull, but can in all instances be picked up on the undersurface of the bones when these are prepared. The posterior arms of the parietals lie partly on the skull roof, but more so on the occipital surface. The distal portions of the posterior arms lie in grooves on the median portions of the squamosals and were capable of moving relative to the squamosals, with the supratemporals intervening to a certain extent. Where the parietals bend over on to the occipital surface of the skull, they are drawn out into a transverse crest which marks the upper limit of the insertion of the neck muscles. This crest turns to follow the upper temporal opening median margin.

Interparietal. The interparietal in *Proterosuchus* has a median process just above the supraoccipital suture. There are small excavations for the presumed insertion of nuchal ligaments just below the crest of the occiput.

Supratemporals. Hitherto undescribed in archosaurs, splint-like but prominent supratemporals can be seen in all the specimens which are well preserved in the temporal region. Laterally they end almost level with the extremities of the opisthotics, but extend beyond the posterior processes of the squamosals.

Postorbitals. The ventral process of each postorbital is grooved on the hind surface to receive the dorsal process of the jugal. The contact with the squamosal is also loose, the posterior process of the postorbital lying in a tapering groove in the squamosal. The contacts with the parietal and postfrontal seem to have been very firm.

The Bones of the Cheek Region (Fig. 2b)

Maxillae. The maxillae have loose, long contacts with the nasals. Below the orbit they are overlapped by the jugals, also apparently loosely. There is no obvious palatal component. Tooth counts vary from 14+ in RC 59 to 29 in BPI 3993. However the number does not necessarily seem to be the same for both sides, and in many cases the state of preservation does not allow for accurate counts to be made.

Jugals. Each jugal is tripartite, with sliding joints for the ventral limb of the postorbital and the posterior ramus of the maxilla. The quadratojugal seems to have rested in a groove on the outer side of the blunt posterior process of the jugal. This joint was also capable of a certain amount of movement.

Squamosals. There is a distinct posterior process overlying the quadrate (Walker, 1968), separated partially from the parietal and opisthotic by the supratemporal. The squamosal thus does not seem to have any firm suture with any of the bones in contact with it. The ventral process overlies the small quadratojugal.

Quadratojugals. These bones are L-shaped, each with a tall dorsal process underlying the squamosal and a shorter anterior process loosely overlying the jugal. The dorsal process forms the anterior boundary to the quadrate foramen. The outer surface of this bone was flat.

Quadrates. These are tall, of typically archosaurian type, capable of rocking in the cup formed by the posterior process of the squamosal, and with a well-developed pterygoid process loosely overlapping the quadrate process of the pterygoid. On the inner face of the pterygoid ramus of the quadrate there is a groove for the reception of the quadrate ramus of the pterygoid. This groove ends in a well-defined wall, and if the end of the quadrate ramus is placed right up to this wall, then the quadrate proper is drawn into a position to make almost a right angle with the skull roof. The condylar portion is waisted one third of the way up, this marking the position of the quadrate foramen. There is a buttress on the posterior surface at the top, immediately under the opisthotic. The articular surfaces are angled so as to force the articulars outward when the jaw was opened. This action also helped to

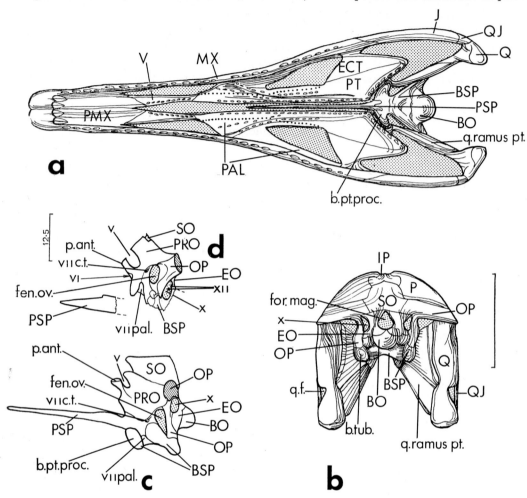

Fig. 3 *Proterosuchus vanhoepeni* (Haughton)
 a. Skull, palatal view.
 b. Skull, occipital view.
 c. Braincase of *Proterosuchus*, lateral view (BPI 3993).
 d. Braincase of *Euparkeria*, lateral view (SAM 7696). Scale-mark is 12·5 mm.
  Scale-mark is 5 cm except for 3d.

keep the retroarticular processes clear of the posterior surface of the quadrate during jaw opening. There is an incipient tympanic notch, similar to that in *Euparkeria* (Ewer, 1965).

The palate (Fig. 3a).
The palate is but little modified from that of a typical eosuchian. The major difference seems to be in the slight arching of the pterygoids in the midline, parallel to the parasphenoid rostrum. There are

strong pterygoid flanges and well developed basipterygoid articulations. The internal nares are elongated backwards similar to those of *Youngina*.

Broom (1903) recognised the general rhynchocephalian nature of the palate of the type of *Proterosuchus fergusi*, but also said that it was reminiscent of that of *Ornithosuchus* (Walker, 1964). As far as the last statement is concerned, there are significant points of detail separating the two genera, though in general Broom was correct in his associations.

Pterygoids. The numbers of teeth indicated on Fig. 3a are a composite reconstruction from specimens BPI 3993 and NM C. 500. There are deep fan-shaped quadrate rami of the pterygoids which overlap the anterior processes of the quadrates.

Epipterygoids. The epipterygoids are composed of two portions, a broad pedestal sitting on the vertical component of the quadrate ramus of the pterygoid, just posterior to the basipterygoid articulation, and a slender process which contacts the parietals. The anterior face of the basal part is notched.

Ectopterygoids. The ectopterygoids are in general as in Broom's reconstruction of the palate (1903) and are relatively large bones overlapping the pterygoid flanges on their ventral surfaces. They seem to have a firm contact with the jugals. On the postero-lateral corner of the ectopterygoid there is a deep excavation which faces partly upward. This seems similar to the excavation noted by Ewer (1965) in *Euparkeria*.

Palatines. As far as can be seen in this material the palatines are almost as large as the pterygoids, forming the surround to the median and lateral sides of the palatine fenestrae. They are toothed in the midline of the skull, and follow the upwardly arched posterior portions of the pterygoids for part of their length, becoming flat anteriorly behind the vomers.

Vomers. These bones seem to be placed far back on the palate, owing to the relatively large premaxillae. The proportions of the internal nares appear to be very similar to *Youngina* or *Prolacerta*, as figured by Romer (1966), and much longer than normally expected in any archosaur. The posterior limits of the internal nares are placed just anterior to the upwardly arched portion of the palate, and would thus be ideally placed to duct the inhaled air into this channel. It is unlikely that the arched portions of the pterygoids supported a fleshy false palate, as there are teeth present both on the median rims of the pterygoids and on the roof portions of both the pterygoids and palatines. The vomers are toothed, each having a single row of functional teeth.

The braincase (Fig. 3b, c and d)

In general the braincase is very similar to that of *Captorhinus* (Price, 1935). Detailed differences exist particularly in the interpretation of the limits of the parasphenoid and prootic, but in general the braincase is very similar to *Sphenodon* (Romer, 1966, fig. 198) and the arrangement of the bones of the braincase and cranial nerves is also identical to those of *Euparkeria* (SAM 7696) (Cruickshank, 1970). There are no ossified laterosphenoids, the prootic being primitive in structure with an open trigeminal notch.

Exoccipitals. The exoccipitals form the sidewalls to the foramen magnum, a pear-shaped opening. They meet on the midline both above and below the foramen magnum, and with the basioccipital form the condyle. A large jugular foramen lies laterally to the exoccipital and is in turn bounded on the outside by the ventral process of the opisthotic.

Basioccipital. The basioccipital is a large bone flooring the posterior part of the braincase. On the occipital surface it is almost entirely separated from the basisphenoid by the ventral processes of the

Opisthotics. While not clear in any of the present material, it is likely that nerve XII emerged through foramina between the basioccipital and exoccipital, as happens in *Euparkeria*.

Basisphenoid. The basisphenoid continues forward from the basioccipital under the middle portion of the braincase, and is exposed on the ventral surface. There is a wide and deep sella turcica with a rather small dorsum sellae, but the relationships of these are not very clear in the specimen available (BPI 3993). Ventrally the basisphenoid is drawn out into a pair of basal tubera which do not contain the foramina ovales as Broili and Schröder (1934) indicate. They do have circular excavations on their posterior surfaces, however, which may have had sterno-mastoideus muscles inserted therein. There are strongly developed basipterygoid articulations.

Parasphenoid. The parasphenoid is characterised by its extremely long rostrum which is grooved on its dorsal surface. It does not overlap the basisphenoid as extensively as in *Captorhinus*.

Supraoccipital. The supraoccipital forms the median part of the upper portion of the braincase, but apparently is all but excluded from the foramen magnum. It extends forward under the parietals to form part of the roof to the braincase.

Opisthotics. The opisthotics have powerful paroccipital processes which fit into sockets in the lateral extremities of the occipital portions of the parietals, and possibly the supratemporals. There are moderately developed ventral processes of the opisthotics which separate the jugular foramen and fenestra ovale. These latter lie at the same level, just under the paroccipital processes. There are no post-temporal fenestrae.

Prootics. In *Proterosuchus* the prootics are primitive in structure and the exit for Nerve V is through an open notch. No participation of the basisphenoid occurs in the pila antotica, as distinct from *Captorhinus*.

The lower jaw (Fig. 2c).

The lower jaw is long and narrow, with a loose symphysis and a prominent upturned retroarticular process. There are large Meckelian fossae, but no mandibular fenestrae. It is apparent that the dentary was capable of a limited fore-and-aft movement relative to the post-dentary bones (Hoffman, 1965, and above p. 91). The tooth-row of the lower jaw is much shorter than that of the maxilla, the anterior teeth on the dentary being able to meet those of the premaxilla only with difficulty, if at all.

Owing to the narrowness of the skull, and the intimate association of the lower and upper jaws in all cases it was not possible to prepare completely any inner surface of the lower jaws.

Dentaries. With the quadrate in an upright position it would be necessary for the dentaries to slide forward an additional 15 mm relative to the post-dentary bones to be able to bring their anterior teeth exactly in apposition with those of the premaxillae. If the quadrate articular end was swung to its farthest back position (e.g. Broili and Schröder, 1934, fig. 1), then the anterior teeth on the dentary would contact the upper jaw level with the posterior limit of the premaxilla. There is a very loose symphyseal joint. The anterior region of the dentary is pierced by a number of foramina.

Surangulars. These form the upper margins of the lower jaws, between the dentaries and the articulars. On the inside of the jaw rami they form the outer walls to the Meckelian fossae.

Articulars. Broili and Schröder (1934, figs 1 and 8) show a relatively small articular with a strip of prearticular (goniale) lying in front of the articular on the outer surface and, on the inner surface, an extensive exposure forming the inner wall to the Meckelian fossa. In NM C. 3016 the corresponding strip of bone is in fact part of the articular. The inner surface is very difficult to interpret, but is probably the same as figured by Broili and Schröder.

The articular is therefore rather different in *Euparkeria* (Ewer, 1965, fig. 2a) when compared with that of *Proterosuchus*.

In addition to the foregoing bones Broili and Schröder describe both splenials and coronoids in their specimen. Nothing is visible of these bones in the present material.

Table 3

Centra Lengths

		SAM 591	SAM K. 140	RC 59	NM C. 500	NM C. 3016	BPI 3993	BPI 4016	BPI 4220
	1								
	2		2·86			2·56	2·50*	2·28	
	3		3·16*			3·05	3·71	2·89	
	4					3·17	3·71		
	5					3·07			
	6					2·68			?2·71
	7					2·28	?3·59		
	8					2·40			
	9					2·23			
	11		?2·53						
	12		?2·46						
	13		?2·46						
	14		?2·68						
2nd caudal	29			?1·67		1·76			
	30			?1·77		1·91			
	31			?1·92*		1·84			
	32			?1·92*		1·90			
	33			?1·88		1·92			
	34			?1·84		1·92			
	35			?1·84*		1·87*			
	36					1·90			
	37					2·00			
	38					2·10			
	39					2·14			
	42					2·20*			
	43					2·20*			

* Estimated value
? Position doubtful

The post-cranial skeleton (Figs 4-10, Plate I, Tables 2, 3 and 4)

The post-cranial skeleton is in general remarkably lepidosaurian in character and has few specifically archosaurian features (Hughes, 1963, p. 230). The pose of the limbs was that of an animal which had poor locomotion on land and was probably clumsy in its movements. This is reflected in the heavily ossified limb girdles and the very primitive nature of the limb bones. There are no cleithrum and suprascapula ossifications, nor a sternum.

The vertebral column (Fig. 4a and c, Plate I)

In specimen NM C. 3016 there is an articulated vertebral column, complete as far as the 16th caudal vertebra. There are 9 cervical, 16 trunk, 2 sacral and 16 caudal vertebrae, making a total of 43 as preserved. Young (1936, fig. 6) reconstructs a tail of one specimen from dissociated caudal vertebrae, showing 27 as a possible complement. While not a reliable reconstruction, the number is realistic for such an animal. Broili and Schröder report a pro-atlas element in *Proterosuchus*, as does Ewer (1965) for *Euparkeria*.

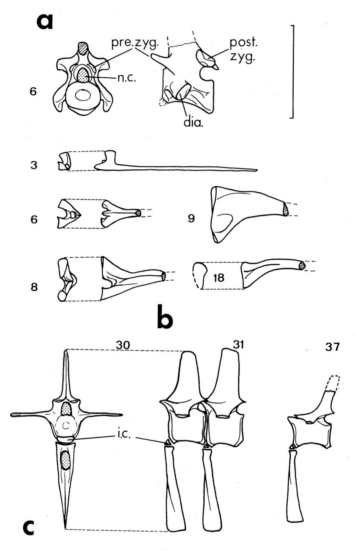

Fig. 4

a. Proterosuchian cervical vertebra from near Harrismith (BPI 4220). Note that there is no parapophysis on the centrum.
b. Ribs.
c. Caudal vertebrae.
b and c are *Proterosuchus vanhoepeni* (Haughton)

The number accompanying each illustration corresponds to the position in the vertebral column. Scalemark is 5 cm.

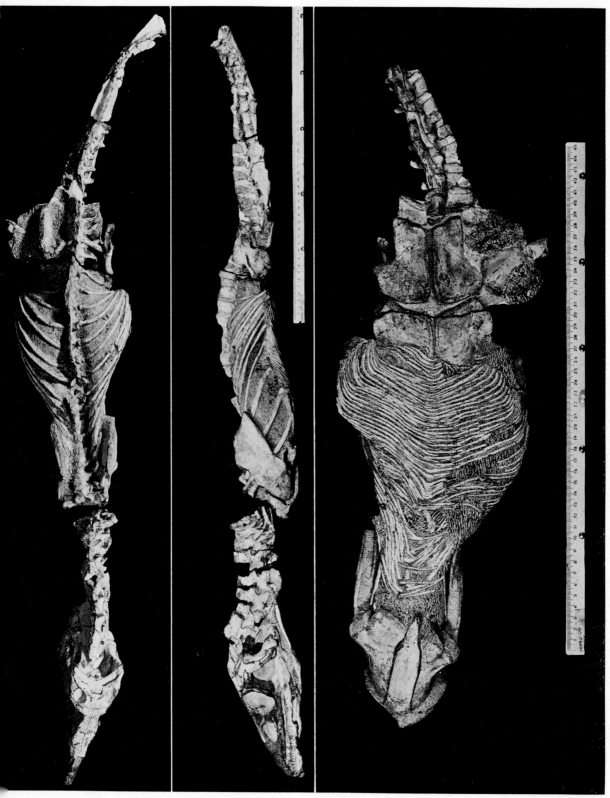

Plate I *Proterosuchus vanhoepeni* (Haughton) NM C.3016. Lateral and dorsal views of the complete skeleton. Ventral view of the rib cage.

Neck region (Fig. 4a). There are intercentra associated with the first 8 vertebrae of NM C. 3016. Between the 8th and the 9th no intercentrum can be seen, and the trunk region of this specimen is obscured by the rib cage. Intercentra are seen again only in the tail. Whereas the ribs of the first eight vertebrae are slender and not very long, that of the ninth is much more robust. At the same time it does not reach completely to the ventral surface of the body, and for this reason it can be assumed that the neck region had nine vertebrae (Hoffstetter and Gasc, 1969, p. 254), a not uncommon count in lepidosaurians, and typical of present-day crocodilians.

The centra of the neck region are elongated and with the count of nine vertebrae, give the animal a very long neck. The centra are amphicoelous, though by no means notochordal and the neurocentric suture is very difficult to see. These centra are only slightly keeled.

Broili and Schröder (1934, figs 11-14) show that both the tuberculum and the capitulum of each cervical rib articulates with the centrum, the intercentrum being relatively small and not involved in the rib articulation at all. In NM C. 3016, however, it is plain that the parapophysis is largely on the intercentrum for the first six vertebrae, and that only on the seventh and subsequent cervical vertebrae does the intercentrum become too small for the parapophysis, which then moves onto the centrum low down.

In *Sphenodon* (Hoffstetter and Gasc, 1969, fig. 28) a similar state of affairs exists, except that the cervical ribs are reduced when compared to *Proterosuchus*, but on the fourth vertebra the capitulum is clearly in contact with the intercentrum and the tuberculum with the centrum at a moderately high position. By the fifth vertebra the diapophysis and parapophysis become joined and lie totally on the centrum. This state of affairs is seen on the 9th centrum in *Proterosuchus*. At the same time that the diapophysis and parapophysis join, the ribs become holocephalous in both *Sphenodon* and *Proterosuchus*.

The neural spines in the neck region are high in all except the atlas, with that of the axis being elongated. On the fourth and subsequent neural spines there are lateral expansions at the tip ("spine tables"), (Ewer, 1965, fig. 19).

The zygapophyses are relatively large, flat and horizontally orientated.

Trunk region. By the tenth vertebra the synapophyses have migrated up the sides of the centrum to form prominent transverse processes level with the base of the neural spine. Progression towards the rear of the body is marked by a decrease in the depth of the synapophyses and a reduction in the length of the transverse processes. The last three presacral ribs are fused to their transverse processes.

The neural spines at the front of the trunk region are high, wide, and shorter than those of both the cervical vertebrae and those of the mid and hindermost trunk vertebrae. The typical thoracic vertebra has a tall, blade-like neural spine with no spine table. The centra in the trunk region are procoelous.

The zygapophyses are similar to those of the neck vertebrae, but become smaller towards the sacrum.

Sacral vertebrae (Fig. 8b, d, e). There are two sacral vertebrae in NM C. 3016, whose neural spines are lower than those preceding them. The sacral ribs are fused to the transverse processes, but not to the inner surfaces of the ilia. The sacral vertebrae are not fixed together, a distinct line of matrix lying between their zygapophyses in NM C. 3016.

Caudal region (Fig. 4c). The caudal region is marked by the presence of strong, but narrow, neural spines; elongated, equal-length centra with long, thin transverse processes; intercentra and long chevron bones. The intercentra become smaller posteriorly; the last separate one is seen lying between the 12th and 13th caudal centra. Chevron bones are present until the 16th caudal vertebra (Plate I, lowermost picture), and presumably were present for many vertebrae behind that.

Ribs (Fig. 4b)

Neck region. The atlas rib seems to have been a simple single-headed rod (Broili and Schröder, 1934, fig. 14), but nothing is seen of it in this material.

The axis rib is double-headed in NM C. 3016, but Broili and Schröder (fig. 14) show a third process running forward. As described above, the capitulum articulates with the intercentrum, and the tuberculum with the diapophysis which is placed low down on the centrum.

The ribs on the third to eighth vertebrae are clearly triple-headed in specimen NM C. 3016, the shaft of the preceding rib lying in the groove formed by the capitulum and the third process of the one behind. This seems to have been to increase the rigidity of the neck in the same way as in modern crocodilians. With the relatively long neck of *Proterosuchus* this would have been important.

The ninth cervical rib is holocephalous, but still retains a vestige of the third head, all three processes lying in almost the same plane. As described above, the shaft of this rib is very much more robust than the preceding ones.

Table 4. Length of Cervical Ribs NM C. 3016

1.	—
2.	9·30
3.	10·33
4.	8·97*
5.	—
6.	6·10*
7.	5·62
8.	—
9.	—

* Estimated value

Trunk region (Plate I). All the ribs of the trunk region, except those of the last three presacral vertebrae are robust and free to move on the synapophyses. The heads become less and less deep and at the same time the trace of the third head becomes smaller and smaller, so that by the ninth trunk vertebra practically no trace of the third head is seen. The last three presacral ribs are firmly ankylosed to their transverse processes and do not meet the ventral surface of the body.

Sacrum (Fig. 8 b, d, e). The sacrum is composed of two vertebrae with the ribs ankylosed to the centra. The ribs are deep dorso-ventrally and they are divided distally, the front one having two processes and the hind one three.

Gastralia (Plate I). The gastralia in general seem to be very similar to those of *Euparkeria*. The anterior series in NM C. 3016 are badly disturbed by the ventral ends of the anterior trunk ribs, but the posterior series are almost undamaged.

The pectoral girdle (Fig. 5)

Scapula (Fig. 5a). Broom (1932) figured a scapula in outer view and while in general this is correct, the specimen examined, NM C. 3016, has a pectoral girdle which is shorter and wider. The glenoid is of typically primitive 'screw shape'. There is a powerful overhang to the glenoid, and the coracoid foramen is situated not far in front of it.

Coracoid (Fig. 5c). These are separate ossifications, and no division into coracoid and post-coracoid elements exists. It is assumed that they met in the midline as reconstructed.

Clavicle (Fig. 5a). The clavicles have tapering dorsal portions which are grooved to fit over the anterior edges of the scapulae, with widened portions curving round the front of the scapulae. Lying between the broad portions of the clavicles and the coracoid ossifications are the upturned lateral extensions of the interclavicle.

Interclavicle (Fig. 5b). This is of a typically primitive type, 'T'-shaped with a thin but broad plate lying on the midline, overlapping the coracoids.

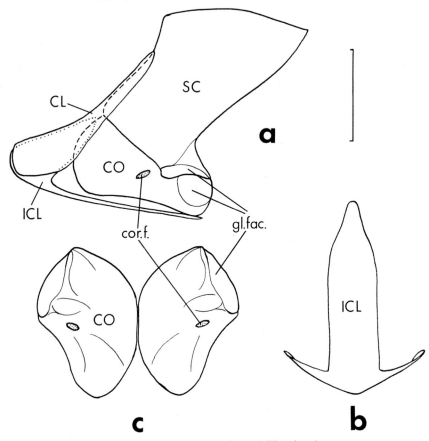

Fig. 5 *Proterosuchus vanhoepeni* (Haughton)
a. Pectoral girdle, lateral view.
b. Interclavicle, ventral view.
c. Coracoids, ventral views.
Dashed line shows outline of scapula and coracoid, behind clavicle and interclavicle. Dotted line shows outline of interclavicle between coracoid and clavicle. Scale-mark is 5 cm.

The forelimb (Figs 6 and 7)

Humerus (Fig. 6a, b and c). The humerus was reconstructed from fragments associated with NM C. 3016, according to the proportions given by Young (1963). It does not differ in any respect from the Chinese material.

Radius (Fig. 6d). This is shorter than the humerus with a slightly concave head. The shaft is oval in section and the distal end slightly expanded.

Ulna (Fig. 6e). There is no olecranon process on the ulna, and it is only slightly longer than the radius. There is a shallow sigmoid notch, and the distal end is more convex than that of the radius.

Carpus (Fig. 7). There is one forefoot among the material, which is incomplete and disarticulated (SAM K. 140). There is no degree of finality in the reconstruction, but it does show that there is no elongation of any of the elements. In fact the reverse could well be true and the carpus may be partly 'reduced'.

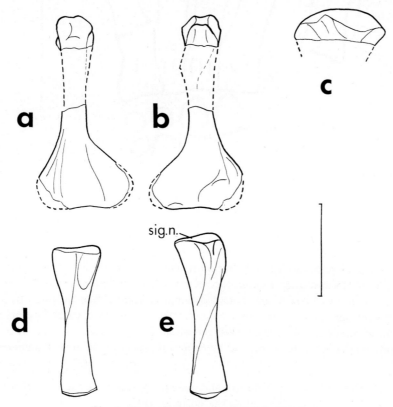

Fig. 6 *Proterosuchus vanhoepeni* (Haughton)
    a. Humerus, dorsal view.
    b. Humerus, ventral view.
    c. Humerus head, anterior view.
    d. Radius, anterior view.
    e. Ulna, anterior view.
    Scale-mark is 5 cm.

Manus (Fig. 7). Four metacarpals (SAM K. 140) were identified, and are assembled in what appears to be a logical order, bearing in mind the probable increase in their lengths laterally. One metacarpal has a phalange in natural articulation. One other phalange is in association with a terminal phalange. The latter are not part of the first digit, because the width of the proximal part of the phalange is very much greater than that of the phalange in articulation with the first metacarpal, and in any case would give a wrong (or unusual) count of phalanges for a normal first digit.

The forefoot is clearly very much smaller than the hind.

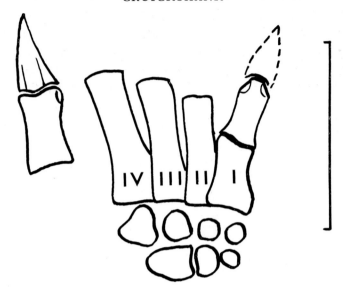

Fig. 7 *Proterosuchus vanhoepeni* (Haughton)
Manus reconstructed.
I-IV Metacarpals.
Scale-mark is 5 cm.

The pelvic girdle (Fig. 8a-e)

Ilium. The ilium is of primitive pattern, but with one peculiar specialisation in that the anterior portion is produced downward and excludes the pubis from the acetabulum. The acetabulum is, therefore, dorso-ventrally elongated and bounded along the front rim of the ilium by a heavy ridge. There are two sacral vertebrae whose transverse processes are not fused to the internal surface of the ilium, but there are well marked facets for their reception. The posterior vertebra has three processes, the foremost process sharing its facet with the posterior process of the bifurcated anterior sacral vertebral process.

Ischium. The ischia are primitively plate-like and extend posteriorly to the level of the prolongation of the ilium. They are slightly thickened on the posterior rims.

Pubis. Apart from its preclusion from the acetabulum, the pubis is very similar to that of *Labidosaurus* (BPI 3838). The pubis is flat in general with an incipient tuberosity on its ventral surface which does not extend beyond the limit of the plate-like portion. At the base of each tuberosity there is a small thyroid fenestra.

The Hind Limb (Figs 9 and 10)

Femur (Fig. 9a-d). The femur is that of a quadrupedal animal. The shaft is almost straight and there is a deep excavation for the pubo-ischio-femoralis group of muscles. The articulations for the tibia and fibula are largely ventral. The proximal end is hollowed, presumably for a cartilaginous cap, and directed partly upward.

Tibia (Fig. 9e and f). This is a stout bone, almost as long as the femur (see Table 2). The head is expanded and has a stout cnemial crest. The distal end is slightly hollowed for the ankle joint and almost circular in section.

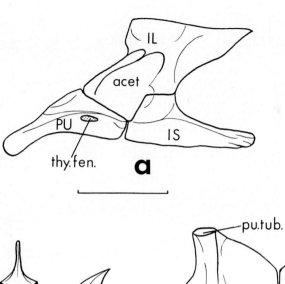
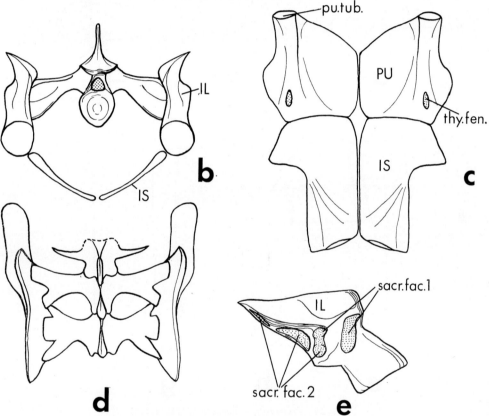

Fig. 8 *Proterosuchus vanhoepeni* (Haughton)

a. Pelvis, lateral view.
b. Section through pelvis immediately in front of sacrum.
c. Pubes and ischia, ventral view.
d. Sacrum, dorsal view.
e. Left ilium, inner view.

Scale-mark is 5 cm.

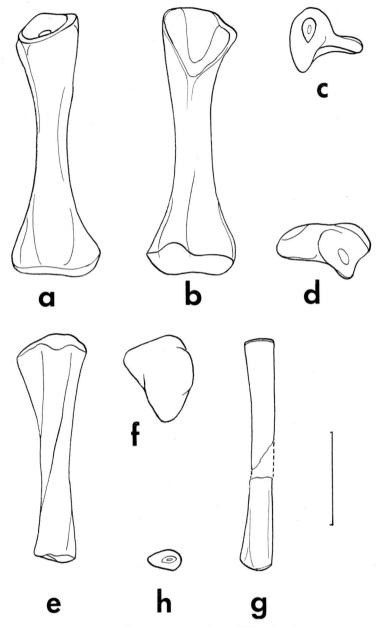

Fig. 9 *Proterosuchus vanhoepeni* (Haughton)
  a. Femur, dorsal view.
  b. Femur, ventral view.
  c. Femur, proximal end.
  d. Femur, distal end.
  e. Tibia, anterior view.
  f. Tibia, proximal end.
  g. Fibula, anterior view.
  h. Fibula, distal end.

Scale-mark is 5 cm.

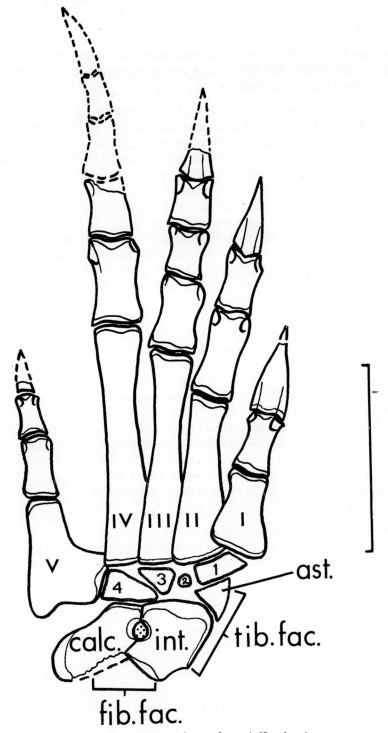

Fig. 10 *Proterosuchus vanhoepeni* (Haughton)
Pes reconstructed
I-V Metatarsals.
1-4 Distal tarsal bones.
Scale-mark is 5 cm.

Fibula (Fig. 9g and h). The fibula is almost the same length as the tibia, very much more slenderly built and oval in cross-section. The distal end is broadly rounded in anterior view to fit the angle between the calcaneum and the intermedium.

Tarsus (Fig. 10). The tarsus comprises three bones in the proximal row and four in the distal. The overall structure is exactly as in the rhynchosaurs described by Hughes (1968), except that the astragalus (tibiale) is smaller and there is a foramen between the calcaneum and the intermedium. There is a lateral tuber on the calcaneum, and this bone fits against the large intermedium by means of the same peg and socket arrangement described by Hughes for the rhynchosaurs. As Hughes emphasises, this pattern of ankle is eosuchian in nature.

The calcaneum (fibulare) in specimen NM C. 3016 is slightly damaged on its proximal surface, so that the articulation with the fibula is not seen with certainty. The lateral portion is drawn out into a tuber which is thinner than the rest of the bone. Medially it thickens and the area in contact with the intermedium is hollowed. Proximally to the main articulation with the intermedium there is a vertical groove, which forms half of the foramen. Distally, the thickened portion articulates with the fourth distal tarsal.

The intermedium is a larger bone than the foregoing, being hollowed on its dorsal surface and with a ventral process or boss medially. The posterior margin of the lateral groove is drawn into a peg to help lock the intermedium to the calcaneum. Distally it articulates with the median part of the fourth, the third and possibly the second distal tarsals. Laterally, the rounded surface is in contact with the astragalus. There are facets for both the tibia and fibula separated by a shallow notch.

The astragalus is essentially a triangular bone, with a postero-ventral projection. It is much smaller than the other two bones just described and functionally was probably part of the intermedium. It articulates distally with the first distal tarsal.

In specimen NM C. 3016 the first distal tarsal bone has been forced on top of the proximal end of its corresponding metatarsal, but its likely position in life is as figured. The second distal tarsal is a small round bone and the third and fourth are triangular in shape, their apices overlapping each other, and the base of the latter slightly overlapping the medial part of the fifth metatarsal. The arrangement described above is very similar to that in the tarsi of the rhynchosaurs described by Hughes.

Pes. The proximal and distal articular surfaces on the metatarsals are not grooved to any extent. The fifth metatarsal is "hooked", the median portion being overlapped to some extent by the fourth metatarsal and the fourth distal tarsal. There is a moderately developed hamate process on the lateral edge of the fifth metatarsal. The phalangeal count is 2 : 3 : 4 : ? 5 : 3. As the phalanges become more distal they develop more strongly grooved and keeled articular ends. The terminal phalanges are laterally compressed and form strongly recurved claws.

DISCUSSION

Palaeoecology

To the localities listed in Table I must be added Harrismith, the site from which the type of *Chasmatosaurus vanhoepeni* Haughton was collected. *Proterosuchus* localities are thus distributed fairly evenly round the South, West and North portions of the *Lystrosaurus* zone. In all these localities the strata producing *Proterosuchus* are low in the succession (Kitching, pers. comm.).

In addition to *Proterosuchus* and *Lystrosaurus* these localities have yielded *Lydekkerina*, *Procolophon*, *Scaloposaurus*, *Nanictosaurus*, *Ictidosuchops*, *Ericiolacerta*, *Glochinodontoides*, *Glochinodon*, *Galesaurus*, *Micrictodon*, *Platycraniella*, *Thrinaxodon*, *Myosaurus* and *Prolacerta*. It should be noted that no fish remains have been recorded from these localities and that fish are not reliably recorded from any *Lystrosaurus* zone locality at all as opposed to the Lower and Upper Beaufort. Likewise amphibian remains are not as common as might be expected from beds laid down under conditions supposedly wetter than the preceding *Cistecephalus* zone (Parrington, 1948).

Kitching (1968, p. 68) notes that the *Lystrosaurus* species which occur alongside *Proterosuchus* in the lower levels of the zone are frequently large, e.g. *L. amphibius*. However, apart from these large species the general size of the members of this fauna is not great. Even *Proterosuchus* itself would not exceed 1·5 metres overall when fully grown. Whereas in *Proterosuchus* the total movement between the lower jaw and quadrate was restricted by the upturned retroarticular process, the ability of the quadrate to rotate backwards on the squamosal would act to increase the gape considerably. *Proterosuchus* was thus quite capable of opening its mouth wide enough to deal with the majority of the fauna, and even to accelerate the lower jaw sufficiently to snap the jaws shut with some force.

In speculating on the mode of life of *Proterosuchus* the following points seem to be important:

1. carnivorous dentition,
2. terminal nares,
3. long neck,
4. plate-like, primitive limb girdles,
5. ventral extension of the ilium,
6. primitive orientation of the limbs,
7. well ossified limbs, carpus and tarsus, and
8. horizontal zygapophyses.

Because of its archosaurian characters, particularly in the neck region, and presumably because of the presupposition of a wet period during *Lystrosaurus* zone times, Broili and Schröder suggested that *Proterosuchus* lived similarly to present-day crocodiles. The position of the external nares and the state of ossification of the limbs, carpus and tarsus would however indicate a terrestrial existence. The deepened ventral part of the ilium, with the strong vertical acetabular buttress lying anterior to the acetabulum, is adapted to take strong forward thrusts from the large hind limbs. This in association with the long neck might indicate that *Proterosuchus* lay in wait for its prey, which was captured by a lunge from a place of concealment. A final pointer towards a terrestrial life for *Proterosuchus* is the presumed vertical hind end of the skull. If the crocodiles are typical of aquatic reptiles, then it would be expected that the back of the skull in all such animals would slope forward. If the lower jaw of *Proterosuchus* is positioned so that the symphyseal teeth on the lower jaw are touching the premaxillary teeth, then the hind end of the skull is vertical.

However, the zygapophyses of the entire vertebral column are largely horizontal, allowing for easy lateral flexure of the body. Crocodiles swim by lateral flexure of the body and tail and it is thus quite possible that *Proterosuchus* was an accomplished swimmer. Charig (1965) has pointed out that the disparity of the fore and hind limb sizes in early archosaurs can be explained as an adaptation for swimming, using the hind limbs as powerful paddles at low speed.

Therefore, although the indications are that *Proterosuchus* was a land-dweller in the first place, feeding on the small-sized *Lystrosaurus* zone fauna, it could probably take to the water if occasion demanded.

The relationships of the Proterosuchia

Reig (1967) has postulated an unusual origin of the archosaurs from the pelycosaurs. His claims are based on the presence, in one pelycosaur, of an antorbital fenestra (Olson, 1965) and the apparent delay in ossification of the posterior coracoid element in some other pelycosaurs (Romer and Price, 1940). He also draws attention to the many superficial similarities between the Pelycosauria and the Proterosuchia. However, these similarities are those which might be seen in two groups of primitive reptiles deriving their ancestry from the same group of cotylosaurs.

On p. 566 Reig states that the Proterosuchia have no otic notch, yet in the material described (Haughton, 1924; Broili and Schröder, 1934; Fig. 1 and Fig. 2b above) there is clearly an emargination of the quadrate of the same sort as in *Euparkeria* (Ewer, 1965). In addition, the occipital condyle can be aligned with the suspensorium, when the quadrate is placed in the upright position. These two points, therefore, deny his claim that the younginids or millerettids are unlikely ancestors of the Proterosuchia.

If the antorbital fenestra in *Varanodon* (Olson, 1965) is genuine, then serious thought must be given to it. However, in therapsids and pelycosaurs there is frequently a depression of the skull bones in front of the eye, which may be accentuated by post-mortem damage, but until the specimens can be actually examined further discussion on this matter will lead to nothing definite.

On his p. 567 Reig enumerates a number of other skull characters to support the primary point of the antorbital fenestra. Among these he claims that the Proterosuchia and the varanospid pelycosaurs have a large lower temporal fenestra and a mandibular fenestra (in *Ophiacodon*, Romer, 1966, fig. 259). *Proterosuchus* has been shown not to have a mandibular fenestra in the accepted sense of the archosaurs' mandibular fenestra, and the possession of the large lower temporal fenestra is not really relevant. It could be said that the dicynodonts have both these characters, but they are not considered to be archosaur ancestors.

Ophiacodont vertebrae are similar to proterosuchian vertebrae and also to captorhinomorph vertebrae (Hughes, 1963), and so comparisons involving the vertebral columns of these animals are not valid.

Likewise, the scapula characters quoted by Reig are those of primitive reptiles, and do not imply a pelycosaur-archosaur relationship. The delay in ossification in the posterior coracoid element in some pelycosaurs may be a valid point, but taken by itself cannot be used to create such a relationship. The ilium of *Proterosuchus* is unique in possessing a ventral process which excludes the pubis from the acetabulum, as opposed to the condition in crocodiles where the ischium, by forward growth, does a similar thing. It is thus unlike that of either any pelycosaur or *Shansisuchus*, except in very general terms. Nor can the proterosuchian pelvis be said to be 'triradiate' in the sense that a semi-bipedal archosaur pelvis is triradiate, e.g. *Euparkeria*.

The humerus of both the proterosuchians and the pelycosaurs is of 'primitive' type, but that of *Proterosuchus* has neither an entepicondylar nor an ectepicondylar foramen. The femora are likewise indicative of a primitive, sprawling gait, and cannot be used to judge relationships.

Finally, as has been shown above, the ankle of *Proterosuchus* is more like that of an eosuchian than any other type, and has no real resemblance to those of pelycosaurs.

To sum up Reig's arguments, his case can be confined to the points concerning the antorbital fenestra and the possible lag in the ossification of the posterior coracoid in some pelycosaurs.

In opposition to Reig's case, it has been shown that the proterosuchians have many eosuchian characters and as these in turn can be readily derived from the millerettids via the younginids, there seems no point in forcing them into a relationship not supported by the mainstream of evidence.

In particular the presence of a tympanic notch, the eosuchian ankle and the loss of the tabular are all characters which tend to separate the proterosuchians from pelycosaurs at almost any level of evolution. Thus, it will be assumed that the evolution of the proterosuchians is from the Eosuchia, and not the synapsids.

Romer has proposed (1956, 1966, 1968) that the archosaurs and lepidosaurs represent separate groups deriving their ancestry independently from the millerosaurs. However, on the grounds of economy of hypotheses, this now ought to be reconsidered in the light of the foregoing descriptions.

As pointed out by Hughes (1963, pp. 221-2) the skull of *Proterosuchus* is archosaurian in general pattern, but it retains a number of primitive characters. These include the presence of a pineal opening and interparietal, postfrontal and supratemporal bones. There are in addition teeth on the palate and, most interesting, on the vomers. The external nares are laterally directed and sub-terminally placed. The internal nares are very long and more like those of *Youngina* than any archosaur. The braincase is of very primitive pattern with no laterosphenoid. There is no reflected hind border on the lower temporal fenestra and in this character the skull is also more like an eosuchian than an archosaur. There is also a strong case to be made for the skull being metakinetic. The lower jaw has no mandibular fenestra. The postcranial skeleton is primitive and, of particular importance, the ankle is eosuchian in design.

Whereas the archosaurian characters of the skull of *Proterosuchus* listed by Hughes are found to a greater or lesser extent in later archosaurs, the presence of an antorbital fenestra is the only one present in *Proterosuchus* which is really diagnostic of archosaurs.

In the vertebral column a number of archosaurian characters are seen, but these too seem to be as much a retention of conservative characters as specifically archosaurian. For example, close resemblances exist between the neck vertebrae of *Proterosuchus* and *Caiman* (Broili and Schröder, 1934, fig. 17) and also *Sphenodon* (Hoffstetter and Gasc, 1969, fig. 28). Likewise the trunk vertebrae show a number of primitive features, notably the web of bone joining the diapophysis and parapophysis, but they have few which are specifically archosaurian. For instance, the parapophysis is never borne entirely on the transverse process and the centra are not keeled. However, there is nothing in the vertebral column of *Proterosuchus* which debars it from being ancestral to the true archosaur state. The sacrum is of interest particularly in the way that the transverse processes are divided distally. The nearest comparison in this respect is to *Clevosaurus*, an Upper Triassic rhynchosaur from the fissure deposits of England (Swinton, 1939). Further information on the sacra of rhynchosaurs is clearly needed.

The limb girdles are primitive in their general structure, with the 'sauropsid' specialisation of the single coracoid ossification and no sternal ossification. The ilium is specialised and of interest in the possession of the ventral process which eliminates the pubis from the acetabulum. This, however, is probably one of the few characters which is applicable at the generic level only.

The extremities are of note inasmuch as the carpus shows no elongation of any of the wrist bones, as might be expected in a crocodile ancestor, and the tarsus is of almost typically eosuchian design (Hughes, 1968).

It thus seems on close examination that *Proterosuchus* retains more features of a conservative nature than it possesses features truly archosaurian. The general structure of the skull is, however, clearly antecedent to that of e.g. *Euparkeria*. Likewise the vertebral column is ancestral to the archosaur type, though retaining several primitive features. Even the crocodilian ankle joint is readily derived from that of an eosuchian by the elimination of the astragalus as a separate ossification, the migration of the fibula on to the intermedium and the development of the incipient peg-and-socket joint between the intermedium and the calcaneum. The characteristic feature of *Proterosuchus*, apart from the antorbital fenestra, is the downturned premaxilla. The structure and relationship of this rostrum is exactly the same as in the Rhynchocephalia as noted by Broili and Schröder. The presumed late descendants of the Proterosuchia, the Phytosauria, also retain this rostrum, with a very much elongated maxillary region (Walker, 1968; Romer, 1966, figs 216 and 217). A further point of interest is the similarity between the bifurcated sacral ribs of *Clevosaurus* and *Proterosuchus*. Hughes (1968) has demonstrated very well the similarity of the eosuchian ankle with that of the rhynchosaur. A similar comparison has also been noted with the ankle of *Proterosuchus*.

Romer (1968, p. 124), quoting Tatarinov (1964), points out that at least one effort has been made to include the Rhynchocephalia within the Eosuchia, and if it is remembered that herbivorous groups are nearly always preceded by a closely related carnivorous group, then it is possible that *Proterosuchus* represents a slightly modified carnivorous rhynchocephalian.

If these points are true, then the Diapsida are a truly monophyletic group, tracing their ancestry back to the Millerosauria through the Eosuchia. There is thus no need to postulate a separate origin for the Lepidosauria and Archosauria as Romer does.

Of the Eosuchia, *Youngina* is probably early enough to be typical of the first true diapsids (Watson, 1957).

The suggested classification of the Proterosuchia is almost identical to that of Hughes (1963, p. 235). The definitions are based on those of Romer (1956).

Subclass Diapsida
Infraclass Archosauria
Order Thecodontia
Suborder Proterosuchia

Family Proterosuchidae— *Proterosuchus* (= *Chasmatosaurus* = *Elaphrosuchus*)
L. Trias. S. Afr., EAs.
Archosaurus U.Perm EEu.

Family Erythrosuchidae—*Erythrosuchus* L.Trias. S. Afr., EAs.
Euparkeria L. Trias. S. Afr.
Shansisuchus L. Trias. EAs.

The affinities of the other forms mentioned by Hughes are so doubtful that they are not included in this classification.

In Romer (1956, pp. 591-592) the Order Thecodontia is well defined, and only two slight modifications are suggested. These are that supratemporals are present in the earliest forms and that in some the ilium may be ventrally extended to exclude the pubis from the acetabulum.

The consideration that the suborder Proterosuchia may be descended from a line of carnivorous Rhynchocephalia leads to a certain confusion as to the distinction between subordinal and familial characters, but the following definitions are suggested tentatively.

Suborder Proterosuchia.

External nares subterminal. Postfrontal and supratemporal present in the most primitive members. Upper temporal fenestra faces dorsally. Parietal foramen present in some. Otic notch moderately developed. Skull metakinetic. Squamosal with process extending over head of quadrate. Broad parietals. Long interpterygoid vacuities, extending forward between vomers. Palate slightly arched posteriorly. Epipterygoid well developed. Paroccipital processes slant diagonally backwards. Post-temporal fenestrae small or absent. Braincase primitive with no laterosphenoid. Teeth sub-thecodont. Teeth present on pterygoids, palatines and vomers. Intercentra present. Centra amphicoelous in the cervical region, procoelous in the trunk. Parapophyses not borne on long transverse processes. Long cervical ribs. Pelvis of primitive type, with plate-like pubes and ischia. Hind limbs longer than fore. No ossified armour.

Family Proterosuchidae.

Proterosuchians little developed beyond the eosuchian grade. Skull elongated, high posteriorly and with characteristic premaxillary rostrum. No external mandibular fenestrae; dentary capable of movement relative to post-dentary bones. Ankle almost identical to that of an eosuchian, containing three bones in the proximal row, four in the distal. Calcaneum with well developed lateral tuber.

Genus 1. *Archosaurus* U. Perm. EEu.

Inadequately known as yet, but with a definite proterosuchian skull table.

Genus 2. *Proterosuchus* L. Trias. EAs., S. Afr.

Typified by premaxillary rostrum, a long neck region and no posttemporal fenestrae. Sacral ribs divided distally and ilium with ventral process excluding pubis from acetabulum. Probable overall length not more than 1·5 m.

Family Erythrosuchidae.

Proterosuchians differing from the Proterosuchidae in lacking the premaxillary rostrum and whose ankles contain only four bones; two proximal and two distal. Calcaneum with tuber of varying sizes. Skull relatively shorter than in the Proterosuchidae.

Genus 1. *Erythrosuchus* (= *Garjainia* = *Vjushkovia*) L. Trias. S. Afr., EAs.

Short neck region. Intermedium (? astragalus) globular, calcaneum with median ventral process and moderate tuber. Large sized animal, two metres or more in length. The actual provenance of this genus will depend on its possession or not of a laterosphenoid ossification. If present, then the affinities are probably more with the pseudosuchia.

Genus 2. *Euparkeria* L. Trias. S. Afr.

Small sized animal with ankle reduced from condition seen in *Erythrosuchus*. Probably incipiently bipedal.

Genus 3. *Shansisuchus* L. Trias. EAs.

Similar in many ways to *Erythrosuchus*, but with two antorbital fenestrae on each side of the face and with an ankle-joint more crocodilian than that of any other proterosuchian.

ACKNOWLEDGEMENTS

This paper is presented as a tribute to Dr F. R. Parrington, F.R.S. who introduced me to Karroo palaeontology. The original MS. was criticised by Professors Romer and Crompton, Drs Hughes, Kemp, Keyser and Joysey and Messrs Gow and Kitching. To Dr S. H. Haughton, F.R.S., I am especially grateful for suggesting the project and his lively interest in its progress. Dr T. H. Barry, the late Drs S. H. Rubidge and A. C. Hoffman gave unstinting loan of specimens in their collections over a long period of time. Mr D. Wolfaardt took the numerous photographs on which the reconstructions are based.

The Murray Fund of the University of Edinburgh provided a grant for the purchase of equipment which has been extensively used in the preparation and photography of the specimens described in this paper. The C.S.I.R. of the Republic of South Africa provided financial assistance both for the running of the Bernard Price Institute for Palaeontological Research (B.P.I.) and my travel, without which this work would not have been finished in time for its presentation at the 18th Symposium of Vertebrate Palaeontology and Comparative Anatomy. The balance of the money came from extremely generous grants to the B.P.I. made by the Research Committee of the Council of the University of the Witwatersrand.

A final word of thanks is due to the painstaking efforts of the preparators in the B.P.I., especially Jonas Morifi and Elias Makapane.

REFERENCES

BRINK, A. S. 1955. Notes on some thecodonts. *Res. Nas. Mus. Bloemfontein*, **1**, 141-148.
BROILI, F. and SCHRÖDER, J. 1934. Beobachungen an Wirbeltieren der Karrooformation. V. Über *Chasmatosaurus vanhoepeni* Haughton. *Sitz. Bayer. Akad. Wiss. München*, (for 1934), 225-264.
BROOM, R. 1903. On a new reptile (*Proterosuchus fergusi*) from the Karroo beds of Tarkastad, South Africa. *Ann. S. Afr. Mus.*, **4**, 159-163.
—— 1932. On some South African pseudosuchians. *Ann. Natal Mus.*, **7**, 55-59.
—— 1946. A new primitive proterosuchid reptile. *Ann. Transv. Mus.*, **20**, 343-346.
CHARIG, A. J. 1965. Stance and gait in the archosaur reptiles. *Advancement of Science*, **22**, 537.
CRUICKSHANK, A. R. I. 1970. Early thecodont braincases. In *Proc. Second International Symposium on Gondwana Stratigraphy and Palaeontology. Cape Town and Johannesburg*. Ed. S. H. Haughton C.S.I.R. Pretoria.
EWER, R. F. 1965. Anatomy of the thecodont reptile *Euparkeria capensis* Broom. *Phil. Trans. R. Soc. Lond.*, (B), **248**, 379-435.
HAUGHTON, S. H. 1924. On a new type of thecodont from the Middle Beaufort Beds. *Ann Transv. Mus.*, **11**, 93-97.
HOFFMAN, A. C. 1965. On the discovery of a new thecodont from the Middle Beaufort Beds. *Res. Nas. Mus. Bloemfontein*, **2**, 33-40.
HOFFSTETTER, R. and GASC, J-P. 1969. Vertebrae and ribs of modern reptiles. In *Biology of the reptilia*. Vol. **1**. C. Gans, A. d'A Bellairs and T. S. Parsons. Eds. Academic Press, London and New York.
HUGHES, B. 1963. The earliest archosaurian reptiles. *S. Afr. Jour. Sci.*, **59**, 221-241.
—— 1968. The tarsus of rhynchocephalian reptiles. *J. Zool. Lond.*, **156**, 457-481.
KITCHING, J. W. 1968. On the *Lystrosaurus*-zone and its fauna, with special reference to some immature Lystrosauridae. *Palaeont. afr.*, **11**, 61-76.
OLSON, E. C. 1965. Chickasha vertebrates. *Oklahoma Geological Survey* Circular **70**, 48-57.
PARRINGTON, F. R. 1948. Labyrinthodonts from South Africa. *Proc. zool. Soc. Lond.*, **118**, 426-445.
PRICE, L. I. 1935. Notes on the braincase of *Captorhinus*. *Proc. Boston. Nat. Hist. Soc.*, **40**, 377-386.
REIG, O. A. 1967. Archosaurian reptiles: a new hypothesis on their origins. *Science*, **157**, 565-568.
ROBINSON, P. L. 1962. Gliding lizards from the Upper Keuper of Great Britain. *Proc. geol. Soc. Lond.*, **1601**, 137-146.
—— 1967. The evolution of the lacertilia. *Colloq. int. Cent. Natn. Rech. Scient.*, **163**, 395-407.
ROMER, A. S. 1956. *Osteology of the reptiles*. Chicago University Press, Chicago.
—— 1966. *Vertebrate paleontology*. 3rd Ed. Chicago University Press, Chicago.
—— 1968. *Notes and comments on vertebrate paleontology*. Chicago University Press, Chicago.
ROMER, A. S. and PRICE, L. I. 1940. Review of the Pelycosauria. *Geol. Soc. Amer. Spec. Paper*, **28**, 1-538.
SWINTON, W. E. 1939. A new Triassic rhynchocephalian from Gloucestershire. *Ann. Mag. nat. Hist.* (11), **4**, 591-594.
TATARINOV, L. P. 1961. Material on the U.S.S.R. pseudosuchians. *Palaeont. Zh. U.S.S.R.* **1**, 117-132.
—— 1964. Lepidosauria. *In: Osnory Paleontolgii*. J. A. Orlov Ed. Part 5. Amphibia, Reptiles, Birds. 439-492. Moscow.
THORNLEY, A. L. 1970. Epidermal remnants of *Proterosuchus vanhoepeni* (Haughton). *Palaeont. afr.*, **13**, 57-60.
WALKER, A. D. 1964. *Ornithosuchus* and the origin of the carnosaurs. *Phil. Trans. R. Soc. Lond.*, (B) **248**, 53-134.
—— 1968. *Proterosuchus, Proterochampsa*, and the origin of phytosaurs and crocodiles. *Geol. Mag.*, **105**, 1-14.
WATSON, D. M. S. 1957. On *Millerosaurus* and the early history of the sauropsid reptiles. *Phil. Trans. R. Soc. Lond.*, (B) **240**, 325-400.
YOUNG, C. C. 1936. On a new *Chasmatosaurus* from Sinkiang. *Bull. Geol. Soc. China*, **15**, 291-311.
—— 1958. On the occurrence of *Chasmatosaurus* from Wahsiang, Shansi. *Vert. Palasiat.*, **2**, 259-262.
—— 1963. Additional remains of *Chasmatosaurus yuani* Young from Sinkiang, China, *Vert. Palasiat.*, **7**, 220-222.
—— 1964. The pseudosuchians in China. *Palaeont. sinica*, **151**, (n.s.c. **19**), 109-204.

ADDENDUM

Roy Chowdhury (pers. comm.) mentions a new proterosuchian from the Yerrapalli fauna. This he describes as being three metres long with a neck about a third of this. An estimate for the length of the neck in NM C. 3016 is 21-23 cm and for the pre-caudal region (excluding the neck) is 37 cm (Plate I). Assuming that the overall length of the specimen when alive was not more than 1·5 metres, then the neck would comprise 15% of the total length, or 35% of the whole precaudal length. Both values seem very high, but the significance of them will be enhanced with extra study. However, this condition in the early proterosuchians may foreshadow that seen in *Tanystropheus*, an enigmatic form from the Swiss Triassic. Apart from the similar primitive post-cranial skeletons, both *Tanystropheus* and *Proterosuchus* have very long necks and fore- and hindfeet of very disproportionate sizes.

ABBREVIATIONS USED IN FIGS 1-10

A	Angular	P	Parietal
ART	Articular	PAL	Palatine
BO	Basioccipital	PMX	Premaxilla
BSP	Basisphenoid	PO	Postorbital
CL	Clavicle	POF	Postfrontal
CO	Coracoid	PRF	Prefrontal
D	Dentary	PRO	Prootic
ECT	Ectopterygoid	PSP	Parasphenoid
EPI	Epipterygoid	PT	Pterygoid
EO	Exoccipital	PU	Pubis
F	Frontal	Q	Quadrate
ICL	Interclavicle	QJ	Quadratojugal
IL	Ilium	SA	Surangular
IP	Interparietal	SC	Scapula
IS	Ischium	SCP	Sclerotic plates
J	Jugal	SO	Supraoccipital
L	Lacrimal	SQ	Squamosal
MX	Maxilla	ST	Supratemporal
N	Nasal	V	Vomer
OP	Opisthotic		

acet.	acetabulum	*pre.zyg.*	prezygapophysis
antorb.fen.	antorbital fenestra	*pt. ramus q.*	pterygoid ramus of quadrate
ast.	astragulus	*pu. tub.*	pubic tuberosity
b.tub.	basal tubera	*q.f.*	quadrate foramen
b.pt. proc.	basispterygoid process	*q. ramus pt.*	quadrate ramus of pterygoid
calc.	calcaneum	*r.art. proc.*	retroarticular process
cor. f.	coracoid foramen	*sacr. fac.*	sacral facets
dia.	diapophysis	*sig. n.*	sigmoid notch
dor. proc.	dorsal process of epipterygoid	*thy. fen.*	thyroid fenestra
fen. ov.	fenestra ovale	*tib. fac.*	tibial facet
fib. fac.	fibula facet	*V 2+3*	notch for cranial nerve V
for. mag.	foramen magnum	*VI*	point of exit for cranial nerve VI
gl. fac.	glenoid facet	*VII pal.*	point of exit for palatal branch of cranial nerve VII
i.c.	intercentrum		
int.	intermedium	*VII c.t.*	point of exit for facial branch of nerve VII
n.c.	neural canal	*X*	jugular foramen and nerves X, XI, and XII
p. ant.	pila antotica		
pin. f.	pineal foramen	*XII*	points of exit of nerve XII in *Euparkeria*
post. zyg.	postzygapophysis		

A. J. CHARIG

The evolution of the archosaur pelvis and hind-limb: an explanation in functional terms

INTRODUCTORY NOTE

This essay deals only with selected aspects of the archosaur pelvis and hind-limb and is greatly simplified in many respects. A full consideration of all aspects of the subject would require far more space than is available here.

To facilitate comparisons all photographs and drawings in lateral view are of the left side, i.e., the animal's head would always be to the left.

LOCOMOTOR TRENDS IN THE ARCHOSAURS

It has long been recognised that the most important characteristic of the archosaurs, to which they owed their dominant position in the land faunas of the Mesozoic, was their development of greatly improved methods of locomotion. This is not disputed, although the gross structural changes of which we know (and which might just as well be termed mechanical changes) were probably no more important in bringing about those improved methods of locomotion than were the unknown histological and physiological changes which almost certainly accompanied them. Further, there can be little doubt that the new methods of locomotion resulted from *two* evolutionary trends. One of these led towards the adoption of a bipedal stance and gait; the other (paralleling a similar trend in mammalian evolution) consisted essentially of the changing of the limb posture from a sprawling position, with the limbs projecting sideways from the body and the belly resting on the ground, to a more upright position, with the limbs supporting the body from beneath and moving in a vertical plane. It has hitherto been believed, however—the story is perpetuated in most of the text-books—that the trend towards bipedality was the more important of the two; it is also believed that it was initiated at the very beginning of archosaur history, at the very beginning of the Mesozoic (i.e., early in Lower Triassic times), and that its prior development was a prerequisite for the initiation of the other trend. A necessary concomitant of this was the belief that all quadrupedal archosaurs must be *secondarily* quadrupedal, having reverted to that condition from bipedality.

The assumption that the trend towards bipedality began at the beginning of archosaur history was challenged for the first time at the 9th Symposium on Vertebrate Palaeontology and Comparative Anatomy, held at University College London in 1961; the programme included a paper entitled 'Stance and gait in the archosaur reptiles'. Another paper on the same topic, developed further, was read in Cambridge at the 1965 meeting of the British Association for the Advancement of Science (Charig, 1965, 1966; Charig et al., 1965). These suggested that the stem group of archosaurs, the

Thecodontia, were *not* bipedal (except perhaps in a few isolated instances), that bipedality originated independently in only *some* of the daughter groups, and that many of the quadrupedal groups were *not* secondarily quadrupedal; that is, they were not reversions from bipedal forbears. Some workers (such as Halstead, 1969) have embraced these ideas more enthusiastically than did their originator; others have accepted them with caution and with certain reservations (Romer, 1966, 1968); yet others seem to have virtually ignored them (Colbert, 1969).

To the suggestion that the Thecodontia were not bipedal are now added the further suggestions that they had improved their limb posture from the sprawling position only partially and only in the more advanced members of the order; that the essential distinction between the Thecodontia (and Crocodilia) on the one hand and the Saurischia and Ornithischia on the other lies in the adoption by the latter two orders of the 'vertical' limb posture; and that the various changes which took place in the dinosaur girdles and limbs, especially in the pelvis, the femur, the cruro-tarsal connexion, the tarsus itself and the pes, are largely explicable in terms of this 'vertical' limb posture and, in some cases only, of bipedality. Indeed, it is possible to go further and to offer at least partial explanations for the remarkable differences between the pelves of saurischian and ornithischian dinosaurs and for the adoption by dinosaurs of both orders of such characters as digitigrady and bipedality. 'Vertical' limb posture, not bipedality, is the primary character; it is always accompanied by digitigrady and *sometimes* by bipedality, both consequent upon its prior adoption and, therefore, regarded as secondary.

In the past it has been decided by various authorities, seemingly on purely arbitrary grounds, that the primary diagnostic character by which members of both dinosaurian orders (Saurischia and Ornithischia) might be distinguished from the parent Thecodontia, in particular the Pseudosuchia, is the presence of a fenestration in the acetabulum. This distinction has gained general acceptance in archosaur taxonomy. By a happy coincidence this character is also ideal in the present circumstances, despite its apparently arbitrary nature, for it seems reasonable to suppose that the acetabular fenestration is an excellent indicator of the 'vertical' posture of the limbs (see below, p. 132). Certain characters, such as the 'crocodiloid' tarsus, are typical concomitants of the unfenestrated acetabulum found in the Pseudosuchia; others, like the inturned head to the femur and digitigrady, are almost invariably associated with the widely fenestrated acetabulum of the dinosaurs.

NEW LIGHT ON THE ORNITHISCHIAN PELVIS

The distinction between the Saurischia and the Ornithischia, as their names imply, is based on the form of the pelvis (Fig. 1).

According to the text-books the most striking characteristic of the ornithischian pelvis is the fact that the pubis has *two* elongated rami. One ramus is directed forwards, rather like the entire pubis of pseudosuchians and saurischians; the other is directed downwards and backwards to lie immediately beneath the ischium, like the pubis of birds. (In many Cretaceous ornithischians the pubis is simpler and is considered to be reduced in various ways.) There are two entirely different interpretations of the homologies of these two rami; each has been regarded as the true pubis and the other as a secondary outgrowth. It is now generally accepted, however, that the posterior ramus is the true pubis, which was originally directed forwards and which has rotated backwards through 90°. The evolutionary history and functional significance of these developments have hitherto been little understood.

Previous studies have been hampered by the fact that virtually all that was known of the structure of the ornithischian pelvis was based on animals of Upper Jurassic age or later. New light has now been shed on the matter by the discovery within the last twenty years of new, earlier ornithischians with good pelvic material. Most notable are a specimen from the Lower Lias of Dorset, probably a small *Scelidosaurus* (entire pelvis shown in Fig. 2; with femur also in Plate VIA), and various ornithischians from the Upper Trias of southern Africa. The pubes of all these otherwise very different animals share one striking characteristic: the anterior ramus is scarcely developed at all.

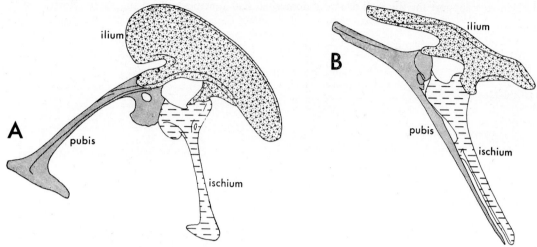

Fig. 1 **A.** Pelvis of saurischian dinosaur (*Ceratosaurus*). [After Gilmore]
B. Pelvis of ornithischian dinosaur (*Thescelosaurus*). [After Romer]
See p. 155 for abbreviations used in Figs 1-11 and Plates.

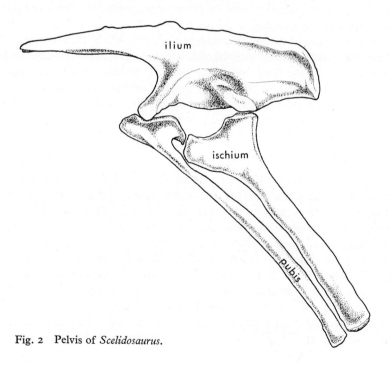

Fig. 2 Pelvis of *Scelidosaurus*.

It therefore seems likely (1) that the ornithischian pubis is primitively *without* a properly developed anterior ramus; (2) that the large anterior ramus of later Ornithopoda, Stegosauria and Ceratopsia was a secondary development; and (3) that Ankylosauria, which do not have an anterior ramus, have not lost it but never possessed it. This confirms the current belief that the posterior ramus represents the true pubis. It also confirms that the anterior ramus of the pubis was not essential to the ornithischian as a support for the wall of the abdomen, as had previously been suggested (Romer, 1927, p. 244).

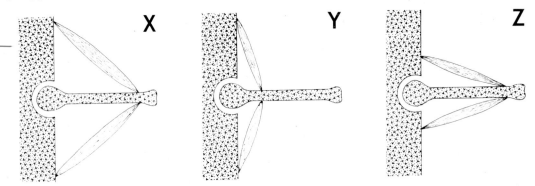

Fig. 3 The origin and insertion of muscles running from a limb-girdle to a propodial.

Hitherto the terms *triradiate* and *tetraradiate* have been applied to the saurischian and ornithischian pelves respectively to describe their essential difference: the typical ornithischian was supposed to have a bifurcated pubis and hence an additional 'prong'. Neither was a particularly good term, for in both cases the ilium was counted as only one prong despite the fact that it projects both forwards and backwards. Further, some confusion existed in that certain authors (e.g. Colbert, 1969, p. 197) regarded the posterior ramus of the ornithischian pubis as part of the same prong as the immediately adjacent ischium, considering instead that the additional prong of the tetraradiate ornithischian pelvis was the anterior process of the ilium; the latter process, however, is well developed also in many saurischians. It is now suggested, in the light of the new information, that the terms *triradiate* and *tetraradiate* be used no longer; better terms would be *propubic* for the saurischian pelvis and *opisthopubic* for the ornithischian. Most other reptile pelves could be described as propubic, the bird pelvis as opisthopubic.

GENERAL NOTES ON LIMB MUSCLES

It should be remembered that all the extrinsic muscles which insert on the hind-limb originate primitively from the pelvic girdle or from the caudal region of the vertebral column. Sometimes, however, the origin of certain protractor muscles may transfer to the ventral surfaces of the centra and transverse processes of the trunk vertebrae; an example of such a muscle is part of the pubo-ischio-femoralis internus of crocodiles (equivalent to the psoas of mammals).

Different positions of origin and insertion of limb musculature relative to the skeletal fulcrum confer different advantages and disadvantages. Fig. 3 shows alternative arrangements for muscles moving a propodial which is articulated with its girdle by a ball-and-socket. In X the origin and insertion of the muscles are more or less equidistant from the fulcrum; in Y and Z they are not. This means that in X, as the propodial approaches the girdle, the origin and insertion will approach each other closely; in Y and Z, by contrast, they will remain well apart. But it is found that in living vertebrates striated muscle does not normally contract by more than about 30% of its relaxed length. It is, therefore, obvious that arrangement X limits the degree to which the propodial can approach the girdle, no matter what may be the distance from fulcrum to origin and insertion. Incidentally, the approach of the propodial right up to the girdle would necessitate the very considerable extension of the opposing muscle.

Arrangement X is also unsatisfactory in that, when origin and insertion are far from the fulcrum, the muscle extends across the broad angle between the body and the propodial as a great sheet which might hamper the limb's general mobility. On the other hand, it is equally unsatisfactory when both origin and insertion are close to the fulcrum, for the muscle must then be short and weak.

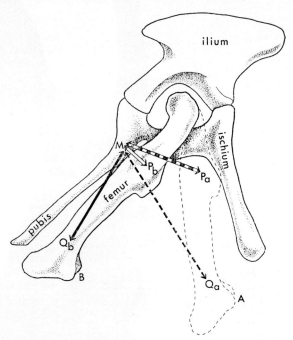

Fig. 4 A femoral protractor in a prosauropod saurischian. M represents its point of origin on the pubis; P and Q represent two alternative points of insertion on the femur. The femur has moved from position A to position B.

None of the difficulties mentioned above applies to arrangements Y or Z. Here, however, there is a different objection. When a muscle pulls upon a limb-bone its rotatory effect (i.e., the magnitude of the turning couple produced) depends upon several factors; one of these is the angle of pull, for, of the force which the muscle exerts, only the component normal to the axis of the moving bone is effective in rotating the latter about the fulcrum. In Y, as the propodial approaches the girdle, the direction of pull of the muscle becomes very oblique to the shaft of the bone so that the rotatory component becomes small and ineffectual; and in Z the same applies throughout the entire arc of rotation. This difficulty could reach significant proportions in arrangement X only if the muscle were over-extended.

Z is open to an even more serious objection. The muscle insertion is at a fixed distance from the fulcrum, whatever the arrangement, and in its movement describes an arc about the latter; thus, when the origin of the muscle is close to the fulcrum (as it is in Z), the distance from origin to insertion cannot vary much and the contraction of the muscle is, in consequence, greatly restricted. This does not apply to Y, where the origin of the muscle is far from the fulcrum.

In Z a large muscular contraction is theoretically necessary in order to produce a relatively small

Plate I COTYLOSAURIA. *Limnoscelis paludis* Williston. Lower Permian, New Mexico. Cast of complete skeleton.

angular displacement, yet in practice the contraction is greatly restricted for the reason just given; the arc of rotation of the propodial must therefore be comparatively narrow. In Y, on the other hand, a small muscular contraction will produce a relatively large angular displacement, so that the arc of rotation of the propodial is not restricted by these factors; but the arrangement is to some extent unsatisfactory in that the force is applied to the lever so close to the fulcrum that it needs to be comparatively large to be effective. Z is essentially a 'low gear' arrangement, producing slow and powerful movement and typical of large heavy animals; Y is a 'high gear' arrangement, producing rapid but weak movement and better suited to small light animals. In these respects X offers a reasonable compromise.

Fig. 4 affords a more concrete example of these matters. The muscle inserting closer to the acetabulum (MP) cannot operate satisfactorily because of its limited contractility; it cannot on its own move the femur from position A to position B, for MP_b is much less than 70% of MP_a. The muscle inserting farther from the acetabulum (MQ), however, has contracted by only 30% in moving the femur from A to B; MQ_b is exactly 70% of MQ_a. At the same time the angle of pull of muscle MP changes very little while that of MQ becomes much more oblique, more nearly parallel to the shaft of the femur.

In short, no arrangement is perfect; the nature of the compromise evolved in each particular case must depend upon the circumstances. A Y-type arrangement would probably be the most suitable for the hind-limbs of a coelurosaur, something between X and Z for those of a sauropod.

LIMB POSTURE AND THE PELVIS

(a) *Primitive amphibians and reptiles* ('*sprawlers*')

In primitive amphibians and reptiles (Fig. 5A) the hind-limb projects sideways from the body at right angles to the longitudinal axis; the thigh lies horizontally (indeed, the knee may well be higher than the hip-joint), the lower leg stands vertically, and the foot is placed flat on the ground. This is shown well by the skeleton of the primitive cotylosaur *Limnoscelis* (Plate I). The plane of the limb flexure is transverse, and the only effect of extension of the right-angled or acute articulation at the knee is to raise the trunk from the ground; to keep it there requires considerable effort.

Locomotion may still be performed to some extent by throwing the trunk and tail into sigmoid waves (Fig. 5D) as in the salamanders and *Limnoscelis*. The ability to do this is retained in lizards, from which have evolved numerous legless forms including the snakes, and in various groups of swimming reptiles.

Locomotion by the limbs involves swinging the femur in a horizontal arc. The ilium is primarily for the attachment of the pelvis to the vertebral column; the forwardly directed pubis provides the origin of the femoral protractor muscles, the backwardly directed ischium (and the base of the tail) those of the retractors, and both lie more horizontally than vertically. The arc of movement of the femur is extensive, even though the muscles cannot pull it as far forwards as the pubis. In all tetrapods, of course, the retractory phase is the 'power stroke' of locomotion; the protractory phase is merely the recovery stroke. It is therefore particularly important that conditions (i.e. location of origin and insertion) should permit the effective operation of the ischiofemoralis and caudifemoralis muscles.

The pelvis and femur of two Quaternary 'sprawling' reptiles—Pleistocene tortoise and living monitor lizard—are shown in Plate II.

(b) *Advanced reptiles and their descendants*

More advanced reptiles and reptilian descendants (birds and mammals) are partly or wholly modified towards the adoption of what may be termed a 'vertical' posture of the limbs (Fig. 5B, C). Where

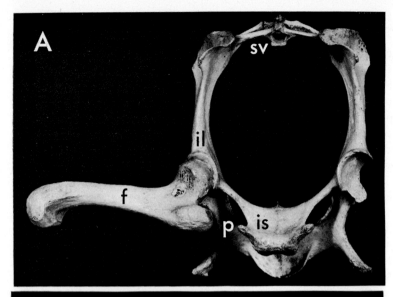

Plate II
A. CHELONIA. *Testudo vosmaeri* Fitzinger. Pleistocene, Rodriguez. Sacrum, pelvis and left femur from behind.

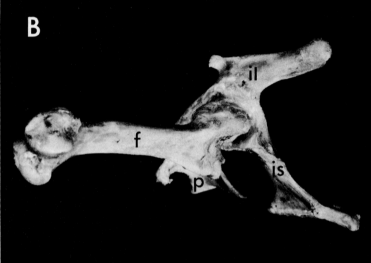

B. SQUAMATA. *Varanus niloticus* (Linnaeus). Living Nile Monitor. Left side of pelvis and left femur in oblique posterolateral view.

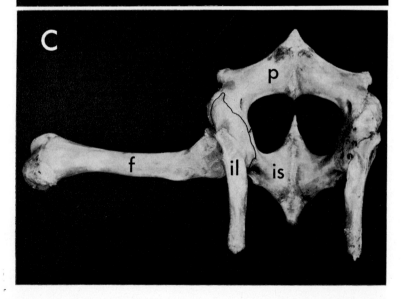

C. Same individual as B. Whole pelvis and left femur from above.

modification is complete the femur has been swung forwards and inwards and the humerus backwards and inwards, so that both limbs are held in a vertical plane beneath the side of the body; the knee and elbow are directed forwards and backwards respectively (Fig. 5F). The limbs now move back and forth, each in a parasagittal plane. The advantages of this 'vertical' limb posture and movement are manifold. The effort which the 'sprawler' requires to drag its belly along the ground is no

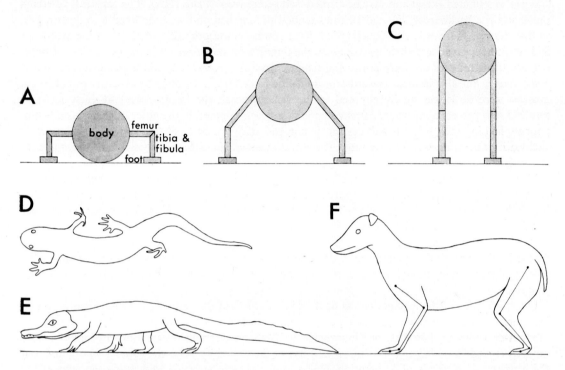

Fig. 5 **A.** A 'sprawler'.
  B. A 'semi-improved' tetrapod with its limbs held as for rapid progression.
  C. A 'fully improved' tetrapod.
  D. A 'sprawler' from above, showing sigmoid lateral undulations.
  E. A 'semi-improved' tetrapod (*Alligator mississippiensis*) during rapid progression. [After Schaeffer]
  F. A 'fully improved' tetrapod in plantigrade pose. [Modified from Gray]

longer needed, yet keeping it off the ground imposes little or no strain on the extensor muscles of the limbs. Because the animal's weight is transmitted to the ground through the foot the latter is less likely to slip when horizontal forces are applied to it during locomotion (see below, p. 148); and in any case, all transverse horizontal forces operating between the feet and the ground and between the limbs and the body are reduced. Most important of all, the potential efficiency of the limb as a locomotor organ is greatly increased, for it can be employed not only as a lever but also as an extensible strut to push the body forwards. (A further advantage for mammals is that the adoption of a 'vertical' limb posture enables some of them—e.g., greyhound—to supplement their other locomotor mechanisms by bounding, i.e. by alternately flexing and extending the back in a dorsoventral direction.)

(i) *'Semi-improved' reptiles.* The most primitive archosaurs, the Thecodontia, were certainly incapable of adopting a 'vertical' limb posture. None of the dinosaur hallmarks—the fenestration in the acetabulum, the strong supra-acetabular crest, or the inturned head to the femur—had yet developed. The pubis and ischium of *Chasmatosaurus*[1], the earliest and most primitive thecodontian, are essentially horizontal plates. All later, more advanced thecodontians could nevertheless modify their stance and gait to a limited extent and may be termed 'semi-improved' (Plate IIIA). The forwardly directed pubis and the backwardly directed ischium tended to lengthen and were directed also downwards, so that they lay more or less obliquely; the fulcrum (acetabulum), the sites of origin of the protractor and retractor muscles and their insertions on the femur were no longer all in the same plane. Consequently those muscles not only pulled the femur forwards (protraction) and backwards (retraction) respectively, but also inwards towards the mid-line (adduction). In other words, the pubofemoral muscles were no longer totally opposed to the ischiofemoral; the force applied by each could be resolved into two components, the first component of each opposed to the other but the second components helping each other to adduct the femur and raise the body off the ground. The proterosuchian *Erythrosuchus* was an early stage, the pseudosuchian *Mandasuchus* (shown in the photograph) was rather later and more typical.

The earliest crocodilians differ little from pseudosuchians in the form of the pelvis. Modern crocodilians (Plate IIIB) still fall essentially into the same category, although several peculiarities in pelvic structure have evolved since Triassic times. They are more dinosaur-like than the pseudosuchians in having a small acetabular fenestration; they are less dinosaur-like in that the supra-acetabular crest is scarcely developed at all. Further, the pubis is excluded from the acetabulum by the ischium; pubis and ischium both descend very steeply; the ischium expands into a large plate below; and the ilium, though massive, is quite short posteriorly and is virtually without an anterior process.

Our knowledge of the locomotion (as of the musculature) of the extinct pseudosuchians is in part inference—from comparative anatomy and simple mechanics—and in part pure conjecture. In modern crocodilians, however, our knowledge is based on fact and observation. Locomotion in a modern crocodilian (*Alligator mississippiensis*) has been studied by Schaeffer (1941), who noted that there were two methods of locomotion. One was for slow progression and was essentially of the 'sprawling' type found in lower reptiles. The other, for rapid progression, was characterised by a partial adduction of the femur which enabled the latter to swing in a plane approximately 50° off the horizontal and kept the body raised from the ground. (In this connexion it is interesting to note that one reliable authority—J. Lucas, personal communication—has himself seen an eight-foot crocodile run *bipedally* for a distance of some thirty yards.) These observations justify the use of the term 'semi-improved' for the locomotion of crocodilians and suggest a similar method for the structurally similar pseudosuchians.

With the body clear of the ground, the hind-limb partly rotated forwards and a highly evolved, flexible tarsus the 'semi-improved' archosaur could walk and run much better than its Permian ancestors and Triassic cousins.

In modern crocodilians, however, as in their more primitive ancestors, protraction of the femur must still be accomplished by the pubofemoral musculature; this is in spite of the fact, mentioned above, that part of the pubo-ischio-femoralis internus has transferred its origin to the undersides of the centra and transverse processes of the trunk vertebrae. The iliofemoral musculature can play no part in protraction, for its site of origin extends no farther forwards than the front edge of the

---

[1] Cruickshank, in his essay in this volume, regards *Chasmatosaurus* as a junior synonym of *Proterosuchus*.

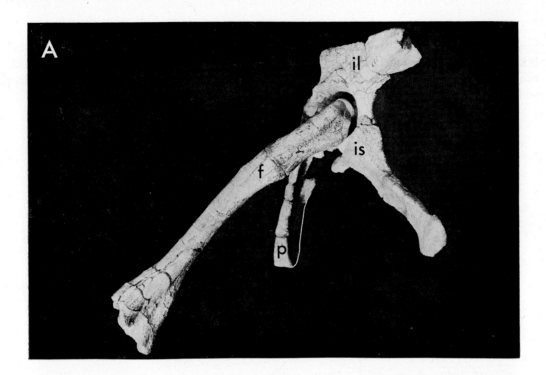

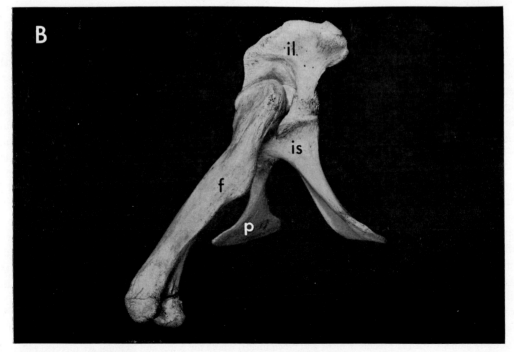

Plate III.
A. PSEUDOSUCHIA. *Mandasuchus tanyauchen* Charig (holotype).
Manda Formation (Middle Trias), Tanganyika.
Left side of pelvis and left femur in oblique posterolateral view.

B. CROCODILIA. *Crocodylus niloticus* Laurenti. Living Nile Crocodile.
Left side of pelvis and left femur in oblique posterolateral view.

acetabular socket (Fig. 6); the muscle originates immediately *dorsal* to the socket and a little behind it and can function only as an adductor. Indeed, there are no muscles arising from the ilium which run backwards to insert on any part of the leg, and there is therefore none originating on that element which could draw the leg forwards (see below, p. 133).

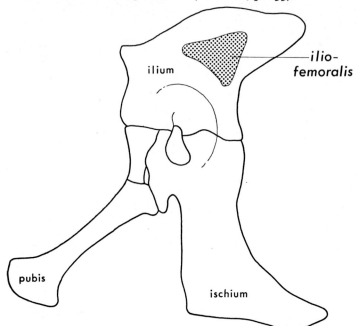

Fig. 6 The origin of the iliofemoralis in *Alligator*. [After Romer]

(ii) *'Fully improved' reptiles.* In more advanced tetrapods such as dinosaurs (and mammals) the rotation of the hind-limbs into a vertical plane beneath the body is relatively complete. This necessitates that the dinosaur have, among other things, an inturned head to the femur (see below, p. 144 and Fig. 8B). Pressure from the head of the femur against the acetabulum, due to the contraction of the muscles running from the pelvis to the leg, is now directed upwards (together with pressure due to the animal's weight), rather than towards the mid-line; in consequence a strong supra-acetabular crest is developed and the centre of the acetabulum fails to ossify. (As pointed out above, one of the major advantages of the 'vertical' limb posture is that it reduces *all* transverse horizontal forces operating between the limbs and the body as well as those between the feet and the ground.) The whole hind-limb has been rotated through 90° so that the plane of flexure is parasagittal instead of transverse; bending of the knee in that plane is therefore necessary, not only to enable the limb to function as an extensible strut, but also to accommodate the differing vertical height of the femur at different positions in its swing. Thus the femur is inclined forwards and a little downwards; there is still a marked flexure at knee and ankle, and the foot also is directed forwards. This would have the effect of bringing the femur very close to the pubis; it would lie a little more laterally, of course, but still very close. The plane of movement is vertical and, if protraction and retraction were still to be brought about by pubofemoralis and ischiofemoralis respectively, the arc of movement would be restricted to the sector between the pubis and the ischium.[1]

[1] In those animals, such as pseudosuchians and crocodilians, where the pubis and ischium are directed obliquely downwards but where the femur is only partly adducted, the arc of rotation of the femur is less restricted because it is not moving in a vertical plane; the more 'vertical' its movement, the greater the restriction.

In many cases the limited contractility of muscle would restrict the arc of movement still further, to only a part of the sector between pubis and ischium (see above, pp. 124-125). No archosaur femur has a trochanter lateral to the head which extends far enough upwards beyond the acetabulum to enable it to act in any significant degree as a counter-lever, i.e. so that a muscle inserted on it would move the main shaft of the femur in a direction *opposite* to its direction of pull. Because of this the archosaur femur cannot be protracted by caudifemoralis, iliofemoralis or ischiofemoralis pulling backwards, nor can it be retracted by a muscle such as the pubo-ischio-femoralis internus pulling forwards. (Nevertheless in some dinosaurs the femur may have swung about an axis situated a little *below* its head; where a comparatively small head fitted into a wide acetabulum, as seems to have been the case in *Massospondylus* (Plate IVA), the movement of the head was more probably a sliding back and forth against the ventral surface of the supra-acetabular crest rather than a rotation in one spot.)

The arc of movement of the femur would be limited by yet another factor in those cases where the origin and insertion of the pubofemoralis were not equidistant from the acetabulum (see p. 125 and Fig. 4). As the femur swung forwards the angle between its shaft and the direction of pull of the pubofemoralis would rapidly diminish so that the rotatory component due to the contraction of that muscle would be small and ineffectual.

In the absence of other appropriate modifications the evolution of the 'vertical' limb posture would thus have presented the dinosaurs with certain difficulties in connexion with the movement of the femur. First, the main femoral protractors and retractors (pubofemoralis and ischiofemoralis respectively) would have changed their function to adduction to a much greater degree than they had done in the pseudosuchians and crocodilians; because the femur was rotated forwards, this applies to the pubofemoralis far more than to the ischiofemoralis, which latter in any case could easily be supplemented in its retractor function by musculature originating on the base of the tail (caudifemoralis). Secondly, if the femur had been protracted as far as was necessary for efficient locomotion it would have been much too close to the site of origin of the very muscles protracting it, i.e. to the pubis. As explained above, it is theoretically possible for a pubofemoral muscle to draw the femur right up to the pubis provided that the origin and insertion of the muscle are not equidistant from the acetabulum; but such an arrangement would be grossly inefficient, for the angle of pull would become more and more unfavourable as the femur approached the pubis. I have therefore named this whole problem, crudely but graphically, the 'femur-knocking-on-the-pubis' problem. Some dinosaurs managed to overcome these difficulties, and in different ways.

At this point it should be mentioned that in the resting crocodile the femur is directed forwards, outwards and a little upwards and the pubis obliquely forwards and downwards; it might therefore be argued that the femur has 'passed' the pubis without involving the animal in any functional difficulties. But it must be remembered (a) that the restriction of the arc of rotation of the femur is only partial in those animals where that element is still directed obliquely outwards (see footnote, p. 132); (b) that in the resting animal the body has sunk under its own weight, so that the femur has been pushed passively beyond the limit of active protraction; (c) that in the crocodile there has been a partial transfer of the origin of some of the femoral protractor muscles to the underside of the vertebral column; and (d) that, in any case, the crocodile may be presumed to be less efficient in its locomotor adaptations than were the dinosaurs.

(iii) *Sauropodomorphs.* The sauropodomorph pelvis differs essentially from that of the pseudosuchians only in that the acetabulum has a large fenestration and a well developed crest above it. The arc of movement of the femur must therefore have been very restricted.

Smaller prosauropods like *Massospondylus* (Plate IVA) were perhaps less cursorial, and certainly

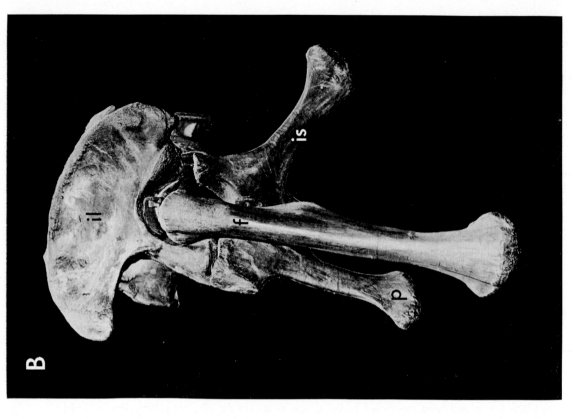

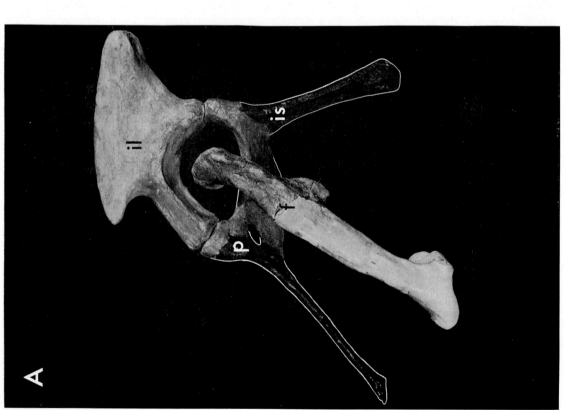

Plate IV.

A. PROSAUROPODA. *Massospondylus harriesi* Broom. Upper Red Beds (Upper Trias), Cape Province, South Africa.
Left side of pelvis and left femur in lateral view (cast of partly restored specimen).

B. SAUROPODA. *Diplodocus carnegii* Hatcher. Morrison Formation (Upper Jurassic), Wyoming.
Left side of pelvis and left femur, with vertebrae, in lateral view (cast).

less bipedal, than is usually believed. The comparatively unspecialised structure of the foot supports this suggestion, and their early extinction may indicate that they could not compete successfully with their more mobile contemporaries. Even the 'vertical' limb posture may not have been fully realised.

The vast sauropods such as *Diplodocus* (Plate IVB) (and perhaps also the larger prosauropods) were presumably slow and lumbering. Their graviportal limbs, which supported their great weight like four stout columns, could never have moved far from a vertical position; indeed, they probably moved like the limbs of large heavy mammals such as elephant (see Gregory, 1912) and rhinoceros, in which the degree of protraction and retraction and the length of a step are restricted and which operate mainly as levers (Gray, 1968, p. 247). Their feet too are relatively short, like the feet of graviportal mammals. But in some forms (notably *Diplodocus*) the pubis is directed so steeply downwards that it is difficult to see how the pubofemoral muscles alone can have protracted the femur sufficiently; the iliofemoralis and iliotibialis may be discounted as accessory protractors (see below, this page), but the possibility remains that part of the pubo-ischiofemoralis internus, inserting near the head of the femur, had transferred its origin from the pubis to the front of the ilium or (as in Recent crocodiles) to the undersides of the last few dorsal vertebrae.

(iv) *Theropods*. In many theropod dinosaurs, those carnivorous reptiles of wholly bipedal habit, the 'femur-knocking-on-the-pubis' problem was solved automatically by their adoption of bipedality. This tilted up the vertebral column and with it the pelvis, so that the pubis and the ischium were once more suitably placed to act as the site of origin of the main femoral protractors and retractors respectively. It is certainly true of the earlier theropods (Fig. 1A; Plates VA, B) that there was no great development of the anterior process of the ilium to act as an alternative site of origin of the protractor musculature, as developed in the ornithischians (see below), nor is there an anterior ramus to a backwardly displaced pubis.

Colbert (1964) drew a distinction between the 'brachyiliac' pelvis of some saurischian dinosaurs (sauropodomorphs) and the 'dolichoiliac' pelvis of others (theropods). The brachyiliac pelvis, with a short anterior process to the ilium and a 'twisted' pubis, is not unlike that found in certain pseudosuchians and is clearly the more primitive of the two types; it is characteristic of the prosauropods, and Colbert stated that it also continues in the sauropods in modified form. By contrast, the dolichoiliac pelvis, with an elongated and expanded anterior process to the ilium and with a slender rod-like pubis, is typical of the theropods. Colbert believed that the expansion of the ilium in the dolichoiliac pelvis was correlated with an increase in the large muscles of the hind-limb, especially the 'extensor' muscles effecting the recovery phase of the stride, and that the developments of the dolichoiliac pelvis as a whole were correlated with the perfection of bipedality.

The root of Colbert's belief was evidently Romer's reconstruction of the pelvic musculature of *Tyrannosaurus* and other saurischian dinosaurs (1923c), which, in turn, could be no more than wise conjecture based on his studies of crocodilian myology (1923b). Even if we accept Romer's reconstruction (and I am in no position to object to it), it is difficult to accept Colbert's contention that the iliofemoralis functioned as one of the main femoral protractors, for, though its site of origin is very large and does extend a short way forward of the acetabulum, it is centred directly *dorsal* to the latter; indeed, a rather greater area of attachment lies *behind* the centre of the socket. The site of origin of the iliotibialis admittedly does extend far forward of the acetabulum, curving downwards anteriorly, but this muscle inserts such a long way down the leg that it could not have protracted the femur very far forwards. What *is* possible, once again, is that part of the pubo-ischiofemoralis internus had transferred its origin from the pubis to the front of the ilium or to the undersides of the last few dorsal

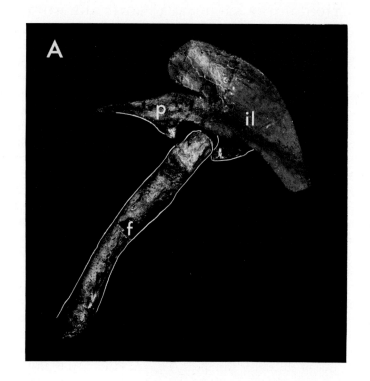
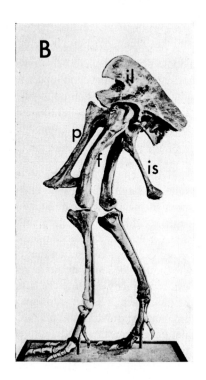
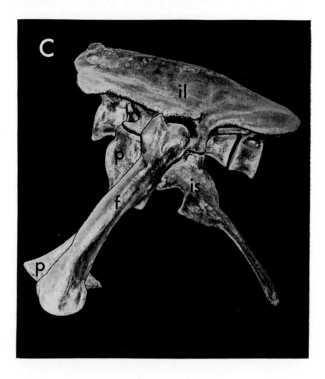

vertebrae. There is no physical evidence of such muscles, but there is ample space for them in, say, *Tyrannosaurus*; and it is undoubtedly true that, inserted on the femur just below its head, they would have been very well placed for protraction. Such muscles, if they existed, could have swung the femur farther forwards, to the pubis or even beyond, irrespective of whether or not the major part of the protractive swing was still accomplished by muscles originating on the pubis itself. There would therefore have been no absolute necessity for the theropod to tilt up its pelvis and trunk; indeed, some theropods may not have done so, especially when moving at speed.

Objection may also be made to Colbert's choice of the Upper Cretaceous carnosaur *Tyrannosaurus* (Plate VC) to illustrate his point. *Tyrannosaurus* represents the culmination of theropod evolution and has an exceptionally long anterior process to the ilium; as mentioned above, the earlier forms (such as the Lower Jurassic *Sarcosaurus*, Plate VA, and the Upper Jurassic *Antrodemus*, Plate VB) have a much shorter process.

I am not entirely convinced that all sauropodomorph and theropod pelves may be so clearly distinguished as 'brachyiliac' and 'dolichoiliac' respectively. Sauropods too have a forward extension of the ilium (see *Diplodocus*, Plate IVB, and compare the carnosaur and sauropod ilia illustrated by Colbert in his fig. 2 on page 9). In all these saurischians the lateral surface of this forward extension faces obliquely outwards and a little upwards (*Tyrannosaurus*) or at least directly outwards (sauropods), whereas in ornithischians (see below, p. 139) the lateral surface is directed steeply downwards. This difference must surely have some relevance to the problem of whether the anterior process of the ilium served as the site of origin of large muscles running to the leg; the ornithischian process appears to be better adapted to that purpose. Romer (1923a) pointed out that the process is not homologous in the two dinosaurian orders: in the saurischians it is an extension of the external iliac surface, in the ornithischians it is an exaggeration of the anterior iliac spine and is more dorsal in position.

It seems to me that the expansion of the ilium in saurischians, both forwards and backwards, is more probably to be correlated with the establishment of a firmer attachment between the pelvis and the vertebral column than with the development of new areas of origin for the extrinsic musculature of the hind-limb, except in so far as these might have been required for the transfer of fibres from the pubis.

It must also be pointed out that the skeleton of *Tyrannosaurus* in the Dinosaur Gallery of the British Museum (Natural History) (Plate VC) was mounted with its body in a far too horizontal position; this was done because it would otherwise have been too tall for the Gallery. Newman, who made the mount, has attempted to rationalise this (1970) by stating that the posture was much more bird-like than is suggested by earlier mounts. There are arguments which tend to favour his belief and others which tend to refute it. In refutation it might be claimed that his analogy with the flightless birds is entirely invalid. As admitted by Newman himself, *Tyrannosaurus* differs from a bird in having a long

---

Plate V.

A. THEROPODA (infra-order indet.). *Sarcosaurus woodi* Andrews (holotype).
Lower Lias (Lower Jurassic), Leicestershire.
Left ilium, pubis and femur in lateral view (the ischium is unknown).

B. CARNOSAURIA. *Antrodemus valens* Leidy.
Morrison Formation (Upper Jurassic), Colorado.
Left side of pelvis and both complete hind-limbs, with vertebrae, in lateral view; as mounted in U.S.N.M. [After Gilmore]

C. CARNOSAURIA. *Tyrannosaurus rex* Osborn.
Lance Formation (Upper Cretaceous), Wyoming.
Left side of pelvis and left femur, with vertebrae, in lateral view (model); as mounted in B.M.(N.H.).

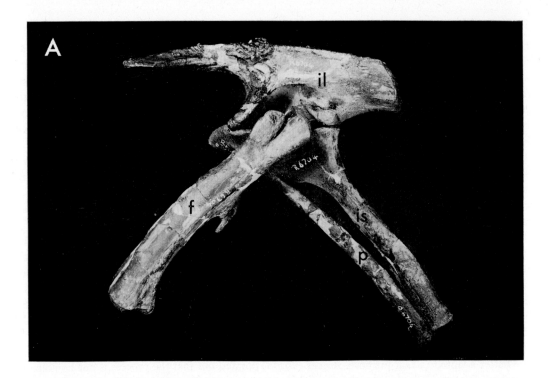

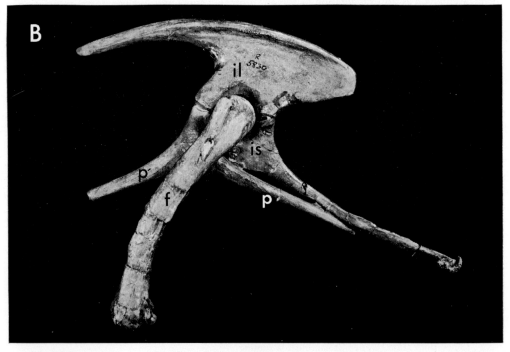

Plate VI.

A. ORNITHISCHIA (sub-order indet.). Cf. "*Scelidosaurus harrisoni* Owen".
Lower Lias (Lower Jurassic), Dorset.
Left side of pelvis and left femur in lateral view.

B. ORNITHOPODA. *Hypsilophodon foxi* Huxley. Wealden (Lower Cretaceous), Isle of Wight.
Left side of pelvis and left femur in lateral view (partly restored).

thigh and a heavy tail (instead of a short thigh and virtually no tail); it differs also in having a propubic pelvis, instead of opisthopubic, and in being an extremely large and massive animal. In view of my comments above, however, no arguments should be based upon the fact that the posture advocated by Newman places the femur very close to the well-developed pubis; in the mount the femur lies immediately lateral to the pubis, partly concealing it, and in a reconstruction of the skeleton drawn by Newman (Halstead, 1969, fig. 35c) the femur is distinctly *anterior* to the pubis.

Any consideration of the function of the pubis in *Tyrannosaurus* and related theropods must also involve its peculiar ventral expansion in those forms, the so-called 'pubic rocker'. This matter was discussed by Marsh (1896, p. 161). It may be that when the animal lay on its belly the 'rocker' rested on the ground, enabling the pubis—through the ilium, sacrum and vertebral column—to support the great weight of the back and prevent it from crushing the viscera.

(v) *Early ornithischians*. The earliest ornithischians, such as *Scelidosaurus*[1] (Plate VIA), appear to have solved the 'femur-knocking-on-the-pubis' problem in an entirely different manner. It may be presumed that at first the ornithischian pelvis evolved in the same way as the sauropodomorph or theropod pelvis, following the trends initiated by the pseudosuchian ancestors of both. The muscles running from the pubis and ischium to the femur became more and more effective as adductors and correspondingly less effective as protractors and retractors respectively; this change of function was more marked in the case of the pubic musculature because of the forward rotation of the knee. But while the theropods swung the pubis further forwards, away from the forwardly directed femur, the ornithischians gradually transferred the function of protraction of the femur to the musculature arising on a different element, the ilium.

The ilium developed an elongated and slightly downwardly curved anterior process; the lateral surface of this faced obliquely outwards and downwards, thus providing a site of origin for the musculature which was far enough forward of the femur to enable that musculature to protract the femur effectively. If the anterior process of the ilium had not been elongated in this manner it would not only have been too small for the origin of a large femoral protractor but also too far back; with the femur in the retracted position it would have been pulling in a direction almost parallel to the bone itself. Perhaps the same (pubofemoralis) muscle shifted its origin, fibre by fibre; more likely it was a different muscle, the iliotibialis, which passed right down over the knee to insert on the tibia and acted simultaneously as an extensor. Under these conditions the pubofemoralis *sensu stricto* no longer played an essential part in the protraction of the femur. The stage would eventually be reached when the iliac musculature pulled the femur as far forwards as the pubis, so that, in the forward part of the swing, the pubofemoralis (if it functioned at all, with the point of insertion so close to the point of origin) would function solely as an adductor. Let us now suppose that the iliac musculature continued to pull the femur forwards, beyond the pubis, so that, in the forward part of the swing, any contraction of the pubofemoralis would result in the *retraction* of the femur. It is difficult to see how such a system could operate at all; for not only would the pubofemoralis muscle be performing entirely different functions according to the position of the femur (protraction, adduction, retraction), but the distance between the points of origin and insertion of the muscle would vary to an impossible degree, being very close together as the femur swung past the pubis and quite remote when the femur was fully retracted. Under those circumstances it was surely advantageous to the animal that the body of the pubis should rotate backwards to lie beneath the ischium in characteristic ornithischian fashion, for

[1] Newman (1968) points out that the lectotype of *Scelidosaurus harrisoni* Owen, the type- and only species of the genus, is the isolated knee-joint of a megalosaurid saurischian and is not an ornithischian at all. The name, however, continues in general use for the complete ornithischian skeleton in the B.M. (N.H.), no. R. 1111, pending an application to the International Commission on Zoological Nomenclature to change the lectotype.

not only did that enable the pubofemoralis to function as an accessory femoral retractor in all positions of the femur but it also placed its origin far enough from its insertion for the muscle to operate effectively.

(vi) *Later ornithischians*. Later, however, many ornithischians developed an anterior ramus to the pubis. Both *Iguanodon* (Plate VIIA) and *Stegosaurus* (Fig. 7A) have 'typical' ornithischian pelves, with a large anterior ramus to the backwardly directed pubis. In the small *Hypsilophodon* (Plate VIB) the anterior ramus is rather less developed. In *Triceratops* (Plate VIIB) the powerful anterior ramus of the pubis is directed downwards as well as forwards, the posterior ramus is so short that it cannot be seen at all in lateral view, and the whole pelvis bears a strong superficial resemblance to that of a

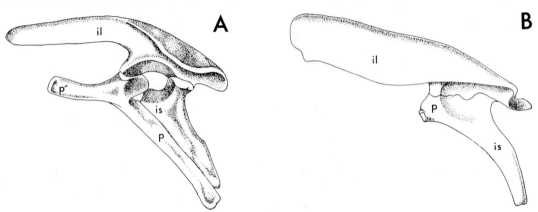

Fig. 7 **A.** Pelvis of *Stegosaurus* (Upper Jurassic). [After Gilmore].
B. Pelvis of *Ankylosaurus* (Upper Cretaceous). [After Romer].

saurischian; in the primitive Protoceratopidae, however, the anterior ramus too is fairly short. In the Ankylosauria (*Ankylosaurus*, Fig. 7B) reduction of the pubis has proceeded much further still; the posterior ramus is vestigial or absent and the anterior ramus is missing altogether, so that the pubis is represented by nothing more than part of the acetabular border. Incidentally, some ankylosaurs of the family Nodosauridae are exceptional among dinosaurs in that the acetabulum is (secondarily) closed.

The reasons for this development of an anterior ramus to the pubis of later ornithopods, stegosaurs and ceratopsians are not at all clear. It might be considered that the ornithopods, having already solved the 'femur-knocking-on-the-pubis' problem by the development of an extended anterior process to the ilium, had then *over*-compensated by becoming bipedal; the consequent tilting of the pelvis placed the anterior process of the ilium *too* far from the femur. But it seems likely that the earliest ornithischians known, which were without an anterior ramus, were already bipedal, and conversely that the stegosaurs and ceratopsians were not. Rather does this development appear to be correlated with large size than with bipedality. Be this as it may, where the anterior ramus to the pubis was present it might well, in its turn, have functioned as the site of origin of the main femoral protractor—in such cases probably the ambiens muscle.

(vii) *Birds*. The pelvis of the bird (Plate VIIIA) has certain similarities to that of the ornithischian dinosaurs (hence the name of the latter order); the most striking of these lies in the fact that in both groups the pubis is directed backwards to lie beneath the ischium. Other important characters of the bird pelvis are the absence of an anterior ramus to the pubis (although there may be a very short

Plate VII.

A. ORNITHOPODA. *Iguanodon bernissartensis* Boulenger. Wealden (Lower Cretaceous), Belgium. Left side of pelvis and left hind-limb, with part of vertebral column, in lateral view (cast).

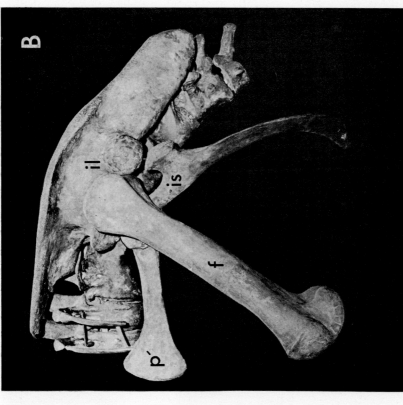

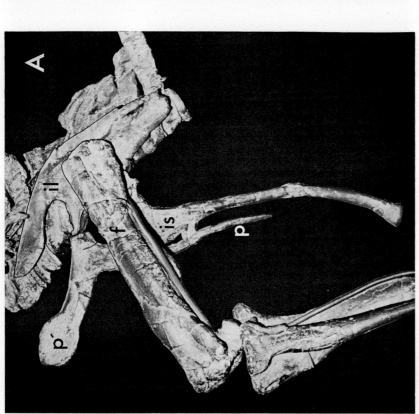

B. CERATOPSIA. *Triceratops prorsus* Marsh. Lance Formation (Upper Cretaceous), Wyoming. Left side of pelvis and left femur, with part of vertebral column, in lateral view (cast).

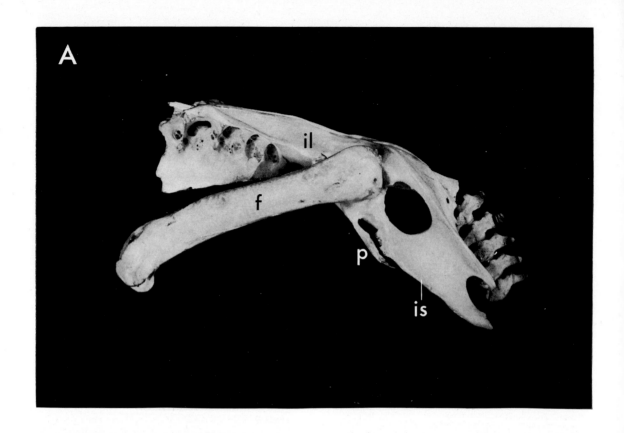
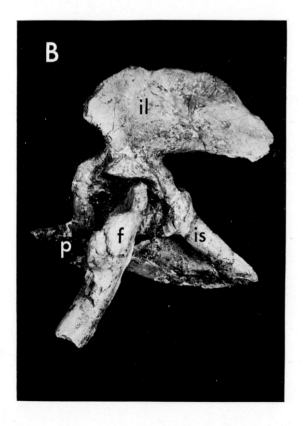
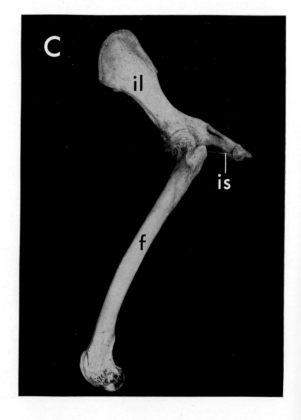

pectineal process in approximately the same position); the absence in most cases of pubic and ischiadic symphyses; and the great elongation of the ilium, firmly attached to the synsacrum.

Some workers (e.g. Galton, 1970) believe that the similarity of the respective pelves suggests a phylogenetic affinity between ornithischians and birds, closer than the affinity of either group with any other order derived from the Pseudosuchia. It seems to me, however, that the earliest known ornithischians (a) occur so soon after their pseudosuchian ancestors, (b) are already so specialised as ornithischians (for example, in their possession of a median predentary bone in the lower jaw) and (c) are already so diverse among themselves that any common ancestry they shared with birds must have been so short-lived as to have been virtually non-existent. This matter will be dealt with more fully elsewhere (Charig, in preparation). In view of this and of the undoubted fundamental differences between the respective pelves it is better to assume that the similarities are due to convergence, the backward rotation of the pubis in the bird having been brought about in a manner analogous to that outlined above for the ornithischians. It therefore seems unlikely that there is any greater affinity between birds and ornithischians than that due to the origin of both from the pseudosuchians.

(viii) *Therapsids and mammals*. Like the dinosaurs, therapsids and mammals showed a trend towards the adoption of a 'vertical' position of the limbs and were therefore confronted by much the same problems. The mammalian solution approximates very broadly to that of the early ornithischians. Seen from the left (Plate VIIIC) the various pelvic elements have 'rotated' anti-clockwise; the ilium extends far forwards and dorsally but not at all back, the pubis is shortened in front, and the pubo-ischium as a whole has moved posteriorly to provide a large area of origin for the powerful retractors of the limb.

These characters appear in the higher therapsids (Plate VIIIB) in incipient form.

(ix) *The sacral attachment*. All 'fully improved' archosaurs are characterised by a stronger attachment of each half of the pelvis to the sacrum, i.e. by an increase in the number of sacral vertebrae (vertebrae of which the ribs attach to the ilia). In lower reptiles the typical sacral count is 2; a few cotylosaurs and *Eunotosaurus* have only one sacral, or one and a partly developed second, and conversely there are a few other primitive forms (generally large, heavy and thoroughly terrestrial reptiles, such as pareiasaurs) which have as many as 4 or even 5. In 'semi-improved' archosaurs the typical number is still 2, although exceptionally there may be 3 (*Ornithosuchus*) or 4 (*Scleromochlus*). In the 'fully improved' dinosaurs this number may be greatly increased, some having as few as 3 sacral vertebrae but others (Ceratopsia) as many as 10 or 11.

The reasons for this are very plain. With the 'sprawling' posture the body rests on the ground; with the 'vertical' posture its weight is suspended from the vertebral column, which is itself supported well off the ground by the limbs through their girdles—especially the pelvic girdle. The latter must therefore be firmly connected to the vertebral column.

Plate VIII.

    A. AVES. *Cathartes aura* (Linnaeus). Living Turkey Vulture.
       Synsacrum and left femur, with part of vertebral column, in lateral view.

    B. THERAPSIDA. *Cynognathus crateronotus* Seeley (holotype).
       Upper Beaufort Series (Lower Trias), Cape Province, South Africa.
       Left side of pelvis and part of left femur in lateral view.

    C. MAMMALIA. *Canis familiaris* Linnaeus. Living Domestic Dog.
       Left side of pelvis and left femur in lateral view.

THE FEMUR AND THE CRUS

Unfortunately space does not permit a detailed consideration of the femur or of the tibia and fibula in the present article. Of particular interest, however, are certain characters which may be used to distinguish 'semi-improved' archosaurs from 'fully improved' or (less frequently) 'sprawlers' from improved types in general; the changes are the natural results of the alterations in limb posture.

In 'semi-improved' types such as pseudosuchians and crocodilians (Fig. 8A) the femur is held horizontally or obliquely and its proximal end is pulled inwards against the acetabular socket by components of the forces exerted by the extrinsic musculature. The shaft of the femur is gently sigmoid, and its proximal end, though turned a little inwards, does not form a distinct 'head'. By contrast, the femur of the 'fully improved' dinosaurs (Fig. 8B) has a more vertical posture and its proximal end is pressed by gravitational forces due to the animal's own weight against the more dorsal part of the acetabulum, i.e. against the downwardly facing surface of the strongly developed supra-acetabular crest. This dinosaur femur possesses a nearly straight shaft (often arched a little dorsally in bipeds) separated by a more or less distinct neck from a very distinct head; the head is turned strongly inwards. As might be expected, the articular surface on the proximal end of the femur is nearly terminal in pseudosuchians and crocodilians but in dinosaurs it faces more upwards than inwards.

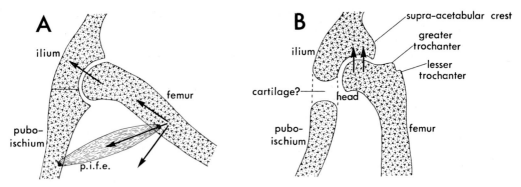

Fig. 8 Diagrammatic transverse sections through the acetabula of:
   A. A 'semi-improved' reptile, e.g. *Crocodylus*.
   B. A 'fully improved' reptile, e.g. *Thescelosaurus*. Arrows show the direction of the pull exerted by the pubio-ischio-femoralis externus, its components, and the direction of thrust of the head of the femur.

The dinosaur femur, both saurischian and ornithischian, is often characterised by the possession of greater and lesser trochanters; these, which are not homologous with the structures of the same name in mammals, lie lateral to the head (the lesser lateral to the greater) and are directed proximally. Unique to 'semi-improved' and 'fully improved' archosaurs is the fourth trochanter, situated ventrally farther down the shaft; it is entirely absent in the primitive sprawling archosaur *Chasmatosaurus* but is represented by a rugose area or a small protuberance in pseudosuchians and crocodilians and by a large powerful crest in many dinosaurs (especially bipedal forms). In large graviportal quadrupeds these trochanters tend to be lacking or developed only weakly.

The major distinction between the tibiae of 'semi-improved' and 'fully improved' archosaurs was pointed out recently by Bonaparte (1969, p. 475). The greatest diameter of the proximal end-surface runs more or less antero-posteriorly in both groups; that of the distal end-surface does likewise in the 'semi-improved' forms, but in the 'fully improved' dinosaurs it runs approximately latero-medially. This gives the tibia of Triassic and some later dinosaurs a 'twisted' appearance.

Bonaparte, in the same place, quotes a personal communication from Attridge as suggesting a functional association between this 'twisted' dinosaur tibia and the characteristic dinosaur tarsus, in which the articulation is mesotarsal and the astragalus and calcaneum are firmly positioned against the tibia and fibula. Indeed, in many dinosaurs the astragalus is locked to the tibia by an ascending process on the former and in some instances also by a descending flange on the back of the latter (Charig *et al.*, 1965). By contrast, the 'straight' tibia of pseudosuchians and crocodilians is associated with a 'crocodiloid' tarsus (see below); there is no firm connexion between crus and tarsus, for the calcaneum is fixed to the metatarsals and moves as part of the pes.

THE TARSUS AND THE PES

(a) *General*

As with the femur and the crus, so may the tarsus and the pes of 'semi-improved' and 'fully improved' archosaurs be distinguished from those of 'sprawlers' and from each other. The foot of the 'sprawler' is simply a flexible base for the leg; the foot of more advanced forms includes a propulsive lever.

The 'sprawler' usually has a simple, unspecialised tarsus. In pseudosuchians (Fig. 9) and crocodilians there is a complex 'crocodiloid' tarsus in which most of the movement takes place at a ball-and-socket articulation between the astragalus and the calcaneum; from a functional viewpoint the astragalus forms part of the crus, the calcaneum, which bears a prominent tuber, part of the pes. In dinosaurs the tarsus is again copmpratively simple, with astragalus and calcaneum more or less firmly fixed to the crus (see above); the articulation is mesotarsal, of the 'roller-bearing' type, and there is no

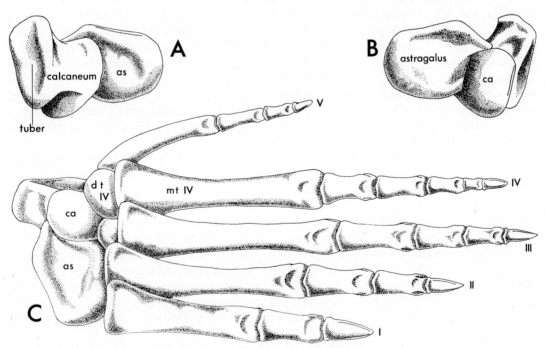

Fig. 9 Left tarsus and pes of the pseudosuchian *Ticinosuchus* (Middle Trias):
  A. Astragalus and calcaneum from behind.
  B. Astragalus and calcaneum from in front.
  C. Entire tarsus and pes in anterodorsal view. [Reconstruction, after Krebs]

calcaneal tuber. Certain aspects of the question of the evolution of the tarsus will be touched upon below (pp. 150, 152), but the whole matter will be dealt with more fully elsewhere (Charig, in preparation).

The foot of the 'sprawler' (Fig. 10A) is also primitive in structure, with each phalangeal series much longer than its metatarsal and with digit IV the longest. Pseudosuchians (Figs 9C and 10C) and crocodilians (Fig. 10D) show a reduction in the length of the digits, for the phalangeal series are generally no longer (or even shorter) than the corresponding metatarsals[1]; they also show the beginning of a trend towards a bilaterally symmetrical foot, with digit IV slightly shorter than III and digit V greatly reduced. The foot of a generalised quadrupedal dinosaur (Fig. 10E) is not unlike that of a pseudosuchian or crocodilian, although the phalangeal series are not usually shorter than their metatarsals; the feet of bipeds (Fig. 10G, H) tend more strongly towards bilateral symmetry about digit III, digit I being reduced as well as digit V and either or both sometimes disappearing altogether.

The structure of the pes, hitherto generally ignored, provides a useful key to the stance and gait of extinct reptiles. It also affords some indication as to whether a suggested phylogenetic relationship was possible or not; for example, it seems to me unlikely that a quadrupedal dinosaur with five well developed digits on each foot could have evolved from an habitual biped, for most known forms of incontrovertibly bipedal habits have only three or four digits on the pes. (See also Charig *et al.*, 1965, pp. 204-205.)

Most of the following discussion is concerned with foot posture and digitigrady.

(b) *'Sprawlers'*

When the limb is held in the primitive 'sprawling' position its propulsive backward thrust is transmitted to the substrate through the foot. (Most of the points mentioned in this paragraph and those following, though here concerned with the hind-limb, apply also to the fore-limb.) The magnitude of the frictional force required to prevent the foot slipping when an almost horizontal backward thrust is applied to it depends upon (a) the coefficient of friction between the plantar surface of the foot and the ground, and (b) the total force applied to the ground normal to its surface, irrespective of the area over which it is applied. Where the animal's body is resting on the ground, and its weight is therefore not transmitted through the limb, the frictional forces are low and would not be increased by increasing the size of the foot.

A large foot, however, does confer other advantages on a 'sprawler'. The primitive tetrapod was not always walking on a completely firm substrate; it must often have found itself on mud or sand, substances which behave like solids when the pressure (force per unit area) applied to them is low but more like liquids when the pressure is high. More simply, the larger the foot the less likely would it be to sink into a soft substrate; this is why people wear snowshoes on snow and avoid the use of stiletto heels on a soggy lawn. The greater part of the length of the foot was therefore split up into divergent toes, primitively united by a web of skin. But, even when the animals had migrated on to firmer substrates and had lost the interdigital webbing, the long radiating toes increased the area of *vertical* contact through which the foot could apply horizontal pressures to superficial irregularities of the ground and enabled it to do so in several different directions. Perhaps this is why, in tetrapods such as lizards which retain the primitive 'sprawling' posture of the limbs, the hind-foot is rarely

---

[1] It has often been suggested that the metatarsals of archosaurs are elongated (e.g. by Schaeffer, 1941, p. 446, specifically of crocodilians; by Romer, 1956, p. 413, of archosaurs in general). Relative to the phalangeal series this is indeed true; but relative to other elements, e.g. the tibia, the metatarsals are not elongated, and it is more logical to consider that the phalanges have been shortened. Only in some specialised archosaurs, e.g. the bipedal *Ornithomimus*, are the metatarsals truly elongated.

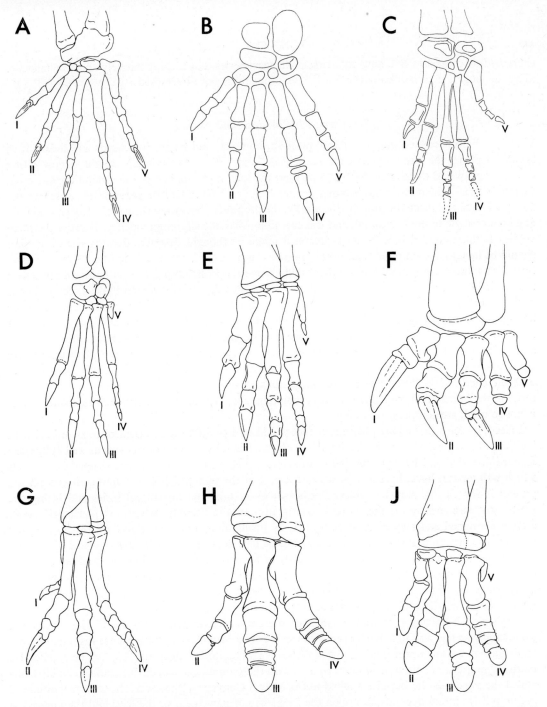

Fig. 10 Left tarsus and pes of various reptiles, in dorsal view, to show the relative lengths of the digits (Not to same scale)

    A. The Recent rhynchocephalian *Sphenodon*. [After Bayer and Schaeffer]
    B. The Lower Triassic therapsid *Thrinaxodon*. [After Broom and Schaeffer]
    C. The Lower Triassic pseudosuchian *Euparkeria*. [After Ewer]
    D. The Recent crocodilian *Crocodylus*. [After Schaeffer and Romer]
    E. The Upper Triassic prosauropod *Ammosaurus*. [After Marsh]
    F. The Upper Jurassic sauropod *Apatosaurus*. [After Hatcher]
    G. The Upper Jurassic carnosaur *Antrodemus*. [After Osborn]
    H. The Upper Cretaceous ornithopod *Anatosaurus*. [After Brown and Romer]
    J. The Upper Cretaceous ceratopsian *Monoclonius*. [After Lull]

shortened; it comprises five long radiating phalangeal series and a comparatively short metatarsus. Why the longest digit is usually the fourth (penultimate) is not understood.

(c) *Forms with 'vertical' limb posture*

When the limb is rotated into a 'vertical' position, however, and moves backwards and forwards in a vertical plane, the circumstances are entirely different. First, the animal's weight is now acting directly downwards through its limb and its foot; the total force applied to the ground normal to its surface is greatly increased; and a much greater horizontal force will be required to overcome the frictional force and make the foot slip. Secondly, the propulsive backward thrust is in any case applied at a greater angle to the horizontal, and this too diminishes the slipping tendency. Because the force applied to the ground is here so much greater it might be thought desirable that the size of the foot should be increased even more than in the 'sprawler' to prevent it from sinking into a soft substrate; but these animals are generally more thoroughly terrestrial in habit and live on hard ground where the possibility of getting bogged down need not be considered. (For this reason I take a sceptical view of the idea that the Mesozoic marshes were frequented by sauropod dinosaurs, weighing up to 70 or 80 tons, yet unsupported by water.) The reverse tendency is therefore evident, and a smaller contact between foot and ground suffices; no advantage is to be gained from having long toes, they may just as well be short. In most terrestrial archosaurs from *Chasmatosaurus* onwards the phalangeal series are of about the same length as their respective metatarsals, a condition which contrasts effectively with the condition in lower reptiles. In this connexion it is interesting to note that the phalangeal series in the human foot, which is used in locomotion, are very short, while those in the hand, which is not so used, are comparatively long.

Indeed, it seems that a long plantigrade foot would be a positive embarrassment to an animal with 'vertical' limbs, for the feet of such animals are characterised by a short contact area as well as by their forward direction and by a greater degree of symmetry about the axis. Most quadrupedal mammals which walk or run have short feet (at least, that part of the foot which comes into contact with the ground is short); this may be achieved either by shortening the phalangeal series and lifting the metatarsus permanently off the ground (digitigrady) or alternatively, where the phalangeal series are not shortened and may even be elongated, by lifting almost the entire foot off the ground and walking on tiptoe (unguligrady). By contrast, bipeds have long feet in order to preserve a reasonably large rectangle of support; the metatarsus tends to be slightly longer in bipedal dinosaurs than in quadrupedal forms, and the toes of the former are usually divergent to enable them to balance more easily when one of their two feet is lifted off the ground. Leaping mammals too have long feet, for reasons which will be explained below (p. 151).

Why should a walking or running quadruped shorten its contact with the ground? The explanation would apply just as well to a quadrupedal dinosaur as to a mammal, for both groups maintained their hind-limbs in a similar posture. It is stated by Gray (1968, pp. 251-252) that ". . . . to be an effective means of propulsion a limb must be in contact with the ground; the longer this contact can be maintained the greater is the impulse applied to the body." Conversely, however, the total magnitude of the propulsive forces depends also upon the frequency of movement of the legs; and the longer the contact with the ground, the lower the frequency of the limb cycle. It is therefore probable that these two factors would largely cancel each other out and might safely be ignored.

It is also noted by Gray (1968, p. 254), however, that "all limbs without exception are digitigrade during the later phases of their retractor movements." He then very lucidly explains the reason for this, the essential point in connexion with the present argument being as follows. When the foot is in contact with the ground and the limb is retracted, the body moves forwards and the degree of flexure of the ankle is increased (see Fig. 11); eventually the body has moved so far forwards that the line of action of its weight, passing vertically downwards through the hip, is directed in front of the anterior end of the metatarsus. Under these circumstances the limb is free to pivot forwards about the

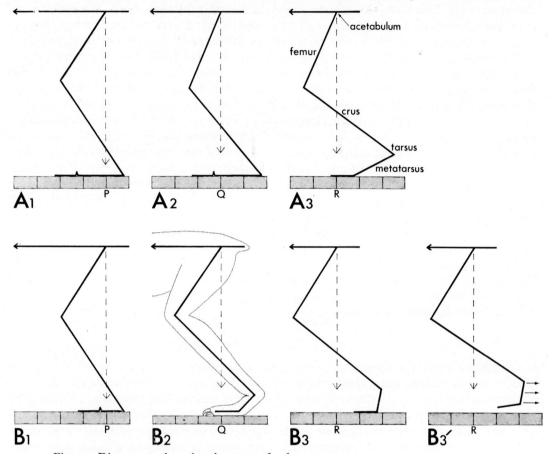

Fig. 11 Diagrams to show the advantages of a short metatarsus.
The A series represents the hind-limb of an animal with a long metatarsus; the B series that of an animal with a short metatarsus but which is otherwise identical. Each moves forwards so that its acetabulum is directly above P (in drawings A1, B1), Q (in A2, B2) and R (in A3, B3). In the A series the metatarsus does not start to rise until the third stage; in the B series it is rising already in the second stage, and by the third stage it has risen much higher. B3′ represents a moment in time immediately following B3 (the acetabulum is shown in the same place) but the backward thrust has now overcome the static friction and the foot has cleared the ground.

distal joints of the metatarsals; and, since the extensor muscles of the ankle are in a state of strain, this it proceeds to do, thereby lifting the metatarsus off the substrate. As Gray remarks, "very great effort by the extensors of the . . . ankles would be required to hold the whole surface of the foot in contact with the ground." It is also evident that a short metatarsus will pivot up like this sooner than a long one, for the animal's body does not have to move so far forwards to produce the same effect.

The limb performs its locomotor function by acting both as a lever and as an extensible strut. The latter effect is achieved by extension of the knee and, in 'improved' tetrapods (where the metatarsus acts as a propulsive lever), by extension of the ankle and of the foot itself; it is with these last two that we are mainly concerned here. Extension of the ankle exerts forces downwards and backwards upon the ground and upwards and forwards upon the animal's body, the latter force being transmitted as a compressional thrust along the line of the tibia. The thrust exerted on the ground is at first directed almost vertically downwards, through the almost horizontal metatarsus, but as the proximal end of the metatarsus and the ankle rise higher off the ground the direction of the thrust becomes progressively more horizontal towards the rear. In the case of walking and running mammals only the horizontal component of this thrust is effective in locomotion, and when that component becomes sufficiently large to overcome the static frictional forces between the foot and the ground the foot slips backwards and clears the ground. With a long metatarsus the ankle has to rise a long way before the horizontal component assumes a significant proportion of the total thrust. With a short metatarsus, however, the direction of the thrust will change more rapidly from downwards to backwards; this will provide forward propulsion sooner and, correspondingly, the foot will be more quickly clear of the ground (Fig. 11, B3′). To put it another way: the longer the foot, the higher will the ankle have to be raised each time to give the forward push to the body and to lift the foot clear of the ground for the protractory phase; thus the whole body will be raised and lowered with every step and the animal will not only be jogging up and down and rolling from side to side but will be expending energy in a useless, wasteful manner. As Gray remarks (1968, p. 262), "From a propulsive standpoint, rhythmical oscillations of the centre of gravity represent a waste of energy".

The total amount of forward propulsion *per step* due to the action of the limb as an extensible strut is obviously greater in an animal with a long metatarsus, provided that the ankle is fully extended. This relatively minor advantage, however, is more than negated by the lower frequency of the gait which is imposed by the possession of a long metatarsus and which reduces the total efficiency of the limb (operating both as a lever and as an extensible strut) *per unit of time*.

(d) *Origin of the 'crocodiloid' tarsus*

In a plantigrade animal the muscles which extend the ankle and those which flex the foot at the metatarso-phalangeal joints are acting against the animal's weight; a long metatarsus would therefore be of advantage in that it would provide more leverage for those muscles. The theoretical considerations adduced above, however, suggest that a long metatarsus would also be *dis*advantageous to a plantigrade animal. An alternative arrangement is therefore developed which provides more leverage for the ankle extensors without undue elongation of the metatarsus: the calcaneum is locked to the metatarsus and extended backwards as a tuber, on which the ankle extensors are inserted instead of on the metatarsus itself. This development is one of the main characteristics of the highly specialised 'crocodiloid' tarsus found in the plantigrade 'semi-improved' reptiles (pseudosuchians, aëtosaurs, phytosaurs, crocodilians), which are generally heavily built.

(e) *Origin of digitigrady*

It is also worthwhile considering the consequences of another alternative solution, that of not bringing the metatarsus down again. It could be kept off the ground permanently and incorporated into the length of the limb, thereby achieving the digitigrade condition and eliminating the need for a calcaneal tuber. The rhythmic oscillations of the centre of gravity would thus be reduced to a minimum; and extension of the ankle, even though limited, could still have some propulsive effect. First, as the leg passed through the early stages of protraction, the ankle would be extended; next, in the later stages of protraction, the ankle would be flexed to lift the toes clear of the ground in the centre of the swing and would then be re-extended; next, as the leg went through the retractory (propulsive) phase, with the foot on the ground, the metatarso-phalangeal joints would be flexed by the forward movement of the body; finally, towards the end of the retractory phase, the ankle would begin to be extended again by the action of its own intrinsic musculature to give the final push before lifting the foot off the ground. Further, once a truly digitigrade condition had been attained, there would be no reason why the metatarsus should not then be elongated, increasing the effective length of the limb as a whole; a continuation of the same evolutionary process in mammals would result in unguligrady. In unguligrade mammals the toes (or toe) also may be elongated, but, even so, the contact of the foot with the ground is very short in the parasagittal direction; the foot has therefore to be lifted very little to clear the ground at the beginning of the protractory phase, and the importance (and size) of the ankle musculature is greatly reduced.

(f) *Observations of limb movements of living animals*

(i) *Crocodilians.* Let us stop theorising and look at the actual movements of the limbs in the animals concerned. Our first and most urgent requirement, a detailed analysis of the limb movements and anatomy of a crocodile, has yet to be carried out; an ideal study would involve both cine-radiography of a moving crocodile and dissection of a dead one. Schaeffer's investigations as published (1941) are insufficient for the present purpose, although he quotes von Huene as having pointed out that "when the crocodile is moving very rapidly, the metatarsals and phalanges appear to snap plantarward, pushing the body forward. This motion . . . is favored by the tuber on the calcaneum, giving the extrinsic flexors of the foot a 'lever grip' " (p. 445).

(ii) *Digitigrade quadrupeds.* Even more unfortunate, all fully digitigrade reptiles are extinct, so that their limb movements can never be studied. Analyses of digitigrade limb movements in quadrupedal walking mammals, however, have been carried out by Manter (1938, on cat) and by Barclay (1953, on dog, goat and sheep). Manter's work showed that the cat's ankle-joint extended as it left the ground (from about 106° to 130° or even 140°), flexed during protraction (to as little as 82°) and then extended again (to as much as 108°) before the foot touched the ground; it remained fixed at about that angle until the metatarso-phalangeal joints rose from the ground again. However, what happens in a digitigrade mammal is not necessarily much indication of what happened in a digitigrade reptile such as a dinosaur and even less of the locomotor pattern in a plantigrade reptile like a crocodile.

(iii) *Bipedal walkers.* A bipedal mammal may run likewise in a digitigrade fashion but must revert to a plantigrade position when walking slowly or at rest; Man is a good example of this. Birds are digitigrade at all times, but preserve a large rectangle of support by the retention of comparatively long toes which diverge widely; presumably the same was true of bipedal dinosaurs to some degree.

(iv) *Leaping mammals.* A leaping mammal, by contrast, cannot be fully digitigrade. Though it still requires a forward (or backward) component to the locomotor thrust of its hind leg, the upward (or downward) component also must be large, sufficiently large to overcome the animal's own weight. This necessitates the sudden, rapid, powerful and complete extension of both knee and ankle from a previous position of complete flexure; and the metatarsus, even at the moment of take-off, cannot be very far out of the horizontal, which requires that it be fairly long.

(g) *Advantages of digitigrady*

Concomitant with the effective incorporation of the metatarsus into the length of the limb in digitigrade animals, and of the proximal phalanges also in unguligrade animals, the ankle becomes a joint within the limb strut itself, rather than the joint between the limb and the foot. The ankle must therefore be strengthened, or at least its flexibility must be limited to the vertical plane and reduced even there.

The assumption of the digitigrade or unguligrade gait, the limitation of the mobility of the ankle and the reduction of the ankle musculature have three beneficial effects upon the animal's locomotion. First, the shortened contact with the ground and the permanent elevation of the heel are better for 'vertical' movement of the limb, as explained above. Secondly, the effective elongation of the limb gives the animal a longer stride. Thirdly, the reduction of the intrinsic musculature of the distal part of the limb moves the centre of gravity of the limb as a whole into a more proximal position, closer to the acetabulum, thereby reducing the moment of inertia of the limb as a whole and the period of its natural oscillation. All these effects make for greater speed.

It is also interesting to note that digitigrady appears to have evolved in dinosaurs and cursorial mammals because a long plantigrade foot would hinder 'vertical' movement of the limbs, not (as is generally believed) in order to lengthen the limb and increase the speed of locomotion. The latter effects were only secondary, for they could have been achieved more simply in other ways.

(h) *Origin of the dinosaur tarsus*

This complete change in the manner of locomotion, from the still rather primitive, plantigrade manner typical of pseudosuchians and crocodilians to the advanced digitigrade manner characteristic of dinosaurs, is of particular interest in its bearing upon the problem of tarsal evolution. It is widely held that the complex 'crocodiloid' tarsus (Fig. 9), found in all but the earliest members of the former category, evolved into the comparatively simple tarsus of dinosaurs with its mesotarsal articulation of the roller-bearing type; this belief is a necessary part of the generally accepted belief that dinosaurs evolved from pseudosuchians. Schaeffer (1941, pp. 442, 446) unquestioningly accepted that the dinosaur tarsus had evolved from the 'crocodiloid' tarsus, and authorities such as Romer seem never to have doubted it. Some workers, however, are now arguing against the possibility of this transition. The first challenger was Krebs (1963), whose main objection was on the grounds that simple structures do not evolve from more elaborate ones; but he has also pointed out (pers. comm.) that no transitional forms are known and that it is in any case difficult to see how the calcaneum could have changed its attachment from pes to crus without a break in functional continuity. It must also be admitted that, whereas all dinosaurs (with a few doubtful exceptions) appear to have lost the calcaneal tuber, all mammals—even the most extreme cursorial forms, digitigrade and unguligrade—have retained it.

There may be something in some of these objections; none of them is insuperable, however, and I find it even more difficult to see how the 'vertical' limb posture of the dinosaurs could have evolved directly from the 'sprawling' type without having first passed through the intermediate, 'semi-improved' stage represented by the pseudosuchians and crocodilians. Perhaps the intermediate stage was represented by a *small* pseudosuchian, as yet unknown, in which—because of its lightness—the development of a complex 'crocodiloid' tarsus with a massive calcaneal tuber was unnecessary.

ARCHOSAUR LOCOMOTION: GENERAL CONCLUSIONS

The sequence of basic locomotor types in the archosaurs was from 'sprawling' to 'semi-improved' to 'fully improved' (or 'vertical'); the change to 'semi-improved' took place during the later part of Lower Triassic times, the change to 'fully improved' towards the end of the Middle Trias. The important features of pelvis and hind-limb correlated with each locomotor type are given below in tabular form (Table 1). The essential distinction between the Pseudosuchia and Crocodilia on the one hand and the Saurischia and Ornithischia on the other lies in the adoption of the 'fully improved' type by the latter orders.

Only *after* an animal had reached the 'vertical' stage could it become an habitual biped. This belief contrasts with the orthodox idea that all archosaurs, from the very beginning of their history, show

Table 1.

	PRIMITIVE 'SPRAWLING' REPTILE	INTERMEDIATE 'SEMI-IMPROVED' ARCHOSAUR	ADVANCED 'FULLY IMPROVED' ARCHOSAUR
SACRUM:			
number of vertebrae	generally 2 (1 or '1½' in a few very primitive forms)	2, occasionally 3	3–11
PELVIS:			
anterior process of ilium	absent	generally weakly developed	generally strongly developed
supra-acetabular crest		weak	strong
acetabulum		imperforate	fenestrated
pubis and ischium	plate-like	somewhat elongated	very elongate
FEMUR:			
position	horizontal	oblique (i.e. partly adducted) when moving fast	vertical (i.e. fully adducted) at all times
head		not distinct, very slightly inturned	distinct, strongly inturned
shaft		gently sigmoid	fairly straight
fourth trochanter	absent	weakly developed	often strongly developed
TIBIA:		not 'twisted'	'twisted'
TARSUS:			
form	simple	'crocodiloid'	simple
articulation		'ball-and-socket' between astragalus and calcaneum	mesotarsal 'roller-bearing'
calcaneal tuber	absent	present	absent
PES:			
posture		plantigrade	digitigrade
phalangeal series	much longer than metatarsals	the same as, or slightly shorter than, metatarsals	the same as, or slightly longer than, metatarsals
trend towards bilateral symmetry	none	slight	strong; very strong in bipeds
longest digit	IV	III	III
reduced digits	none	IV slightly reduced V strongly reduced	I reduced or absent in bipeds IV slightly reduced V strongly reduced, may be absent in bipeds

bipedal tendencies. For example, Romer wrote in 1956 (p. 365) that there was an "archosaur trend toward bipedalism, and a resultant forward turning and semierect pose of the hind limb"; he also stated (*op. cit.*, p. 413) that "Among archosaurs the trend toward bipedalism has generally resulted in the development of a large and strongly built pes of digitigrade pose and with elongate metatarsals." I submit that Romer, like many others in the past who did not then have the information now available, was putting the cart before the horse.

Much of what has been expounded above is largely theoretical, deriving little support from observation or experiment, and it may well prove to be wrong in detail. It nevertheless offers, for the first time, a logical explanation of such peculiarities as the simplification of the dinosaur tarsus and the form of the ornithischian pelvis. I hope that my ideas, whether correct or not, will stimulate discussion leading to an eventual solution of all the problems involved.

SUMMARY

The ornithischian pelvis is primitively without an anterior ramus to the pubis; the posterior ramus represents the true pubis, rotated backwards to lie beneath the ischium. All archosaurs may be classified into three categories according to the characters of the pelvis and hind-limb, which fall into three well-defined suites indicating different types of stance and gait: 'sprawlers' (the earliest Proterosuchia), 'semi-improved' (other Thecodontia and Crocodilia) and 'fully improved' or 'vertical' (dinosaurs and birds). The 'vertical' type is always accompanied by digitigrady and only sometimes by bipedality; the latter character is a *secondary* development, consequent upon the prior adoption of the 'vertical' limb posture. The forward rotation of the femur in dinosaurs and birds would have brought the femur too close to the pubis for the effective operation of the pubofemoral protractors; this difficulty was overcome in different ways by theropods (through their bipedality) and by ornithischians and birds (through the transfer of protractor function to the iliac musculature, with the subsequent development of the opisthopubic pelvis). The possession of a short metatarsus is advantageous to an animal with a 'vertical' gait. The relative lengths of the toes afford some indication of the type of gait employed. The primary functions of digitigrady are to enable the foot to clear the ground more quickly, to promote stability and to conserve energy, *not* to increase the effective length of the limb. The elaboration of the 'crocodiloid' tarsus and the simplification of the dinosaur tarsus are to meet the functional requirements of the 'semi-improved' and 'fully improved' types of locomotion respectively.

ACKNOWLEDGEMENTS

First I must acknowledge the debt I owe to two former members of the Zoology Department at Cambridge: my interest in animal locomotion was originally stimulated by Professor Sir James Gray, and I was led into the field of vertebrate palaeontology, especially reptile palaeontology, by Dr F. R. Parrington. The great influence of their teaching and supervision upon my career, both in my student days and afterwards, is very evident in the present paper, although they cannot be held responsible for its shortcomings.

My thanks are also due to Dr David Hardwick of Imperial College and Dr Richard Jefferies of the British Museum (Natural History), with whom I discussed the mechanical aspects of my theories. Many other people, too numerous to mention, contributed useful ideas and criticisms.

REFERENCES

BARCLAY, O. R. 1953. Some aspects of the mechanics of mammalian locomotion. *J. exp. Biol.*, **30**, 116-120.
BONAPARTE, J. F. 1969. Comments on early saurischians. *Zool. J. Linn. Soc.*, **48**, 471-480.

CHARIG, A. J. 1965. Stance and gait in the archosaur reptiles. *Liaison Rep. Commonw. geol. Liaison Off.*, **86**, 18-19. [Abstract]
—— 1966. Stance and gait in the archosaur reptiles. *Advmt Sci., Lond.*, **22** (103), 537. [Abstract]
—— In preparation. New light on the pelvis of ornithischian dinosaurs.
CHARIG, A. J., ATTRIDGE, J. and CROMPTON, A. W. 1965. On the origin of the sauropods and the classification of the Saurischia. *Proc. Linn. Soc. Lond.*, **176**, 197-221.
COLBERT, E. H. 1964. Relationships of the saurischian dinosaurs. *Am. Mus. Novit.*, **2181**, 1-24.
COLBERT, E. H. 1969. *Evolution of the vertebrates*, 2nd ed., John Wiley & Sons, New York, London, Sydney & Toronto.
GALTON, P. M. 1970. Ornithischian dinosaurs and the origin of birds. *Evolution, Lancaster, Pa.*, **24**, 448-462.
GRAY, J. 1968. *Animal locomotion*. Weidenfeld & Nicolson, London.
GREGORY, W. K. 1912. Notes on the principles of quadrupedal locomotion and on the mechanism of the limbs in hoofed animals. *Ann. N.Y. Acad. Sci.*, **22**, 267-294.
HALSTEAD, L. B. 1969. *The pattern of vertebrate evolution*. Oliver & Boyd, Edinburgh.
KREBS, B. 1963. Bau und Funktion des Tarsus eines Pseudosuchiers aus der Trias des Monte San Giorgio (Kanton Tessin, Schweiz). *Paläont. Z.*, **37**, 88-95.
MANTER, J. T. 1938. The dynamics of quadrupedal walking. *J. exp. Biol.*, **15**, 522-540.
MARSH, O. C. 1896. The dinosaurs of North America. *16th Ann. Rep. U.S. Geol. Surv.*, 133-414.
NEWMAN, B. H. 1968. The Jurassic dinosaur *Scelidosaurus* Owen. *Palaeontology*, **11**, 40-43.
—— 1970. Stance and gait in the flesh-eating dinosaur *Tyrannosaurus*. *Biol. J. Linn. Soc.*, **2**, 119-123.
ROMER, A. S. 1923a. The ilium in dinosaurs and birds. *Bull. Am. Mus. nat. Hist.*, **48** (art. 5), 141-145.
—— 1923b. Crocodilian pelvic muscles and their avian and reptilian homologues. *Bull. Am. Mus. nat. Hist.*, **48** (art. 15), 533-552.
—— 1923c. The pelvic musculature of saurischian dinosaurs. *Bull. Am. Mus. nat. Hist.*, **48** (art. 19), 605-617.
—— 1927. The pelvic musculature of ornithischian dinosaurs. *Acta zool., Stockh.*, **8**, 225-275.
—— 1956. *Osteology of the reptiles*. University of Chicago Press, Chicago.
—— 1966. *Vertebrate paleontology*, 3rd ed. University of Chicago Press, Chicago & London.
—— 1968. *Notes and comments on vertebrate paleontology*. University of Chicago Press, Chicago & London.
SCHAEFFER, B. 1941. The morphological and functional evolution of the tarsus in amphibians and reptiles. *Bull. Am. Mus. nat. Hist.*, **78** (art. 6), 395-472.

ABBREVIATIONS USED IN FIGURES AND PLATES

as	astragalus	*mt*	metatarsal
ca	calcaneum	*p*	pubis
dt	distal tarsal	*p'*	anterior ramus of pubis
f	femur	*p.i.f.e.*	pubo-ischio-femoralis externus
il	ilium	*sv*	sacral vertebra
is	ischium		

A. d'A. BELLAIRS

Comments on the evolution and affinities of snakes

In recent years there has been a revival of interest in the evolution of snakes. New fossil material has been described and important papers dealing with the structure of snakes, and of Squamata in general, have appeared. This volume, produced in honour of Dr F. R. Parrington who has contributed so much to our knowledge of fossil reptiles, seems an appropriate place for a survey of current information. Certain groups of snake-like reptiles will also be considered.

THE MAJOR GROUPS OF SQUAMATA

The Squamata, by far the most successful group of modern reptiles, contains nearly 6,000 species, divided about equally between the lizards and the snakes. The group as a whole appears to have only a single obvious diagnostic character which is universally present; paired, eversible copulatory organs (hemipenes) in the male. Other characteristic features are the elaboration of the organs of Jacobson, which are of great importance as sense organs in most lizards and all snakes, and the development of various types of skull kinesis. There is also a tendency in many groups towards reduction of the limbs and elongation of the body which is, of course, most fully realised in the snakes. It is noteworthy that this tendency is apparently restricted to the lepidosaurian reptiles; it has never been evident in either the synapsid or archosaurian lineages, despite the fact that these have both, in their time, occupied dominant positions and have undergone wide adaptive radiation.

The Squamata are usually classified as a single order containing two suborders, the Sauria or Lacertilia (lizards) and the Serpentes or Ophidia (snakes). Some workers believe, however, that the amphisbaenids, a group of burrowing worm-like reptiles with degenerate eyes, should no longer remain as an infraorder of the Sauria but should be placed in a suborder (Amphisbaenia) of their own. In many ways these animals resemble certain burrowing members of the typically saurian families such as the Scincidae, and also the snakes. For example, the limbs are drastically reduced, the temporal arches are absent, and the sides of the cranial cavity are walled in by extensive downgrowths of the frontal and parietal bones. On the other hand they show a number of unique features such as the reduction of the right lung (instead of the left, as in snakes and snake-like lizards generally; see Fig. 2), the remarkable arrangement of the scales in rings round the body, the heavy ossification of the anterior sphenoid region of the skull, and the structure of the extra-stapes (extra-columella) which extends forwards along the outer side of the lower jaw. Whatever taxonomic status they are given, the amphisbaenids are clearly an isolated group; as fossils they are first known from the Eocene, but it is fair to assume that they originated at a much earlier date. Within the limits of fossorial specialisation they have been remarkably successful and recent forms number about 23 genera and 130 species.

Much of our knowledge of these interesting creatures is due to Carl Gans (see Gans, 1968, 1969, for reviews).

Existing lizards (excluding the amphisbaenids) fall into some fifteen to twenty families, depending on whether a 'lumping' or a 'splitting' classification is used, and these can conveniently be grouped into four infraorders: Gekkota (gecko and pygopodid families), Iguania (iguanids, agamids, chamaeleons), Scincomorpha (skinks, lacertids, teiids and related forms), and Diploglossa or Anguimorpha (anguoids and related forms, and the varanoid or platynotid lizards which include the monitors, heloderms and *Lanthanotus*). Some additional groups consisting entirely of fossil forms are also recognised (Romer, 1956; Hoffstetter, 1962).

In the classification of snakes, many herpetologists have not employed taxa of higher than familial rank. Some workers, however, have tried to segregate the various families into larger groups, as Underwood has done in his important monograph on the classification of snakes (1967). He divides the snakes into three infraorders, the Scolecophidia, Henophidia and Caenophidia. The first two of these may be loosely termed the 'lower snakes' since many of them show characters which are clearly of a lizard-like and therefore primitive nature. The most notable of these is the retention of vestiges of the pelvic girdle and even, in many cases, of the hind limbs (see Gasc, 1966, for a descriptive account).

The infraorder Scolecophidia comprises the Typhlopidae and Leptotyphlopidae, two families of worm-like snakes widely distributed in various warm countries and highly adapted for burrowing life. The relationship of these families with each other and with the more typical snakes has been a matter for debate (McDowell, 1967b). Underwood, however, has brought forward good evidence that they form a natural group of archaic snakes; he regards the Typhlopidae, which is the more successful and diversified of the two, as being in general particularly primitive.

The infraorder Henophidia, which corresponds with the superfamily Booidea in Romer's classification (1956), is a rather diverse assemblage of snakes which also show primitive characters. The Boidae is a fairly large family which can be broken down into various sub-familial but supra-generic groups. It contains the well known giant constrictors as well as a substantial number of small, secretive or burrowing types such as the sand boas (*Eryx*). The Acrochordidae, a small family of water snakes from the oriental and North Australian regions, have usually been classified with the Colubridae but are placed in the Henophidia by Underwood on the basis of certain characters of the arterial and skeletal systems. Three further groups traditionally regarded as families, the Aniliidae, Xenopeltidae and Uropeltidae, are included in this infraorder; they are all small, secretive or fossorial in habit, and sub-tropical or tropical in distribution. The Aniliidae (= Ilysiidae) is represented by *Anilius* (= *Ilysia*) in South America, a form with a type of coral snake coloration (red and black rings) and by *Cylindrophis* and *Anomochilus* in Asia. This is a group of great interest, and one which I am inclined to regard as perhaps being not far from the primitive ophidian stock. *Xenopeltis unicolor*, the single member of the Xenopeltidae, also shows a number of primitive features (such as the relatively large size of the left lung), but seems to approximate more closely to the 'higher' snakes. This oriental species is known as the sunbeam snake because of the iridescent sheen of its scales. The Uropeltidae is a bigger group than either of the two previous ones and its members are more highly specialised for burrowing life. They show some aberrant features such as an unusual type of occipito-vertebral joint. Romer (1956) reduced these three last-named groups to sub-familial status as the Aniliinae, Xenopeltinae and Uropeltinae, lumping them together in the single family Aniliidae. I think that this procedure has certain advantages, though it has been reasonably criticised by Dowling (1959), and is not followed by Underwood (1967).

There remain the 'higher' or advanced snakes of the infraorder Caenophidia. This consists of the

Colubridae (of most classifications) and of the poisonous snakes belonging to the families Elapidae (cobras, kraits and related forms, most of the poisonous snakes of Australia), the Hydrophiidae (sea snakes) and the Viperidae (vipers and pit-vipers, including rattlesnakes). The Colubridae is a vast unwieldy assemblage containing both non-venomous and more or less (usually less) venomous, back-fanged species; Underwood (1967) has made a gallant attempt to break it down into a number of smaller groups. The Hydrophiidae is clearly related to the Elapidae and some workers regard it as only a specialised elapid subfamily (Dowling, 1959; Underwood, 1967). I feel, myself, that this is a pity since the sea snakes are a fairly important and well-defined group, and their distinctive mode of life might be held to justify the retention of their familial status.

The renewal of interest in the problems of classification of the snakes is most welcome, but it is not entirely without its drawbacks: instability of the nomenclature and a proliferation of names. Such changes may be confusing, especially to workers who have had no training in systematic zoology but who often make important contributions to herpetology, especially in such fields as venomology, biochemistry and ultrastructure. New systems of taxonomy must, of course, reflect advances in our knowledge of the interrelationships of animal groups, and it is certainly desirable that all distinguishable groups *should* be distinguished. Nevertheless, as Simpson (1945) has noted, it is not necessary that all should enter formal classification and receive names.

THE MAIN CHARACTERS OF SNAKES

The snakes differ from all other known groups of limbless Squamata in the abundance of their species and the wide range of their adaptive radiation; indeed the group appears to be still undergoing diversification at the present time. Its success seems to be due, in large measure, to modifications of the vertebral column and trunk musculature which increase the efficiency of limbless locomotion and of the jaw apparatus, allowing the engulfment of relatively large prey; many snakes have also evolved special killing techniques, utilising constriction or venom.

Most or all snakes possess the following structural features; some of these have previously been listed by Dowling (1959) and Underwood (1967). Others could doubtless be added and I have omitted reference to the muscular system, a very complex area which has been studied by Haas (1962, and many previous papers), and by Gasc (1967). The term *always* where given below implies 'always in so far as is known'. The anatomy of limbless lizards is not so well known as that of snakes, and further study may show that some of the distinctions between snakes and lizards listed below are invalid.

1. *Always.* Rigid, bony brain-case formed in front by downgrowths of the frontals and parietals which meet the parasphenoid (parabasisphenoid) below (Fig. 1C). Both temporal arches absent. More or less similar conditions are found in amphisbaenids, and in burrowing lizards such as *Acontias* and *Anniella* (Fig. 1B), though in the latter the parietals do not extend so far ventrally nor the parasphenoid so far forwards (Brock, 1941; Bellairs, 1949; El-Toubi et al., 1965).

2. *Always.* Ophthalmic nerves and nerves to eye-muscles enclosed by parietal downgrowths and entering orbit usually through same foramen as optic nerve. In other Squamata with these downgrowths the nerves are not enclosed.

3. The chondrocranium is *generally* platytrabic and never forms a typical interorbital septum such as is present in most other Sauropsida. These features are of great interest and are probably related to changes in the relative proportions of the skull, eyes and nose which have occurred during the evolution of snakes.

In most lizards the orbits are separated by a high, thin sheet of cartilage, often containing

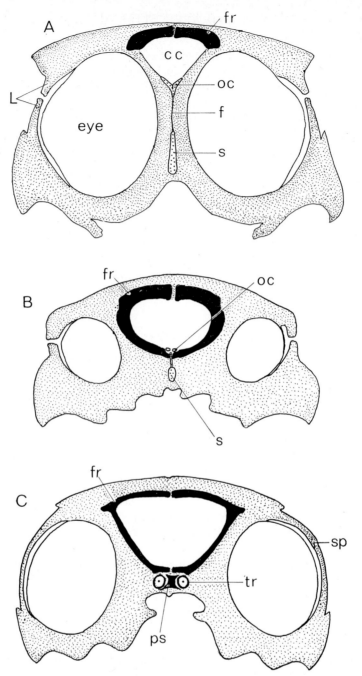

Fig. 1 Diagrams showing cross-sections of head at level of middle of orbits to show tropitrabic and platytrabic conditions. **A**, typical lizard with well developed interorbital septum (tropitrabic); **B**, small-eyed, burrowing lizard with frontal downgrowths and reduced interorbital septum (modified tropitrabic, based on *Anniella*); **C**, typical snake (platytrabic).

cc, cranial cavity. *f*, fenestra in interorbital septum. *fr*, frontal bone. *L*, eyelids. *oc*, part of anterior orbital cartilage (planum supraseptale in **A**). *ps*, parasphenoid. *s*, interorbital septum. *sp*, spectacle. *tr*, trabecula. Other bones not shown.

membranous fenestrae, which is called the interorbital septum; this is the typical tropitrabic condition (Fig. 1A). The septum is developed in the embryo from parts of the trabeculae cranii and orbital cartilages which fuse in the midline (Kamal, 1965). In small-eyed, burrowing lizards such as *Anniella* the septum is very low and the orbital cartilages are reduced; the cranial cavity lies partly between the orbits (Fig. 1B). Conditions in amphisbaenids are hard to interpret without embryonic material, but again, the bones regarded as ossified anterior orbital cartilages (orbitosphenoids) fuse ventrally in the midline and lie dorsally to a rod of cartilage which apparently represents the fused trabeculae (trabecula communis) (Bellairs, 1949).

In snakes the orbital cartilages are absent or vestigial, and the trabeculae generally remain paired in the orbital region (the platytrabic condition); each trabecular cartilage lies in a groove along the side of the parasphenoid and is separated from its fellow by this bone (Fig. 1C). Further forwards the trabeculae unite with each other and become continuous with the nasal septum. Exceptional conditions are: (i) in the scolecophidian snakes, which have minute eyes, the trabeculae are fused into a single rod at the level of the orbits (McDowell, 1967b). This departure from the typical ophidian platytrabic state is perhaps associated, in some cases at least, with a kind of telescoping of the nasal and orbital regions (Haas, 1964). (ii) in the colubrid snake *Psammophis* the frontal downgrowths are raised high above the parasphenoid and the space between is filled by a membranous septum. This condition, which amounts to the formation of a new, ophidian type of 'interorbital septum' is approached in certain other colubrids (Underwood, 1967 and personal communication).

4. *Usually*. Highly mobile jaw skeleton and quadrate, and mobile facial region permitting complex types of cranial kinesis and independent movements of the dentigerous bones on the two sides of the head and, especially, of the lower jaw (Albright and Nelson, 1959; Gans, 1961; Boltt and Ewer, 1964; Frazzetta, 1966).

5. *Always*. Epipterygoid absent. The bone is also absent or reduced in some lizards, both surface-living and burrowing, and in all amphisbaenids.

6. *Usually*. Numerous sharp, recurved teeth on jaws and palate; maxillary teeth may be modified as poison fangs. Salivary glands well developed; one or more glands in the upper jaw often specialised for venom secretion (Taub, 1966; Gabe and Saint Girons, 1969).

Characters 4, 5 and 6 are related. The jaw bones of the Leptotyphlopidae, Aniliinae and Uropeltinae seem to be a good deal less mobile than those of more typical snakes. The upper jaw of typhlopids is mobile, but in many forms shows an unusual type of kinesis; the two halves of the lower jaw are quite firmly united (List, 1966). The dentition is much reduced in the Scolecophidia and in a few other snakes also.

The jaw bones of lizards are more firmly attached to the skull than those of snakes and the main types of saurian kinesis involve flexion at the fronto-parietal (mesokinetic) and/or parieto-occipital (metakinetic) regions which do not appear to be flexible in snakes. The epipterygoid is important in typical saurian kinesis. The palate of most lizards is toothless, though some have teeth on the pterygoids and a few have toothed palatines.

7. *Nearly always*. Exoccipital bones meeting over foramen magnum, excluding supraoccipital (Underwood, 1967).

8. *Always*. Visual cells of retina without the refracting organelle known as the paraboloid which is almost universally present in lizards with differentiated visual cells (Underwood, 1967). (i) *All* snakes lack a scleral cartilage and scleral ossicles; in almost all other Squamata these structures are present. (ii) The eyes of *all* snakes are covered by a transparent spectacle and there is no nictitating membrane. The spectacle of certain burrowing snakes such as *Typhlops* is not a separate structure but merely a transparent window in the scale covering the orbit. Until embryological evidence is

available, it seems best to regard this as homologous with the spectacle of other forms, which is developed in the embryo from the fusion of the upper and lower eyelid primordia. Some lizards possess a spectacle of apparently similar type.

The lachrymal apparatus of snakes also shows certain differences from that in the majority of lizards; as in Squamata generally the lachrymal system appears to be functionally associated with the organ of Jacobson (see Bellairs and Underwood, 1951). Other significant ophthalmological differences between lizards and snakes are described by Walls (1942) and Underwood (1970).

9. Tongue *always* long, deeply forked and very protrusible, well adapted for carrying scent particles to the organ of Jacobson. The tongues of other Squamata are more or less protrusible and, in the majority, are notched or forked at the tip; in chamaeleons the protrusible tongue is specially modified for grasping prey. The only lizards, however, with apparently snake-like tongues are the monitors (*Varanus*) and perhaps the tegus (*Tupinambis*) of the family Teiidae. Accounts of the tongues of snake-like teiids would be of interest.

10. *Always.* Tympanic membranes and Eustachian tubes absent. Stapes, probably without extra-stapes, approximated to quadrate. Comparable though probably not identical conditions are found in many lizards of various types, both surface-dwellers and burrowers. Amphisbaenids lack tympanic membranes and Eustachian tubes but the stapedial apparatus shows modifications of its own (Toerien, 1963; Baird, 1970).

11. Vertebrae *usually* between 160 and 400 in number with zygosphenal articulations; 160 seems to be about the maximum number in lizards and amphisbaenids. Zygosphenal joints are found in certain lizards, including *Iguana* and the snake-like teiid *Bachia* (Hoffstetter and Gasc, 1969).

12. *Almost always.* One or more pairs of forked ribs movably attached to the vertebrae of the cloacal region and surrounding the lymph hearts, followed by several pairs of fused forked ribs (lymphapophyses). Similar conditions occur in some amphisbaenids; in some other snake-like lizards fused forked ribs only are present.

13. Pectoral girdle *always* absent; pelvis *usually* absent, sometimes vestigial. Vestiges of hind limb (e.g. femur) present in Leptotyphlopidae, Aniliinae (Romer's classification) and some Boidae. It is uncertain whether any other Squamata lack all traces of either of the limb girdles.

14. Heart situated well behind the head in *most* (probably all) snakes (Fig. 2C), with thymus bodies just anterior to it. Both common carotid arteries *always* present in Scolecophidia and Henophidia; left one only present in Caenophidia. In most of the snake-like lizards studied the heart lies relatively further forwards and nearer the head (Fig. 2A); the thymus bodies lie just behind the head and further in front of the heart than in snakes. In some amphisbaenids at least, however, the heart is about a quarter of the way down the body (Fig. 2B).

It is interesting that the position of the heart does not necessarily afford a guide to the length of the neck in limbless reptiles; thus in *Amphisbaena* the heart is about opposite the 24th-26th vertebrae, but the vestigial shoulder girdle is at the level of the 3rd vertebra (Hoffstetter and Gasc, 1969).

15. Left lung *always* reduced or absent. In a few Henophidia the left lung is quite large (50-80% the length of the right; Underwood, 1967). The left lung is reduced in many snake-like lizards, apart from amphisbaenids where the right one is smaller (Fig. 2).

16. *Always.* Kidneys placed well in front of cloaca and somewhat asymmetrical, the right being further forwards than the left. In lizards even slight asymmetry of the kidneys is rare.

17. *Almost always.* Gall bladder situated some distance behind liver; in lizards it is contiguous with or near the liver.

18. *Often.* Ventral scales (gastrosteges) enlarged and extending the full width of the belly. This is

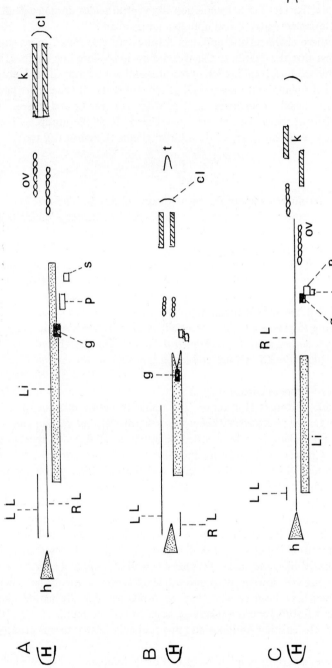

Fig. 2 Diagrams showing positions of main viscera in females of **A**, limbless lizard, the slow-worm *Anguis*; **B**, the amphisbaenid *Diplometopon*; **C**, the snake *Naja* (cobra), after Bergman (1962) and George and Shah (1965). In many other snakes the heart is situated at a relatively more posterior level. A and B drawn from preserved specimens; the spleen was not certainly identified in B. When undamaged the tail of the slow-worm (not indicated in **A**) may be longer than the body.

cl, cloaca. *g*, gall-bladder. *h*, heart. *H*, head. *k*, kidney. *Li*, liver (lobation not shown). *L.L*, left lung. *ov*, ovary. *p*, pancreas. *RL*, right lung. *s*, spleen. *t*, tail tip.

characteristic of Caenophidia, though absent in some (e.g. sea snakes). The ventral scales are not enlarged in scolecophidians and only slightly, if at all, enlarged in some henophidians. Even in the big boas and pythons which practise rectilinear locomotion the ventral scales do not extend the full width of the belly. No other Squamata have typical ophidian gastrosteges.

The fact that a number of these characteristic ophidian features are also found in certain lizards has led some workers to suggest that the lizards and snakes almost grade into one another, and that the snakes can only be distinguished by a combination of features and not by any single feature alone (Dowling, 1959). Underwood has pointed out, however, that some of these characters are peculiar or almost peculiar to snakes. Those listed under items 2, 3, 4, 8, 8(i), 12, 13, 14 and 16 seem to be of special significance and sufficient in themselves to provide a diagnosis of the group. A few others not listed here, such as the microscopic structure of the adrenal glands (Underwood, 1967), may also be important.

EARLY HISTORY

A great deal has been learnt in recent years about the early history of lizards. The first known forms showing the typical saurian modifications of the temporal region, such as a movable quadrate and loss of the lower temporal arch, are from the Upper Triassic. They include the remarkable *Kuehneosaurus* and *Icarosaurus* with elongated ribs which appear to have supported a gliding membrane like that of the modern *Draco* (Robinson, 1967; Romer, 1968). Most of the existing infraorders and some of the existing families had appeared before the end of the Cretaceous; during this period the platynotid or varanoid lizards attained their remarkable success as water reptiles, culminating in the huge marine mosasaurs. It seems probable that the lizards flourished during almost the whole of the Mesozoic and that the group was beginning to realise its wide potentialities even in late Triassic times. Snake-like, essentially terrestrial lizards do not seem to have been definitely identified before the Tertiary although some of the families which contain existing limbless forms such as the Anguidae are known from the Cretaceous.

Most of the remains of Mesozoic snakes consist only of vertebrae and ribs and much of our knowledge of them is due to the work of Robert Hoffstetter. The ophidian nature of some of the earlier snake-like reptiles such as the Lower Cretaceous *Pachyophis* is equivocal, but *Lapparentophis* from Lower Cretaceous deposits in the Sahara is now regarded definitely as a snake. It appears to have been a precursor of the Upper Cretaceous *Simoliophis*, though certain features of the latter, such as the curious thickening (pachyostosis) of the vertebral centra and the reduction of the costal facets, suggest that it lived in the sea (Hoffstetter, 1959, 1962, 1968; Romer, 1968).

Upper Cretaceous and early Tertiary formations have yielded remains of boid snakes such as the large *Madtsoia* and *Gigantophis*, and also of certain marine forms which are placed in the extinct families Palaeophidae and Archaeophidae. Perhaps the most interesting snake fossil is the 6-foot *Dinilysia* from the Upper Cretaceous of Patagonia. Its skull is comparatively well preserved and shows a combination of primitive, or lizard-like, and snake-like characters. Both temporal arches are absent, the brain is enclosed by fronto-parietal downgrowths and the skull kinesis seems to have been of a snake-like type. The animal may have been aquatic since its orbits are dorsally placed like those of the anaconda. Despite this last feature, and the relatively large size of the reptile, *Dinilysia* shows some significant resemblances to the existing Aniliinae and was probably related to this group (Estes, 1966) (see Addenda).

The fossil record therefore gives a very incomplete picture of early ophidian history. We know that snakes were in existence in the Lower Cretaceous, and that the Henophidia, represented by *Dinilysia* and boa-like types, and perhaps also by forms such as *Palaeophis*, (if these can be referred to the henophidian group), had become well established by Eocene times; the earliest known scolecophidians date from the same period. Aglyphous colubrid snakes are first known in the Upper Oligocene, and poisonous snakes of elapid and viperid type during the Miocene.

BURROWING ANCESTRY

It is generally accepted that the snakes arose from lizards of some kind but two main problems remain to be considered: (a) the general adaptive type of saurian ancestors from which the snakes arose, and (b) the group of lizards to which the ancestors of snakes are most likely to have been related.

What may be termed the 'burrowing theory' of snake origins, so brilliantly argued by Walls (1942), and discussed by Bellairs and Underwood (1951), has on the whole been favourably received (Romer, 1956, 1968). It is suggested that the immediate ancestors of snakes were not surface-dwelling lizards, but were creatures which had become highly modified for burrowing life. Their eyes had supposedly become degenerate (or at least simplified), but subsequently underwent re-elaboration in those snakes, (the great majority), which returned to life above ground. Characters such as the rigid cranial box, the loss of the remaining temporal arch, of the interorbital septum, of the middle ear cavity and of the parietal eye, the importance of olfaction by the nose and Jacobson's organ, and of course many features of the eye itself, together with the fossorial habits of many of the 'lower snakes' seem to point in this direction. Furthermore, although the central nervous connections of the optic system are basically similar in lizards and snakes, the comparatively small size of the optic tectum in certain boas which live above ground may be significant (Senn, 1966). If these rather primitive snakes had originated directly from large-eyed, surface-dwelling lizards, one might expect that their optic tecta would be as relatively large as those of typical caenophidians.

Underwood's important work on the complex patterns of the visual cells of Squamata can be interpreted along the lines of the 'burrowing theory' but does not provide clear-cut evidence for it. Underwood suggests that the ancestral Squamata possessed a duplex type of retina, containing both rods and cones; presumably this was retained by the earlier snakes even though their eyes became reduced in association with nocturnal and burrowing habits. The duplex retina of living Boidae most closely approaches the primitive ophidian condition and the visual cell patterns of other snakes, including the Scolecophidia which have rods only, is thought to be derivable fom it. Since lizards typically possess pure cone retinas, (exceptions such as geckos can be explained), Underwood suggests that the Serpentes must have diverged from the Sauria at an early stage, before any of the groups of Recent lizards had arisen (Underwood, 1967, 1970).

Underwood (1967) has made the interesting suggestion that the highly mobile jaws of snakes were acquired before or during their period of subterranean apprenticeship rather than after it, as has been previously supposed. Along with Hoffstetter (1968) I feel that this hypothesis raises serious difficulties. The mobile-jawed Typhlopidae certainly possess many primitive features, but it is hard not to believe that they are also highly aberrant. The curious type of upper jaw kinesis in *Typhlops*, with the maxillary tooth row disposed almost transversely across the skull seems a striking aberration, unlikely to lead on to more typical ophidian conditions. One would guess that it is associated with some special method of raking in wriggling invertebrate prey, perhaps assisted by peristaltic movements of the oesophagus (Haas, 1964).

The skull of a primitive henophidian such as *Cylindrophis* (Aniliinae) with its fairly rigid jaw skeleton seems a much more likely starting point for a reptile aspiring to higher ophidian status. The boid condition could probably be derived from it, without radical alteration, by a loosening of the facial and upper jaw skeleton and elongation of the supratemporal and quadrate so that the jaw joint was shifted behind the skull, increasing the length of the gape and the mobility of the jaws.

The typhlopids seem to have found movable upper jaws of value in their particular burrowing economy. One would imagine, however, that selection pressure for kinesis would in general be strongest in snakes beginning their careers above ground, with opportunities for capturing relatively large prey. According to the 'burrowing theory' the rigid cranial box in which parietal as well as frontal down-growths participate originated as a fossorial adaptation; it could have been pre-adaptive also, as a protection for the fore-brain from large prey, protesting against being swallowed alive.

Berman and Regal (1967) have pointed out that the loss of the middle-ear cavity and tympanic membrane must have been essential for the evolution of a highly mobile quadrate, irrespective of the type of habitat in which the ancestral snakes lived. One could argue, however, that the initial loss of these structures was associated with burrowing habits and the diminished importance of air-borne auditory stimuli; although loss of the tympanic system in lizards is by no means always correlated with burrowing, it is certainly prevalent in forms which are highly specialised for subterranean life. After this loss had occurred in the early snakes, the path to increased streptostyly would have been opened up; like the rigidity of the cranium the loss of the middle ear structures could have been pre-adaptive for the acquisition of new feeding habits.

Although snakes are not completely deaf to air-borne sounds their hearing is comparatively poor, sensitivity being limited to a narrow range of sound frequencies. In burrowing lizards the acoustic papilla in the inner ear is reduced, but in most burrowing snakes including the scolecophidians and Aniliidae it is better developed than that of other snake types. Good development of the papilla is also characteristic of 'lower snakes' in general, irrespective of habitat; the fact that some non-burrowing boids have long, large papillae may indicate that they were derived from burrowing ancestors, or even that the ancestral snakes were all burrowers (Miller, 1968). Study of the inner ear, as of the visual cells, also points to an early separation of the Serpentes from the saurian stock.

It is implicit in the 'burrowing theory' that many basic features of ophidian organisation originated as burrowing specialisations, although these are often masked by secondary adaptations to life above ground. The types of snakes regarded as being most primitive would therefore be those which show burrowing modification most clearly, but which also possess non-adaptive archaic characters such as vestiges of the hind limbs. If one does not accept this theory, as Dowling (1959) has pointed out, it is logical to regard the Boidae as the most primitive existing family since some of its members possess unquestionably archaic features but show no obvious signs of burrowing adaptation. The primitive nature of the boid visual cells has already been noted, but Underwood interprets it along 'burrowing theory' lines.

THE PROBLEM OF PLATYNOTID AFFINITY AND LANTHANOTUS

It is quite possible that the snakes evolved from some unknown stock of early lizards and are not closely related to any of the recognised saurian groups, surviving or extinct. This would be in accordance with Underwood's views (1957b, 1970) on the early divergence of the Serpentes, and also perhaps with the observations on the inner ear structures previously cited.

On the other hand there has for many years been an idea that the snakes have a special relationship with the platynotid or varanoid lizards and it has been suggested that they are descended from dolichosaurs, a family of platynotids known mainly from the Lower Cretaceous. These small reptiles (3 feet or less) were snake-like in form with long slender necks and bodies, zygosphenal joints on their vertebrae and moderately reduced limbs, the posterior pair being the larger. Bellairs and Underwood (1951) examined the evidence for this view and concluded that the case for dolichosaurian, or indeed for general platynotid affinities with the snakes was not proven. Apart from showing certain unsnake-like anatomical features, the dolichosaurs are thought to have been shore-living or marine, and origin from such ancestors is difficult to reconcile with the structure of the ophidian eye, the focal point of Walls' 'burrowing theory'.

In 1954, however, McDowell and Bogert produced a lengthy and stimulating article dealing, *inter alia*, with *Lanthanotus borneensis*, a hitherto little-known lizard from Sarawak. They confirmed its position in a family of its own and established its membership of the platynotid group, along with the monitors and heloderms. They believed that it was even more closely related to the extinct, Lower Cretaceous platynotids known as aigialosaurs, amphibious monitor-like lizards from which the dolichosaurs and mosasaurs were probably derived. McDowell and Bogert then went on to suggest that *Lanthanotus* also shows significant resemblances to snakes and might even be regarded as a 'structural ancestor' of these reptiles, linking the early platynotids with the ancestors of Serpentes.

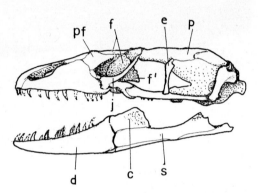

FIG. 3 Skull of *Lanthanotus borneensis*, after McDowell and Bogert (1954). The hinge across the outer side or the lower jaw, mainly between the dentary and coronoid + surangular, is evident. There is a corresponding, but less pronounced hinge on the inner side of the jaw.

c, coronoid. *d*, dentary. *e*, epipterygoid. *f*, frontal. *f'*, frontal downgrowth. *j*, jugal. *p*, parietal. *pf*, prefrontal. *s*, surangular.

Lanthanotus, the 'earless' monitor as it is popularly known, is a long-bodied lizard rather more than a foot in length (maximum recorded size 43 cm) with a blunt muzzle, rather small eyes, short but pentadactyl limbs, and prominent rows of tubercles along its back (see Plate I). It is not so rare as was formerly believed but is nocturnal and cryptic, being sometimes found in holes underground. Like other elongated lizards it moves partly by means of serpentine undulations and shows a marked liking for water. In captivity it is sluggish; it can be fed on raw turtle eggs or on strips of originally frozen plaice (B. Harrisson, 1961; T. Harrisson, 1966; Mertens, 1966). A specimen in the Bronx Zoo eats earthworms and frozen squid, and catches live goldfish in its bowl (Wayne King, personal communication.)

Among the various features of *Lanthanotus* which McDowell and Bogert regarded as significantly snake-like are (1) loss of upper temporal arch (the postorbital arch is present), (2) absence of pineal foramen, (3) presence of frontal downgrowths separating the orbits, (though not of well developed parietal downgrowths), (4) transverse hinge across lower jaw, (5) recurved form of teeth which show alternating type of replacement, (6) teeth on pterygoids and (very few) on palatines, (7) well forked, protrusible tongue (which does not, however, look nearly as snake-like as that of *Varanus*), (8) window in lower eyelid, possibly foreshadowing evolution of spectacle, and (9) absence of fracture planes for autotomy in tail vertebrae. It should be added that (3), (4), (5) and (9) are features of Platynota in general.

McDowell has described the middle ear of *Lanthanotus* in a later paper (1967a) and concludes that it resembles the condition in mosasaurs, and also that 'The snake ear can easily be derived from that of the *Lanthanotus*-mosasaur group through loss of the extra-columella and tympanic cavity'. Estes (1966) mentions that there are certain palatal resemblances between *Lanthanotus* and the fossil snake *Dinilysia*. A further point which may perhaps support an association of the snakes and the Platynota

in general is Werner's observation (1959) that *Varanus gouldii* is the only lizard (out of 40 species studied) with a chromosome complement similar to that of snakes (see Addenda).

Lanthanotus shows a number of unsnake-like characters, of which the most obvious are the following: (i) the relatively slight degree of limb reduction, (ii) the suspension of the quadrate, (iii) the well developed epipterygoid, (iv) the presence of scleral ossicles, and (v) of osteoderms. Furthermore, from the figures available (Fig. 3 here), I doubt whether the brain-case makes any real approach to the platytrabic condition, though serial sections are needed to settle this point. It seems probable that there is a cartilaginous interorbital septum beneath and behind the frontal downgrowths, as there is in *Varanus*.

Despite these considerations, the 'burrowing theory' of snake origins would not, I think, be greatly strained by the acceptance of special relationship between *Lanthanotus* and the ancestors of snakes. The 'earless' monitor is not markedly specialised for burrowing, but it does appear to have fossorial inclinations, employing its head and neck in digging rather than its fore-limbs (Harrisson, 1961). It is also thought that vision plays a small part in its life (Mertens, 1966).

The Mesozoic radiation of the Platynota may have been wider than is generally envisaged; the group is an ancient one with a probable representative known from the Upper Jurassic (Romer, 1968). It is well known that certain groups of lizards such as the skinks show a whole spectrum of body elongation and limb reduction, in some cases correlated with degrees of specialisation for burrowing life. The skulls of surface-dwelling skinks are typically saurian in such features as the development of the interorbital septum; in burrowing skinks such as *Acontias meleagris*, on the other hand, the septum is reduced and a bony cranial box is formed resembling that of snakes (Brock, 1941; El-Toubi *et al.*, 1965). It is quite possible that a comparable spectrum had been evolved among the extinct relatives of *Lanthanotus*. The hallmarks of burrowing adaptation superimposed on the basic platynotid organisation might provide a much closer approach to ophidian structure than *Lanthanotus* does itself.

Very little is known about the 'earless' monitor in the wild, but its habit of remaining for long periods in water has been stressed by those who have kept it in captivity. These aquatic proclivities seem to be in accord with a general trend among the early Platynota, but are not evident from its appearance apart from the dorsally placed nostrils; for example, the tail is not laterally compressed. There are, however, other types of aquatic adaptation besides that of the free-swimming surface-dweller. Is it conceivable that the ancestors of snakes were small-eyed mud-dwellers, combining burrowing with amphibious habits? Some of the present-day caecilians (Gymnophiona) might offer interesting parallels; these animals burrow in moist earth or vegetation near water, though only one small group of them is known to be aquatic in the adult state.

I feel, myself, that McDowell and Bogert have made a good case for the affinities of *Lanthanotus*, and by inference of the Platynota in general, with the snakes, at least in so far as the range of characters which they consider is concerned. Their theory has the merit of providing a positive solution to the baffling problem of the origin of Serpentes and is worthy of serious and continued attention.

Underwood, however, has remained unconvinced that *Lanthanotus* or other Platynota (or indeed any other of the surviving groups of lizards) occupy a key position in relation to the origin of snakes. He argued in a critical review (1957a) that many of the allegedly snake-like characters of the 'earless'

Plate I

Top. The Australian snake-like lizard *Pygopus nigriceps* beginning defensive display, and mimicking a venomous snake. The reptile strikes with the mouth closed and usually past the enemy; the neck is flattened and the throat expanded. From Bustard (1968).

Bottom. The 'earless' monitor, *Lanthanotus borneensis*. (Photo by Tom Harrisson).

monitor can be matched in other groups of lizards, while others again may be of doubtful phylogenetic significance. He suggests that the snake-like form of the hinge across the lower jaw may be a parallel resemblance evolved independently in Platynota and snakes in relation to the habit of feeding on large prey.

Underwood's work on the visual cells, however, seems to provide the most serious objection to the concept of *Lanthanotus* as a structural ancestor of snakes. If the yet undescribed retina of this lizard should prove to show the typically saurian pure-cone condition, like that found in *Varanus*, and even in the small-eyed, nocturnal platynotid *Heloderma*, derivation of the duplex condition, (which is thought to be primitive for snakes), would seem unlikely. One would have to suppose either that the rod elements persisted in the early platynotid stock, or else that they had been evolved *de novo* among the ancestors of snakes.

SNAKE MIMICS

In criticising over-ready acceptance of the Platynota as ophidian ancestors, Underwood (1957b) has drawn attention to the remarkably snake-like character of the pygopodid lizards. The family Pygopodidae is restricted to Australasia and consists entirely of forms with elongated bodies; some species are practically limbless externally, but in others there are appreciable flap-like vestiges of the hind limbs. As in snakes the eyes are covered by a transparent spectacle. Some forms such as *Pygopus* and *Lialis* grow about two feet long and are able to prey on smaller lizards; they are often found under logs but live mainly on the surface of the ground. *Pygopus nigriceps* mimics poisonous snakes in its threat behaviour, expanding its throat and striking at an aggressor (Bustard, 1968; see Plate I here). Another genus, *Aprasia*, consists of six-inch, worm-like forms, modified for burrowing; these show a closer approximation to snakes in cranial structure. Parietal, as well as frontal downgrowths are present and the interorbital septum is reduced to a rod-like, partly ossified trabecula communis; the anterior orbital cartilages are absent. These features are surprising in view of the fact that the eyes do not appear to be drastically reduced; moreover, there is no scleral cartilage, at least in the specimen which Underwood examined, although ossicles are present. Reduction of the middle ear has proceeded even further than in snakes, for *Aprasia* is the only lizard known in which the stapes has virtually disappeared.

It seems now established that the pygopodids are related to, (and probably derived from), the geckos, and the two families are placed together in the infraorder Gekkota. Underwood (1957b) also believes that the worm-like subterranean *Dibamus*, an enigmatic form from the Far East and New Guinea, has affinities with this group. The Gekkota has never been regarded as closely related to, or possibly ancestral to, the snakes and Underwood does not suggest that it should be. Geckos and pygopodids alike, however, do possess snake-like features: loss of the temporal arcades and pineal foramen, and presence of the spectacle eye-covering, (except in a few geckos). There are unsnake-like characters too: the tongue is broad and only notched at the tip, the tail has fracture planes and usually has good powers of regeneration and the heart is situated quite near the head.

CONCLUSIONS

There is good evidence that the Serpentes diverged from the Sauria at an early date, perhaps during the Jurassic. This is supported by the existence of some seemingly deep-seated differences between existing snakes and lizards. Moreover, the structural changes associated with elongation of the body such as asymmetry of certain viscera, appear to be more marked in snakes than in snake-like lizards.

The theory that the snakes have evolved from burrowing ancestors still seems to provide the most satisfactory explanation for many of their peculiarities. It rests on circumstantial evidence however, and should not become accepted as a dogma.

The combination of characters described by McDowell and Bogert in the platynotid lizard *Lanthanotus* suggests to me that there is a genuine phylogenetic relationship between this reptile and the snakes. The resemblances seem perhaps the more significant since they are not associated with extreme burrowing habits or with drastic reduction of the limbs. There are, however, important objections to this view, which remains controversial. As Underwood (1957a) has emphasised, there is still a need for workers in various fields to investigate every line of evidence bearing on the origin of snakes, and they should not be hampered by preconceived ideas.

It is disappointing that the origin of snakes still remains so obscure and that any discussion of the problem involves so much speculation. New palaeontological discoveries are eagerly awaited; in the meanwhile students of existing forms may be stimulated to further investigation of the complexities of squamate evolution.

ACKNOWLEDGEMENTS

I am most grateful to Dr Garth Underwood for his helpful comments on this article, to Dr H. R. Bustard and Mr Tom Harrisson for photographs, and to Mrs Lesley Sheppard for the dissection of *Diplometopon*.

REFERENCES

ALBRIGHT, R. G. and NELSON, E. M. 1959. Cranial kinetics of the generalised colubrid snake *Elaphe obsoleta quadrivittata*. I. Descriptive morphology. II. Functional morphology. *J. Morph.*, **105**, 193-240: 241-292.
BAIRD, I. L. 1970. The anatomy of the reptilian ear. In *Biology of the Reptilia* (Ed. C. Gans and T. S. Parsons), **2**, 193-275. Academic Press, London and New York.
BELLAIRS, A. d'A. 1949. The anterior brain-case and interorbital septum of Sauropsida, with a consideration of the origin of snakes. *J. Linn. Soc. (Zool.)*, **41**, 482-512.
BELLAIRS, A. d'A. and UNDERWOOD, G. 1951. The origin of snakes. *Biol. Rev.*, **26**, 193-237.
BERGMAN, R. A. M. 1962. Die Anatomie der Elapinae. *Z. wiss. Zool.*, **167**, 291-337.
BERMAN, D. S. and REGAL, P. J. 1967. The loss of the ophidian middle ear. *Evolution*, **21**, 641-643.
BOLTT, R. E. and EWER, R. F. 1964. The functional anatomy of the head of the puff adder, *Bitis arietans* (Merr.). *J. Morph.*, **114**, 83-106.
BROCK, G. T. 1941. The skull of *Acontias meleagris*, with a study of the affinities between lizards and snakes. *J. Linn. Soc. (Zool.)*, **41**, 71-88.
BUSTARD, H. R. 1968. *Pygopus nigriceps* (Fischer): a lizard mimicking a venomous snake. *Br. J. Herpetol.*, **4**, 22-24.
DOWLING, H. G. 1959. Classification of the Serpentes: a critical review. *Copeia* (for 1959), 38-52.
EL-TOUBI, M. R., KAMAL, A. M. and HAMMOUDA, H. G. 1965. The origin of Ophidia in the light of the developmental study of the skull. *Z. zool. Syst. Evolutionsforsch.*, **3**, 94-102.
ESTES, R. 1966. Anatomy and relationships of the primitive fossil snake *Dinilysia*. *Yrb. Am. phil. Soc.* (for 1966), 334-336.
FRAZZETTA, T. H. 1966. Studies on the morphology and function of the skull in the Boidae (Serpentes). Part II. Morphology and function of the jaw apparatus in *Python sebae* and *Python molurus*. *J. Morph.*, **118**, 217-296.
GABE, M. and SAINT GIRONS, H. 1969. Données histologiques sur les glandes salivaires des Lépidosauriens. *Mém. Mus. natn. Hist. nat., Paris*, **58A**, 1-112.
GANS, C. 1961. The feeding mechanism of snakes and its possible evolution. *Am. Zool.*, **1**, 217-227.
—— 1968. Relative success of divergent pathways in amphisbaenian specialisation. *Am. Nat.*, **102**, 345-362.
—— 1969. Amphisbaenians—reptiles specialised for a burrowing existence. *Endeavour*, **28**, 146-151.
GASC, J. P. 1966. Les rapports anatomiques du membre pelvien vestigial chez les squamates serpentiformes. [*Anguis fragilis* and *Python sebae*]. *Bull. Mus. natn. Hist. nat., Paris*, **37**, 916-925: **38**, 99-110.
—— 1967. Introduction à l'étude de la musculature axiale des squamates serpentiformes. *Mém. Mus. natn. Hist. nat., Paris*, **48A**, 69-125.
GEORGE, J. C. and SHAH, R. V. 1965. Evolution of air sacs in Sauropsida. *J. Anim. Morph. Physiol.*, **12**, 255-263.
HAAS, G. 1962. Remarques concernant les relations phylogéniques des diverses familles d'ophidiens fondées sur la différenciation de la musculature mandibulaire. In *Problèmes actuels de paléontologie (Evolution des vértebrés)*. *Colloq. int. Cent. nat. Rech. sci.*, **104**, 215-241.

HAAS, G. 1964. Anatomical observations on the head of *Liotyphlops albirostris* (Typhlopidae, Ophidia). *Acta zool. Stockh.*, **45**, 1-62.
HARRISSON, B. 1961. *Lanthanotus borneensis*—habits and observations. *Sarawak Mus. J.*, **10**, 286-292.
HARRISSON, T. 1966. A record size *Lanthanotus* alive (1966): casual notes. *Sarawak Mus. J.*, **14**, 323-334.
HOFFSTETTER, R. 1959. Un serpent terrestre dans le Crétacé inférieur du Sahara. *Bull. Soc. géol. France.* (7 ser.), **1**, 897-902.
—— 1962. Revue des récentes acquisitions concernant l'histoire et la systématique des squamates. In *Problèmes actuels de paléontologie (Evolution des vertébrés)*. *Colloq. int. Cent. nat. Rech. Sci.*, **104**, 243-279.
—— 1968. (Review of) A contribution to the classification of snakes. By G. Underwood. *Copeia* (for 1968), 201-213.
HOFFSTETTER, R. and GASC, J. P. 1969. Vertebrae and ribs of modern reptiles. In *Biology of the Reptilia* (Ed. C. Gans et al.), **1**, 201-310. Academic Press, London and New York.
KAMAL, A. E. M. 1965. The origin of the interorbital septum of Lacertilia. *Proc. Egypt. Acad. Sci.*, **18**, 70-72.
LIST, J. C. 1966. Comparative osteology of the snake families Typhlopidae and Leptotyphlopidae. *Illinois Biological Monographs*, No. 36. University of Illinois Press, Urbana and London.
MCDOWELL, S. B. 1967a. The extracolumella and tympanic cavity of the 'earless' monitor lizard, *Lanthanotus borneensis*. *Copeia* (for 1967), 154-159.
—— 1967b. Osteology of the Typhlopidae and Leptotyphlopidae: a critical review. [Review of List (1966) cited above]. *Copeia* (for 1967), 686-692.
MCDOWELL, S. B. and BOGERT, C. M. 1954. The systematic position of *Lanthanotus* and the affinities of the anguinomorphan lizards. *Bull. Am. Mus. nat. Hist.*, **105**, 1-142.
MERTENS, R. 1966. The keeping of Borneo earless monitors (*Lanthanotus borneensis*) in captivity. *Sarawak Mus. J.*, **14**, 320-322.
MILLER, M. R. 1968. The cochlear duct of snakes. *Proc. Cal. Acad. Sci.*, **35**, 425-476.
ROBINSON, P. L. 1967. The evolution of the Lacertilia. In *Problèmes actuels de paléontologie (évolution des vertébrés)*. *Colloq. int. Cent. nat. Rech. Sci.*, **163**, 395-407.
ROMER, A. S. 1956. *Osteology of the reptiles*. University of Chicago Press. Chicago.
—— 1968. *Notes and comments on vertebrate paleontology*. University of Chicago Press, Chicago.
SENN, D. G. 1966. Über das optische system im Gehirn squamater Reptilien. *Acta Anat. Suppl.* 52 = 1 ad vol. **65**, 1-87.
SIMPSON, G. G. 1945. The principles of classification and a classification of mammals. *Bull. Am. Mus. nat. Hist.*, **85**, i-xvi, 1-350.
TAUB, A. M. 1966. Ophidian cephalic glands. *J. Morph.*, **118**, 529-542.
TOERIEN, M. J. 1963. The sound-conducting systems of lizards without tympanic membranes. *Evolution*, **17**, 540-547.
UNDERWOOD, G. 1957a. *Lanthanotus* and the anguinomorphan lizards: a critical review. *Copeia* (for 1957), 20-30.
—— 1957b. On lizards of the family Pygopodidae. A contribution to the morphology and phylogeny of the Squamata. *J. Morph.*, **100**, 207-268.
—— 1967. *A contribution to the classification of snakes*. British Museum (Natural History), London.
—— 1970. The eye. In *Biology of the reptilia* (Ed. C. Gans and T. S. Parsons), **2**, 1-97. Academic Press, London and New York.
WALLS, G. L. 1942. The vertebrate eye and its adaptive radiation. *Bull. Cranbrook Inst. Sci.*, No. 19, Ann Arbor, Michigan.
WERNER, Y. L. 1959. Chromosomes of primitive snakes from Israel. *Bull. Res. Council of Israel*, **8B**, 197-198.

ADDENDA

ESTES, R., FRAZZETTA, T. H. and WILLIAMS, E. E. 1970. Studies on the fossil snake *Dinilysia patagonica* Woodward Part I. Cranial morphology. *Bull. Mus. comp. Zool. Harv.*, **140**, 25-74.

This important study confirms resemblances with the Aniliinae but reveals some divergent features. The findings neither support nor refute the theory of platynotid-snake affinities.

GORMAN, G. C. and GRESS, F. 1970. Chromosome cytology of four boid snakes and a varanid lizard, with comments on the cytosystematics of primitive snakes. *Herpetologica*, **26**, 308-317.

This new study shows that *Varanus* has a nearly typical saurian karyotope with no firm indication of special affinity with snakes.

C. B. COX

A new digging dicynodont from the Upper Permian of Tanzania

INTRODUCTION

The discovery by Parrington of a specimen of *Endothiodon* in 1930, while passing through the south-western corner of what is now Tanzania, indicated that a Karroo fauna was present in the area. He was the first to make an extended stay in the region expressly to collect fossil vertebrates, but he was soon followed by Nowack, most of whose material was later described by von Huene. Among this were some dicynodonts which von Huene (1942) ascribed to the well known South African genus *Cistecephalus*[1].

As well as seven skulls, this material included vertebrae, ribs, both girdles and the humerus. Most of these were damaged and incomplete, and only those surfaces which lay superficially in the nodules had been exposed. The specimens had been collected in what is now known as the Kawinga Formation (Charig, 1963), which is similar in age to the *Cistecephalus* Zone of South Africa. Material from the Formation is very inviting to the palaeontologist, for it is frequently little, if at all, distorted and is preserved in a matrix which can readily be prepared with acetic or formic acid, leaving beautifully clean bone which can be studied in minute detail. It therefore seemed worthwhile to try to obtain more information from von Huene's specimens by using this technique. Two of the skulls, a lower jaw and all the post-cranial material were very kindly loaned for this purpose in 1961 by Dr F. Westphal of the Geologisches-Paläontologisches Museum, Tübingen, and a further four skulls were made available recently.

Study of the post-cranial material showed that the scapulo-coracoid, humerus and ulna were very unlike those of most dicynodonts. Their distinctness provoked a functional analysis, to try to explain the reasons for the observed specializations. As will be seen, adaptation to digging may be a satisfactory solution to this problem. It has now also become clear that the form under discussion does not, after all, belong to *Cistecephalus*, but represents a new genus. It is therefore named *Kawingasaurus fossilis*—the digging Kawinga reptile. The cranial morphology is figured and described here in sufficient detail to establish the validity of the new genus. In view of the almost perfect condition of the specimens,

[1] Though originally named *Kistecephalus* by Owen (1876), the spelling was later arbitrarily altered to *Cistecephalus* by Lydekker (1890). This emendation has, for the last 80 years, been accepted by all major workers on the dicynodonts (Broom, Watson, von Huene, Broili and Schröder, Haughton, van Hoepen, Brink, Toerien and others). Because the genus was selected by Broom (1906) as one of the six genera used to characterise the zones of the South African Karroo deposits, the name has been widely used in both palaeontological and geological publications. It would therefore be inconvenient now to revert to the original spelling, and application is therefore being made to the International Commission on Zoological Nomenclature to have the name *Kistecephalus* suppressed under the Commission's plenary powers.

it is hoped to describe the skull more thoroughly at a later date, after preparation of the remaining skulls.

Genus *Kawingasaurus nov.*

Generic diagnosis Small toothless dicynodont. Skull c. 4 cm long and 5 cm broad. Skull tapers anteriorly in dorsal and lateral view. Flattened, laterally expanded snout and small, antero-dorsally directed orbits. Wide intertemporal and interorbital bars. Pineal foramen absent. Zygomatic arch slender. Interpterygoid vacuity closed. Otic region inflated. Stapedial foramen present. Ventral region of scapulo-coracoid twisted so that glenoid is directed laterally. Cleithrum present. Spine down outer surface of scapula near its anterior edge. Coracoid region reduced. Proximal and distal surfaces of humerus expanded and at an angle of $85°$ to one another. Humerus bears prominent postero-internal process proximally and large epicondyles distally. Condyles for radius and ulna separate from one another. Ulna bears large olecranon process.

Type species *Kawingasaurus fossilis nov.*

Holotype of *K. fossilis* No. K52 in the Geologisches-Paläontologisches Museum, Tübingen, consisting of a skull.

Paratype Nos. K56 (a skull) and K55 (five skulls plus post-cranial remains) in the above mentioned Museum.

Horizon and Locality From the Upper Permian Kawinga Formation, at a level 78 m above its base, from near the village of Kingori, Songea District, Tanzania. (For a detailed plan of the locality, see Nowack, 1937, fig. 5).

OSTEOLOGY

Skull. Skulls K52 and K55a are of identical size. Most of the features shown in the figures are derived from specimen K52, but the details of the palate are derived from specimen K55a, and the details of the snout and external nares are derived from specimens K55b and K56.

The greatest breadth of the skull, which is across the posterior tips of the squamosals, is just under 5 cm; the length in the midline is 4·0 cm. The bones of the skull are very dense and thin, being translucent in some areas.

In dorsal view (Fig. 1A) the very wide intertemporal bar narrows anteriorly. Few sutures are visible in this region, but it appears to be composed mainly of very wide frontals and parietals. The pineal foramen is absent. The postorbital extends back to meet the squamosal at the posterior end of the intertemporal bar. The interorbital region, also, is relatively wide, the orbits are rather small and are directed antero-dorsally. The nasals extend back as a rectangular projection between the frontals. Just in front of the orbits, the snout suddenly expands laterally, and is unusually flattened. The external nares lie fairly close to the midline, separated by a rounded ridge.

The skull also tapers anteriorly in lateral view (Fig. 1B), its greatest height being at the level of meeting of the skull roof and occiput. Though these two regions meet at right angles to one another, there is a rounded transition between them. The occiput slopes downwards and backwards, and the region of the foramen magnum and occipital condyle is visible in lateral view. The otic region of the braincase is inflated and projects both anteriorly and posteriorly. The squamosal extends very far anteriorly, meeting the maxilla under the anterior half of the orbit. The zygomatic arch is very slender and frail. There is an extensive ossification of the median region of the skull, including not only the sphenethmoid between the orbits, but also extending more posteriorly and ventrally between the anterior parts of the temporal fossae to meet the dorsal surface of the palate.

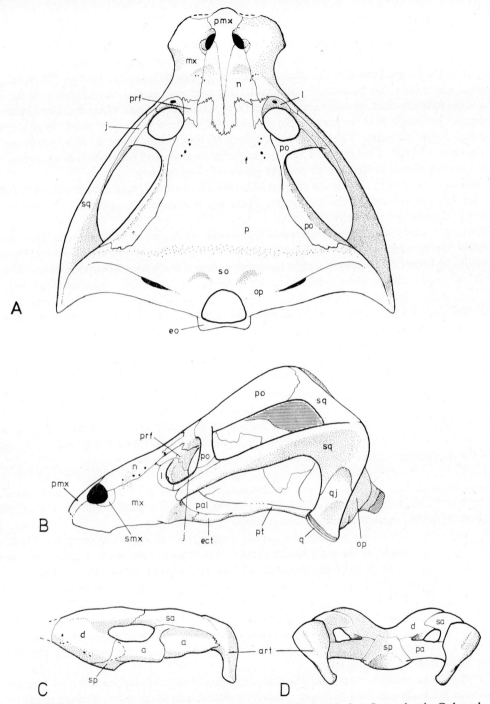

Fig. 1 *Kawingasaurus fossilis*, skull in **A**, dorsal view and **B**, lateral view. Lower jaw in **C**, lateral view and **D**, posterior view. (×2).

Abbreviations: *a*, angular; *art*, articular; *d*, dentary; *ect*, ectopterygoid; *eo*, exoccipital; *j*, jugal; *l*, lacrimal; *mx*, maxilla; *n*, nasal; *op*, opisthotic; *p*, parietal; *pa*, prearticular; *pal*, palatine; *pmx*, premaxilla; *po*, postorbital; *prf*, prefrontal; *pt*, pterygoid; *q*, quadrate; *qj*, quadratojugal; *sa*, surangular; *smx*, septomaxilla; *so*, supraoccipital; *sp*, splenial; *sq*, squamosal.

In ventral view (Fig. 2A) the whole palatal region is unusually flat, except for a low median ridge on the premaxilla. There is no sign of any teeth. The posterior median part of the premaxilla is clasped by the vomer. Lateral to the internal naris, the premaxilla meets the palatine. There is a small foramen where these two bones meet the maxilla. The pterygoid extends far anteriorly on the lateral side of the palate to meet the ectopterygoid, and there is a larger foramen where these two bones meet a posterior projection from the maxilla. No trace of interpterygoid vacuities is visible in either ventral or dorsal view of the palate. The basisphenoid tubera are low, but the fenestra ovalis is large and filled by the enlarged head of the stapes. There is a stapedial foramen and the postero-distal corner of the stapes bears a narrow process which projects posteriorly and dorsally.

The anterior end of the lower jaw (Figs 1C-D, 2B-C) is missing. There is a large fenestra between the dentary, the angular and the surangular. There is a prominent retro-articular process, which is directed ventrally and internally (Fig. 1D).

The post-cranial material was contained in four blocks. As described by von Huene (1942) the first included four vertebrae and two ribs, together with a lower jaw. The second contained four vertebrae, three ribs and other rib fragments, a right humerus, a clavicle and a fragmentary left pelvis. The third block contained a left scapulo-coracoid and a left humerus, and the fourth contained a right humerus. Acid preparation has revealed that the bone which von Huene supposed to be a clavicle is a right ulna, and that all the humeri visible to von Huene were left humeri. Furthermore, an almost undamaged right humerus lay concealed within the fourth block until revealed by acid preparation.

Von Huene's material thus provides an almost complete scapulo-coracoid and a complete humerus and ulna. All these bones are beautifully preserved and are now completely free of matrix. It is therefore possible to study them in detail, to investigate the types of movement possible at each joint and to attempt to assess the functional reasons for the divergence of these bones from the normal dicynodont pattern. This in turn indicates to which particular mode of life *Kawingasaurus* was adapted. The osteology of each element will first be described and figured. The abnormalities of these elements will then be noted and the functional anatomy of the whole pectoral region will be discussed, together with the implications of this as to the mode of life of *Kawingasaurus*.

Scapulo-coracoid (Fig. 3). Though the bone figured is the almost perfect left scapulo-coracoid, this has been reversed in the figures to appear as that of the right side in order to facilitate integration of this bone with the humerus and ulna, which belong to the right side.

No clear sutures are visible between the scapula, precoracoid and coracoid. The glenoid is directed outwards, at right angles to the plane of the surface of the blade of the scapula, but more ventrally. Its surface is well finished. In profile the outline of the glenoid forms about 160° of arc.

The scapula is elongate and narrow. The blade is very constricted above the glenoid and gradually expands dorsally. The posterior border is concave in lateral view, because the postero-dorsal corner of the blade is prolonged posteriorly. The postero-dorsal region faces slightly more posteriorly than the rest of the blade; this fact can be seen especially well in dorsal view (Fig. 3C). As a result, the line of meeting of this region and the rest of the blade is marked by a low ridge running up the outer side of the blade from its posterior edge.

A very pronounced spine runs down the external surface of the scapular blade near its anterior edge. This ridge gradually decreases in height towards its ventral end, where it twists round to run into the anterior edge of the preglenoid region of the scapula. There is no sign of an acromion process arising from this ridge; though its summit is slightly damaged in one or two places, none of the resulting broken surfaces are sufficiently extensive to suggest that they mark the position at which an acromion process had broken off.

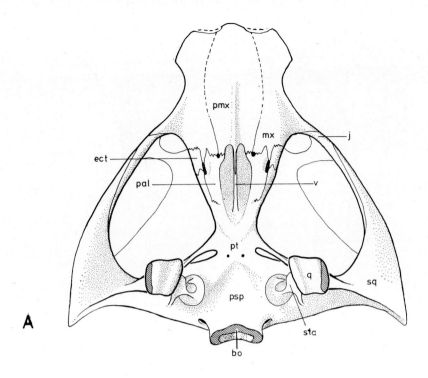

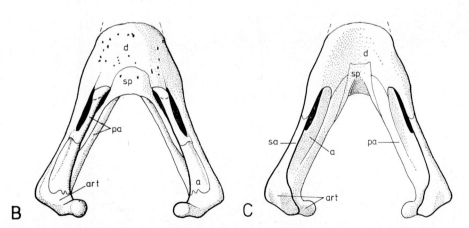

Fig. 2 *Kawingasaurus fossilis*. **A**, ventral view of skull. Lower jaw in **B**, ventral view and **C**, dorsal view. (×2).

Abbreviations: *bo*, basioccipital; *psp*, parasphenoid-basisphenoid complex; *sta*, stapes (other abbreviations as in Fig. 1).

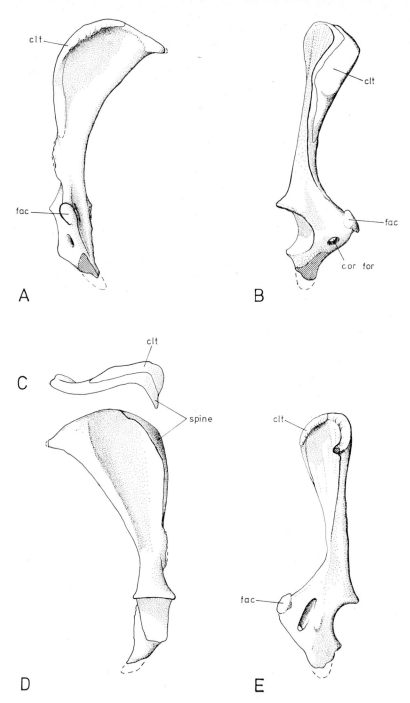

Fig. 3 *Kawingasaurus fossilis*, specimen K55, right scapulo-coracoid. **A**, medial view; **B**, anterior view; **C**, view of dorsal edge of blade; **D**, lateral view; **E**, posterior view. (×2).

Abbreviations: *clt*, cleithrum; *cor for*, coracoid foramen; *fac*, facet for ? clavicle.

Apart from a strip near its summit, the anterior surface of the scapular spine is covered by the cleithrum. This bone is preserved as far as a point nearly half-way down the scapula towards the glenoid, and markings on the scapula show that it originally extended slightly further ventrally. Dorsally it extends half-way along the dorsal edge of the scapula. *Kawingasaurus* is one of the few dicynodonts in which a cleithrum is known. The others are an unidentified specimen in the American Museum of Natural History (catalogue no. 5302) described by Broom (1915), and *Diictodon sesoma* and *Diictodontoides skaios* (Watson, 1960). Like *Kawingasaurus*, these are all small dicynodonts; the cleithrum is certainly absent in the larger Triassic dicynodonts.

As well as covering the anterior surface of the scapular spine, the cleithrum also projects internally beyond the medial surface of the scapula so that a low ridge projects inwards from the antero-dorsal edge and the dorsal part of the anterior edge of the scapula. Only the anterior edge of the blade is curved to follow the outline of the rib cage, the remainder being flat.

Internally, the ventral end of the scapula bears a deep groove which leads into the coracoid foramen. Anterior to this, on the anterior corner of the scapulo-coracoid, a flattened knob projects slightly from the surrounding bone. Though there may be slight distortion in this region, this appears to have done no more than slightly exaggerate what seems to be a real feature of the area.

Though part of its outer surface is damaged and its posterior region is missing, it is clear that the precoracoid-coracoid region was extremely small.

Humerus (Fig. 4). The humerus is extremely twisted, the plane of the distal expansion being at an angle of about 85° to that of the proximal expansion (Fig. 4D). It appears short and massive, since both the proximal and the distal ends are expanded to a width of 20 mm while the length of the bone is only 25 mm. Its articular surfaces are distinct and well ossified. The proximal head faces dorsally, somewhat proximally, and at an angle of about 125° to the long axis of the bone. It is convex across its long axis, but almost flat along this axis; it becomes slightly narrower anteriorly. The condyles for the ulna and for the radius are both convex and are quite distinct from one another. That for the ulna faces distally while that for the radius is on the underside of the distal expansion of the humerus and faces anteriorly and somewhat ventrally. The entepicondylar canal is of normal dimensions.

The proximal dorsal surface of the humerus (Fig. 4G) is dominated by the proximal condyle. Anterior to this there is a large delto-pectoral crest, but this does not extend far distally. Distal to the proximal condyle, a high narrow ridge runs onto the dorsal part of the ectepicondyle. The postero-medial corner of the humerus is enlarged into a prominent process composed of a rather narrow dorsal surface, an extensive posterior surface and a very extensive ventral surface; the end of this process is widest dorsally and narrows ventrally.

The proximal ventral surface of the humerus (Fig. 4B) is wide and concave. It includes all the ventral aspect of the delto-pectoral crest and of the prominent postero-medial process mentioned above. The proximal condyle hardly intrudes onto this surface.

The distal part of the humerus is very wide (Figs 4A, F) since the ectepicondyle and the entepicondyle extend for a considerable distance on either side of the condyles for the radius and ulna. Large areas for muscle attachment are present on these epicondyles, that on the entepicondyle facing anteriorly and distally (Fig. 4A, C), while that on the ectepicondyle faces posteriorly (Fig. 4F).

Ulna (Fig. 5). The ulna is dominated by the large olecranon process, which makes up approximately one-third of the total length of the bone. (Though the distal end of the ulna is damaged, only a little of the bone is apparently missing.) No suture is visible between the olecranon process and the shaft

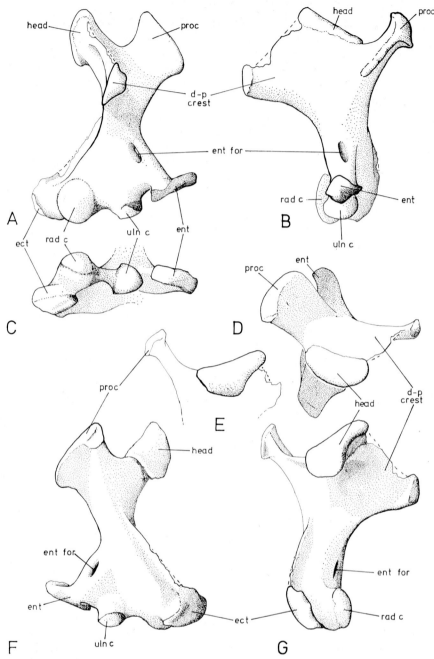

Fig. 4 *Kawingasaurus fossilis*, specimen K55, right humerus. **A**, anterior view; **B**, ventral view; **C**, view of distal end; **D**, proximal view; **E**, view of proximal condyle at right angles to its surface; **F**, posterior view; **G**, dorsal view. Abbreviations: *d-p crest*, delto-pectoral crest; *ect*, ectepicondyle; *ent*, entepicondyle; *ent for*, entepicondylar foramen; *proc*, postero-internal process; *rad c*, condyle for radius; *uln c*, condyle for ulna. (×2).

of the bone. The facet for the ulnar condyle of the humerus is deep, with an overhanging dorsal edge. The facet is slightly elongate, its long axis being directed slightly antero-ventral to the long axis of the ulna itself. Antero-ventral to it lies a distinct, slightly convex facet for the side of the radius.

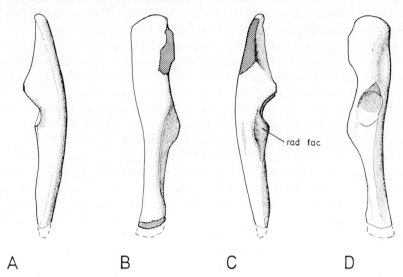

Fig. 5 *Kawingasaurus fossilis*, specimen K55, right ulna. **A**, posterior view; **B**, lateral view; **C**, anterior view; **D**, medial view. (× 2).
Abbreviation: *rad fac*, facet for side of head of radius.

TAXONOMIC POSITION

The broad intertemporal region of *Kawingasaurus* is found in many of the small Permian dicynodonts and may be a primitive feature. Other primitive features include the fact that the postorbital extends far posteriorly, and the persistence of the cleithrum and stapedial foramen. On the other hand, the lack of any teeth is an advanced feature which, together with such specialisations as the complete closure of the interpterygoid vacuity and the primitive retention of the stapedial foramen, is found also in *Cistecephalus* (Broili & Schröder, 1935). It seems, therefore, likely that *Kawingasaurus* is fairly closely related to that genus. However, it is distinguished from it by the following features: its smaller size, the relatively small orbits, the tapering shape of the skull in both dorsal and lateral view, the slenderness of the zygomatic arches and the anterior extension of the squamosal to meet the maxilla. It is therefore clearly a distinct, new genus.

FUNCTIONAL MORPHOLOGY OF THE PECTORAL GIRDLE AND FORELIMB

The right humerus figured is the same size as the left humerus found articulated with the left scapulo-coracoid. The ulna was originally associated with a humerus of slightly smaller size (less than 5% smaller) but this difference is not great enough to invalidate its use in composite restorations of the osteology and myology of the front limb.

The most unusual feature of the scapulo-coracoid of *Kawingasaurus* is the great twisting of the whole bone. In most dicynodonts, if the scapula is viewed at right angles to the plane of the outer surface of the blade, then the glenoid is directed posteriorly and only slightly laterally, so that little of its surface is visible. In *Kawingasaurus*, if similarly viewed, the whole of the glenoid is visible, as it is directed laterally. The spine along the anterior part of the outer edge of the scapula is also unusual; the only other dicynodonts known to possess this are *Zambiasaurus* (Cox, 1969), *Stahleckeria* (von Huene, 1935), *Tetragonias* (Cruickshank, 1967) and *Kingoria* (Cox, 1959). The scapulo-coracoid of *Kawingasaurus* also differs from that of most other dicynodonts in the posterior projection of the postero-dorsal corner of the scapula, in the presence of an internal ridge near the front edge of the blade, in the lack of an acromion process and in the very small size of the coracoid region.

In the humerus, the delto-pectoral crest and the epicondyles are unusually well developed, while the postero-internal region of the humerus is prolonged into a unique process. Due to these features, both the proximal and the distal ends of the bone are very wide in proportion to the length of the bone. The delto-pectoral crest does not extend as far distally as in most dicynodonts. The bone is more twisted than is usual and the proximal condyle does not lie transverse to the main axis of the bone. The condyles for the radius and the ulna are separate from one another, not confluent as in most dicynodonts.

The ulna is unusual in the very great development of the olecranon process.

From all these features it is clear that the pectoral girdle and forelimb of *Kawingasaurus* was extremely specialized. These osteological specialisations must obviously have been correlated with an unusual musculature and with an unusual function of the limb, and an attempt is now made to analyse this. Any discussion of the myology of this region in fossil reptiles is inevitably based on Romer's classic papers (1922, 1944) which provide a reasonable morphological and developmental basis for the subdivision of this musculature and for its restoration in extinct reptiles.

The most fundamental modification in *Kawingasaurus* is probably the rotation of the whole glenoid region antero-laterally, so that it is directed at right angles to the plane of the surface of the blade. This must have altered the posture of the humerus itself, which must have projected laterally instead of postero-laterally. It must also have altered the direction of the thrust of the humerus relative to the girdle as a whole. In normal dicynodonts this thrust is directed antero-medially, and the shoulder girdle is braced to withstand this by the clavicle, which curves anteriorly and medially from the acromion process of the scapula to meet the median interclavicle. Such an arrangement would not have been appropriate to withstand the medially directed thrust of the humerus in *Kawingasaurus*, and it is not therefore surprising to find that there is no sign of a normal acromion process. The most effective way of bracing the scapulo-coracoid of *Kawingasaurus* would have been for the clavicle to buttress against it internally at the level of the glenoid. This region of the scapulo-coracoid has rotated with the glenoid so that it is directed medially instead of anteriorly, and it may be significant that, as described above, there is a distinct flattened facet at the inner corner of this region. This facet would be ideally situated to be the surface against which a short, stout clavicle articulated, bracing the scapulo-coracoid away from the interclavicle, and it will be interesting to see whether future *Kawingasaurus* material will prove or disprove this hypothesis.

The morphology of the scapulo-coracoid, then, suggests that there was a powerful, inwardly-directed thrust along the humerus. The source of this is immediately apparent when the unusually large size of the olecranon process of the ulna is recalled. This process provides insertion for the triceps musculature, which extends the forearm. The presence of an unusually powerful triceps muscle is in turn supported by other osteological features of the humerus. The area of the humerus just posterior to the proximal condyle is normally occupied by a lateral head of the triceps. This region of the humerus is uniquely developed in *Kawingasaurus* into a large postero-internal process, the outer surface of which would have provided a large area of origin for this part of the triceps (Fig. 8D). The more medial part of the triceps normally originates from the distal posterior surface of the humerus. This area, too, is expanded by the presence of a ridge which runs along the antero-dorsal edge of the bone to the ectepicondyle. The scapular head of the triceps presumably originated in normal fashion from just above the glenoid. The triceps as a whole is responsible for extending the forearm. The compression force resulting from this muscle action would have been transmitted antero-medially along the humerus to the glenoid. Any tendency for the head of the humerus to move forward within the glenoid could have been resisted by the teres minor muscle. This muscle normally originates from just above the anterior region of the glenoid and inserts on the dorsal part of the humerus, just anterior to the head. There is a definite hollow in this region in *Kawingasaurus*, which may mark the insertion of such a muscle.

The recovery action, flexion of the forearm, would have been caused by the biceps and brachialis muscles. The biceps normally originates on the coracoid below the glenoid, and the brachialis from the distal ventral surface of the humerus, the two muscles inserting jointly on the inner side of the proximal end of the radius and ulna.

The short, broad humerus with large areas for muscular attachment and the large olecranon process of the ulna are adaptations commonly found in animals in which it is essential to develop the maximum power of limb movement rather than the maximum speed of limb movement. They are found particularly well developed in animals which use the arm for digging. The strong postero-lateral extension of the forearm, which would result from contraction of the large triceps muscle, similarly seems best interpreted as a digging stroke. This is supported in a general way by the resemblance between the shape of the humerus of *Kawingasaurus* and that of the monotremes, which use the arm for digging. However, the detailed movements must have been somewhat different in *Kawingasaurus*, for the head of the humerus is not crescentic like that of monotremes, in which it is pulled backwards and forwards through the glenoid (Howell, 1937) and rotates around its long axis (Jenkins, 1970). Another difference lies in the fact that the articular surfaces for the radius and ulna are confluent in monotremes, but are well separated in *Kawingasaurus*.

Unfortunately, since the hand is not preserved, the morphology of this region is unknown. However, the normal dicynodont manus is rather wide and short, with a phalangeal count of 2, 3, 3, 3, 3. It would therefore be preadapted to the type of digging activity that the modifications of the scapula, humerus and ulna so strongly suggest. The wide, shallow snout, lacking any sharp cutting edges, suggests that *Kawingasaurus* was a carnivore, digging up and eating small invertebrates, perhaps in the mudflats which seem to have been an extensive feature of the African Karroo landscape (Plumstead, 1963).

The adaptations of the scapulo-coracoid and forelimb described and interpreted above have inevitably affected the mechanics of normal locomotion. As will be seen, they have had little effect on upward movement at the shoulder joint, some effect on downward movement at that joint, and considerable effect on walking movements at the elbow. The movements permitted by the osteological nature of the joints will be described first, and the relevant musculature will then be restored as far as is possible.

Viewed from a point at right angles to the centre of its surface (Fig. 4E), the head of the humerus becomes narrower anteriorly. It is fairly flat along a plane parallel with its inner edge, but convex at right angles to this plane. The glenoid is a cylindrical facet, slightly wider at its dorsal end than at its ventral end. However, the surface is slightly twisted so that the ventral part faces dorsally and slightly anteriorly, while the dorsal part faces ventrally and slightly posteriorly. If the associated humerus and scapulo-coracoid are compared, it is noticeable that the surface of the glenoid makes up a much greater arc than does the head of the humerus. As a result, the latter does not at any time contact the whole surface of the glenoid. The humerus does not, therefore, simply hinge within the glenoid, but its area of contact with the glenoid moves dorsally while the shaft of the bone swings ventrally, and *vice versa*. Experiment with these bones shows that, if the humerus is placed at the extreme upward end of its possible arc of movement, its distal end faces downwards and forwards at an angle of about 45° to the vertical (Fig. 6A). Because of the twist in the glenoid surface, by the time the humerus reaches the extreme downward end of its movement, its distal end faces more ventrally, at an angle of about 15° to the vertical.

Since the plane of movement of the humerus at the glenoid joint is mainly up and down parallel with the long axis of the blade of the scapula, its movement would have provided the minimum of antero-posterior movement of the limb if the scapula were positioned vertically in the body. Instead, the scapula must have slanted antero-dorsally, as suggested in other dicynodonts by Watson (1917, 1960). Due to the 85° twist of the humerus, the forelimb would then be directed antero-ventrally towards the ground and the distal end of the radius would have lain internal to the distal end of the ulna. (Such a pose of the radius is supported by the fact that the condyle for that bone on the humerus is directed somewhat posteriorly to the ventral plane of the ventral surface of the humerus).

The stride would, then, have resulted partly from a downward, backward and outward movement of the humerus at the shoulder, accompanied by an alteration in the plane of its distal end, and partly from a backward swing of the forearm at the elbow. These movements are shown schematically in Fig. 6. The musculature concerned with movements at these two joints will now be considered.

The earliest phase of the downward movement of the humerus is normally due to the action of the pectoralis muscle, which would have inserted on the summit and ventral surface of the powerfully developed delto-pectoral crest, and originated from the ventral surface of the sternum and interclavicle. These regions are flat in dicynodonts, and would appear to provide a surface only poorly orientated for the origin of laterally directed muscles. A similar situation exists in bats, where powerful pectoralis muscles originate from a sternum which bears no median bony keel. However, the anterior end of the sternum in bats bears a median ventral bony process, which provides an anterior brace for a series of median ligaments, to the sides of which the pectoralis muscles are attached (Norberg, 1970). An analogous process is found near the anterior end of the interclavicle in some dicynodonts, e.g. *Dinodontosaurus* (Cox, 1965), *Lystrosaurus* and *Tetragonias* (Cruickshank, 1967). It seems very likely that a median ligament, providing an origin for the pectoralis muscle, existed in dicynodonts also.

The second muscle which normally acts to lower the humerus is the coraco-brachialis, which originates from the posterior region of the coracoid. The longus component of this muscle inserts somewhat distally on the underside of the humerus. The area of origin of the brevis component of the coraco-brachialis has been brought close below its area of insertion (the proximal ventral surface of the humerus), as a result of the lateral twist which has affected the whole of the ventral region of the scapulo-coracoid. The coraco-brachialis probably acted mainly during the earlier phase of downward movement of the humerus. The later phase was probably due to the subcoraco-scapularis muscle, which normally originates on the lower part of the inner side of the scapula and inserts on the internal surface of the postero-internal region of the humerus. Because this region of the humerus projects as an unusual process to a level well medial to that of the glenoid, muscles pulling this region upwards would cause the humerus to move downwards. Furthermore, due to the posterior position of this process, contraction of the subcoraco-scapularis muscle would also produce the slight clockwise rotation of the humerus which the shape of the glenoid would in any case impose. The downward and backward swing of the humerus is the propulsive phase of its movement, and so takes place while bearing the weight of the body. It was, therefore, probably important that the humerus should be actively pulled through this rotation, rather than being made to rotate solely by the shape of the glenoid, since this latter method would be liable to lead to rapid wear of the articular surfaces.

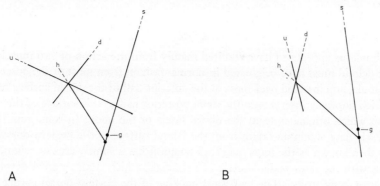

A, B, during the forward swing of the arm. The axis of the scapula (*s*) slopes antero-dorsally. The longitudinal axis of the humerus (*h*) is directed upwards, backwards and outwards, and its head lies below the centre of the glenoid (*g*). The transverse axis of the distal end of the humerus (*d*) faces antero-ventrally, and the ulna (*u*) is similarly directed.

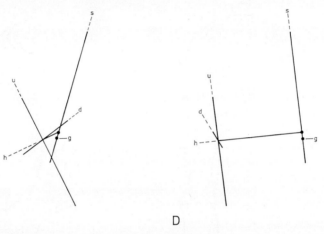

C, D, when the hand has just been placed upon the ground. The long axis of the humerus is now directed downwards, backwards and outwards, and its head lies above the centre of the glenoid. The distal end of the humerus, and that of the ulna, are now directed more ventrally.

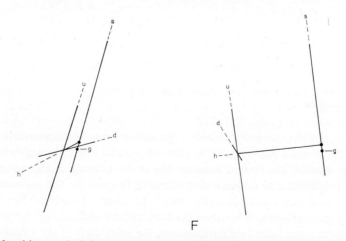

E, F, when the hand has reached the most posterior position, at the end of the stride. The distal end of the humerus has continued to rotate slightly, and the ulna has now swung posteriorly.

Fig. 6 Schematic representation of movements of the right forelimb in lateral view (**A, C, E**) and anterior view (**B, D, F**).

Raising the humerus appears to have resulted mainly from the action of two muscles (Figs. 7, 8). Anteriorly, the deltoid must have originated in normal fashion from most of the outer surface of the blade of the scapula, and inserted over most of the anterior half of the dorsal surface of the humerus. The muscle which normally inserts onto the more posterior half of this surface of the humerus is the latissimus dorsi, which originates from the dorsal fascia of the body. In some reptiles the anterior part of this muscle may obtain an origin from the lateral surface of the postero-dorsal corner of the scapula, and is then known as the teres major. (Though it has a similar area of origin, this muscle is not homologous with the teres major muscle of mammals, which is an external development of the subscapularis, see Cheng, 1955.) The backward hooking of the postero-dorsal corner of the scapula in *Kawingasaurus*, which is separated by a low ridge from the rest of the outer surface of the blade, strongly suggests that such a teres major muscle had become distinct.

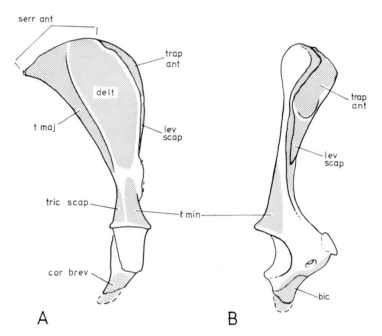

Fig. 7 Areas of attachment of muscles to the scapula. **A**, lateral view; **B**, anterior view.

Abbreviations of the names of muscles: *bic*, biceps; *delt*, deltoid; *lev scap*, levator scapulae superficialis; *serr ant*, serratus anterior superficialis; *t maj*, teres major; *t min*, teres minor; *trap ant*, trapezius anterior; *tric scap*, triceps scapularis.

Another muscle which normally assists in the movement of the humerus is the supracoracoideus, inserted onto the proximal part of the delto-pectoral crest. However, the region of origin of this muscle, anterior to the glenoid, has rotated inwards due to the twisting of the lower part of the scapulo-coracoid, and it is difficult to find any clear course or function for the supracoracoideus muscle. The origin of this muscle often extends more dorsally in advanced mammal-like reptiles until, split by the developing scapular spine, it becomes the infraspinatus and supraspinatus muscles of mammals. Though a scapular spine exists in *Kawingasaurus*, the orientation of the anterior surface of this spine does not suggest that the supracoracoideus had extended onto this region. Instead, this surface more likely provided insertion for the trapezius anterior and levator scapulae superficialis muscles, originating on the posterior surface of the skull and from the ligamentum nuchae. These muscles would brace the dorsal end of the scapula anteriorly, while a similar posterior brace would be produced by the serratus anterior superficialis muscle originating from the spines of the dorsal vertebrae and from the dorsal surface of the ribs, and inserting on the dorsal edge of the scapula.

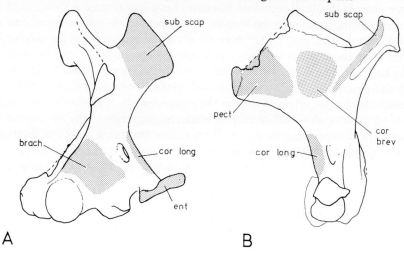

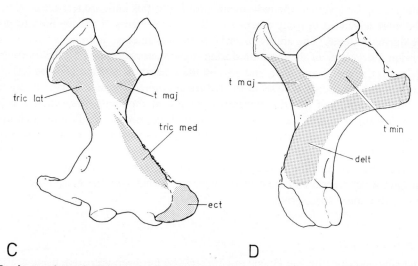

Fig. 8 Areas of attachment of muscles to the humerus. **A**, anterior view; **B**, ventral view; **C**, posterior view; **D**, dorsal view.

Abbreviations of the names of muscles: *brach*, brachialis; *cor brev*, coracobrachialis brevis; *cor long*, coracobrachialis longus; *delt*, deltoid; *ect*, muscles attached to the ectepicondyle; *ent*, muscles attached to the entepicondyle; *pect*, pectoralis; *sub scap*, subcoracoscapularis; *t maj*, teres major; *t min*, teres minor; *tric lat*, triceps lateralis; *tric med*, triceps medialis.

The flexion-extension digging action of the forearm, due to the action of the triceps musculature, has already been considered, but the walking movements at this joint, and the associated musculature, may finally be analysed. Though the hand was probably directed forwards and slightly inwards, so that extension of the elbow would have provided a slight anterior component of movement, the majority of the stride must have resulted from an antero-posterior movement of the radius and ulna nearly parallel with the plane of the distal end of the humerus. Such a movement is shown to be possible by actual experiment, moving the ulna on the humerus. Furthermore, the occurrence of such movements at the elbow are strongly suggested by the character of the epicondyles. The entepicondyle, and to a lesser extent the ectepicondyle, are well developed, and the entepicondyle projects well posterior to the condyle for the ulna. The muscles attached to these epicondyles are those which swing the forearm anteriorly and posteriorly. The backward, power, movement of the forearm at the elbow would have been caused by muscles originating from the entepicondyle. The size of this region in itself suggests that these muscles were well developed, while the projection of the entepicondyle posterior to the condyle for the ulna provided the muscles with a more favourable lever arm for its action upon the forearm. The forward, recovery, stroke of the forearm would have been caused by the muscles which originate from the ectepicondyle, on the dorsal surface of which is a large area for muscle attachment.

In other dicynodonts, in which the humerus was directed more posteriorly, the stride appears to have been due to the movement of the humerus, the distal end of which moved posteriorly and ventrally (Watson, 1917). The relationship between the forearm and the humerus in such forms changed very little, so that the downward movement of the humerus raised the body while its posterior movement drove it forward. There was probably a slight flexure of the forearm, which served to bring the body (and its centre of gravity) over toward the side where the stride was taking place. This must have been accompanied by a slight twisting of the radius and ulna. Apart from these movements, the bones of the forearm moved very little and did not contribute to the stride. Because the condyles for the radius and ulna are close together, the movements possible at this joint are restricted. *Kawingasaurus* appears to be more advanced in using anterior-posterior movements of the forelimb to provide part of the stride during walking. The ability to make these movements, and also to make considerable flexion-extension movements of the forelimb during digging, are permitted by the unusual convexity of the condyles for the radius and ulna, and also by the separation of these two condyles.

The validity of any functional analysis of the nature attempted above is inevitably uncertain. The most firmly founded aspect of it is the movement possible at the joints, since these can be demonstrated by actual manipulation. Though the morphology of the shoulder girdle and forelimb is equally well established, the reasons suggested for the unusual features which they show are purely interpretive. Finally, though the positions and functions of those muscles which had retained their normal relationships are probably well founded by comparison with those of living reptiles, the functions of those which have been radically altered by the adaptive changes in the morphology of the whole region are inevitably more speculative.

SUMMARY

Material originally described by von Huene as *Cistecephalus* has been prepared, using acetic acid, and is re-described. It is found to belong to a new genus and species, which is called *Kawingasaurus fossilis*. The skull, scapulo-coracoid, humerus and ulna are described; all show unusual features. The functional anatomy of these post-cranial elements is analysed and their musculature restored. It is concluded that *Kawingasaurus* was probably adapted to a digging mode of life.

ACKNOWLEDGEMENTS

My first thanks must go to Dr F. Westphal of the Geologisches-Paläontologisches Museum, Tübingen, for making this work possible by lending the specimens. I am grateful to Mr Peter Hutchinson for helpful, critical discussions and for drawing the figures, and to the Central Research Fund of the University of London for a grant to pay for his work. I am also grateful to Dr A. Keyser of the Geological Survey of South Africa for allowing me to see his M.Sc. thesis on *Cistecephalus*.

REFERENCES

BROILI, F. and SCHRÖDER, J. 1935. Beobachtungen an Wirbeltieren der Karroo-formation. 6. Über dem Schädel von *Cistecephalus* Owen. *Sber. bayer. Akad. Wiss.*, **1935**, 1-20.

BROOM, R. 1906. On the Permian and Triassic faunas of South Africa. *Geol. Mag.*, **3**, 29-30.

—— 1915. Catalogue of types and figured specimens of fossil vertebrates in the American Museum of Natural History. 2. Permian, Triassic and Jurassic reptiles of South Africa. *Bull. Am. Mus. nat. Hist.*, **25**, 105-164.

CHARIG, A. J. 1963. Stratigraphical nomenclature in the Songea Series of Tanganyika. *Rec. geol. Surv. Tanganyika*, **10**, (for 1960), 47-53.

CHENG, C-C. 1955. The development of the shoulder region of the opossum, *Didelphys virginiana*, with special reference to the musculature. *J. Morph.*, **97**, 415-472.

COX, C. B. 1959. On the anatomy of a new dicynodont genus with evidence of the position of the tympanum. *Proc. zool. Soc. Lond.*, **132**, 321-367.

—— 1965. New Triassic dicynodonts from South America, their origins and relationships. *Phil. Trans. R. Soc.*, **248**(B), 457-516.

—— 1969. Two new dicynodonts from the Triassic Ntawere Formation, Zambia. *Bull. Br. Mus. nat. Hist. (Geol.)*, **17**, 255-294.

CRUICKSHANK, A. R. I. 1967. A new dicynodont genus from the Manda Formation of Tanzania (Tanganyika). *J. Zool., Lond.*, **153**, 163-208.

HOWELL, A. B. 1937. Morphogenesis of the shoulder architecture. Part V. Monotremata. *Q. Rev. Biol.*, **12**, 191-205.

HUENE, F. VON 1935. *Die fossilen Reptilien des südamerikanischen Gondwanalandes an der Zeitenwende.* I, Anomodontia. Tübingen: Heine.

—— 1942. Die Anomodontier des Ruhuhu-Gebietes in der Tübinger Sammlung. *Palaeontographica*, **94**, 154-184.

JENKINS, F. A. 1970. Limb movements in a monotreme (*Tachyglossus aculeatus*): a cineradiographic analysis. *Science*, **168**, 1473-1475.

LYDEKKER, R. 1890. *Catalogue of the fossil Reptilia and Amphibia in the British Museum.* Vol. IV. London: British Museum (Natural History).

NOWACK, E. 1937. Zur Kenntnis der Karruformation in Ruhuhu-Graben (D.O.A.). *Neues Jb. Min. Geol. Paläont.*, **78** (B), 380-412.

NORBERG, U. M. 1970. Functional osteology and myology of the wing of *Plecotus auritus* Linnaeus (Chiroptera). *Ark. Zool.*, **22**, 483-543.

OWEN, R. 1876. *Description of the fossil Reptilia of South Africa in the collection of the British Museum.* London: British Museum (Natural History).

PLUMSTEAD, E. P. 1963. The influence of plants and environment on the developing animal life of Karroo times. *S. Afr. J. Sci.*, **59**, 147-152.

ROMER, A. S. 1922. The locomotor apparatus of certain primitive and mammal-like reptiles. *Bull. Am. Mus. nat. Hist.*, **46**, 517-606.

—— 1944. The development of tetrapod limb musculature—the shoulder region of *Lacerta*. *J. Morph.*, **74**, 1-41.

WATSON, D. M. S. 1917. The evolution of the tetrapod shoulder girdle and fore-limb. *J. Anat.*, **52**, 1-63.

—— 1960. The anomodont skeleton. *Trans. zool. Soc. Lond.*, **29**, 131-208.

Ch. H. MENDREZ

On the skull of *Regisaurus jacobi,* a new genus and species of Bauriamorpha Watson and Romer 1956 (=Scaloposauria Boonstra 1953),[1] from the *Lystrosaurus*-zone of South Africa

INTRODUCTION

The following study concerns the skull of a new genus and species of mammal-like reptile from South Africa.

Genus *REGISAURUS* nov.

Derivation of name: In honour of Dr F. Rex Parrington, F.R.S., who has contributed so much to our knowledge of Karroo reptiles.

Diagnosis: Regisaurus is distinguished from the related genera *Ictidosuchops, Ictidosuchoides* and *Silphoictidoides* by a large orbit limited posteriorly by a complete postorbital arcade, an intertemporal region of medium breadth surmounted by a small parietal crest but practically lacking a pineal foramen, a premaxilla long internasally, a long and very high lacrimal, a jugal very short anteriorly, a periotic region constituted by the well separated prootic and opisthotic, extensive participation of the parietal in the occiput, a broad interparietal, a high tabular, an occipital condyle exclusively formed by the basioccipital, a primitive secondary palate with a premaxilla-maxilla connection along the vomer, and a high dentigerous tuberosity bearing a great number of pterygoid teeth. The dental formula is 6 I, 1 C, 10 PC.

Type-species: R. jacobi nov.

Species R. *JACOBI* nov.

Derivation of name: In honour of Mr James W. Kitching, who collected the material.

Material: Only the holotype.

Holotype: Skull, posterior half of left dentary and part of postcranial skeleton (F. R. Parrington's field number F.R.P. 1964/27).

Horizon: Lystrosaurus-zone (Lower Trias).[2].

Locality: Zeekoeigat, Venterstad, Bethulie, Cape Province, Republic of South Africa.

Diagnosis: R. jacobi is the only known species.

[1] For the purpose of this paper, the author will use the terms Therocephalia and Bauriamorpha in the sense of Watson and Romer (1956). Therocephalia Watson and Romer (1956) = Pristerosauria Boonstra (1953), and Bauriamorpha Watson and Romer (1956) = Scaloposauria Boonstra (1953).

[2] Twenty feet above the top of the *Daptocephalus*-zone (verbal communication, J. W. Kitching, 1.9.1970). The *Daptocephalus*-zone corresponds to the upper part of the classical *Cistecephalus*-zone (J. W. Kitching, in press).

DESCRIPTION

The skull, of medium size, is triangular with an elongated snout, with large and oblique nares and relatively large orbits, nearly as long as the temporal fossae. These last are comparatively small and are slightly wider than they are long.

The chief measurements of *Regisaurus* are:

Total length of the skull	119 mm
Maximum width	73 mm
Minimum intertemporal width	11 mm
Minimum interorbital width	17 mm
Distance between the anterior border of the premaxilla and the anterior orbital border	59 mm
Distance between the anterior border of the premaxilla and the anterior border of the temporal fossa	85 mm
Breadth of the snout at the level of its maximal constriction	27 mm
Height of the snout at the same level	24 mm
Maximum height of the occiput (reconstructed)	approx. 30 mm
Maximum width of the occiput	71 mm
Total length of the maxillary dentition	32 mm
Total length of the postcanine tooth row	24 mm

Figures: The skull being well preserved, the drawings are only slightly corrected to allow for the flattening of the parietal region and for part of the asymmetry.

I—SKULL

DERMAL BONES

1. *Dorsal surface* (Fig. 1 and Plate I, 1)

In dorsal view, the skull shows the general arrangement of the Therocephalia and Bauriamorpha (Fig. 1). It is characterised by a wide orbit, a rather short temporal fossa and weak zygomatic and postorbital arches. The snout is elongated and very pointed. Posteriorly its breadth increases slowly and regularly, except for an extremely slight but sudden constriction behind the level of the canines.

The two united *premaxillae* (*Pmx*) are well preserved and show the nasal process, convex forward, penetrating deeply between the nasals (*Na*). The lateral process, or maxillary process, of each premaxilla (*pr.mx. Pmx*, Fig. 2) is overlapped for some distance by the maxilla (*Mx*). Anteriorly, near the root of the nasal process, the dorsal premaxillary foramen (*f.pmx.d.*, Fig. 2) appears to be double on the left side, but on the right side, a small groove, exposed probably because of weathering, runs between the locations of those two foramina. The posterior premaxillary foramen (*f.pmx.p.*) is visible dorsally (Fig. 5).

The teeth are lost, but there are six alveoli in each premaxilla, as in most ictidosuchids[1].

The *septomaxilla* (*Smx*) is a triangular bone, well developed on the snout surface and quite broad for its length.

There are two septomaxillary foramina, instead of only one as usually described in mammal-like reptiles[2]:

---

[1] "ictidosuchid", used provisionally here in the meaning given by Haughton and Brink (1954).
[2] This feature has also been observed in *Akidnognathus* by Brink (1960, page 156 figs B, C, D and page 164) and in *Moschorhinus* and *Whaitsia* by the author.

Plate I *Regisaurus jacobi*—*Lystrosaurus*-zone of Zeekoeigat, Venterstad, Bethulie (South Africa). F. R. Parrington field No.F.R.P./1964/27.

Fig. 1 Dorsal view of the skull (approx. × 1).
Fig. 2 Ventral view of the skull (approx. × 1).

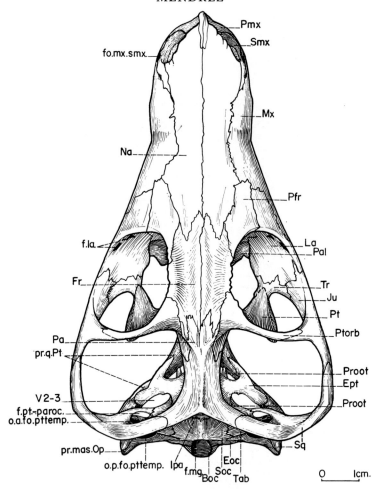

Fig. 1 *Regisaurus jacobi*—Dorsal view of the skull (×1). *Boc*, basioccipital; *Eoc*, exoccipital; *Ept*, epipterygoid; *f. la.*, lacrimal foramen; *f.mg.*, foramen magnum; *fo. mx.-smx.*, maxillo-septomaxillary fossa; *f. pt.-paroc.*, pterygo-paroccipital foramen; *Fr*, frontal; *Ipa*, interparietal; *Ju*, jugal; *La*, lacrimal; *Mx*, maxilla; *Na*, nasal; *o.a.fo.pttemp.*, anterior opening of the posttemporal fossa; *o.p.fo.pttemp.*, posterior opening of the posttemporal fossa; *Pa*, parietal; *Pal*, palatine; *Pfr*, prefrontal; *Pmx*, premaxilla; *pr.mas.Op*, mastoid process of the opisthotic; *Proot*, prootic: *pr.q.Pt*, quadrate process of the pterygoid; *Pt*, pterygoid; *Ptorb*, postorbital; *Smx*, septomaxilla; *Soc*, supraoccipital; *Sq*, squamosal; *Tab*, tabular; *Tr*, transversum; *V 2-3*, notch in the epipterygoid for the maxillary and mandibular branches of the trigeminal nerve.

—The anterior one (*f.smx.a.*, Fig. 5), is situated under an overhang of the septum, not far from the suture between premaxilla and maxilla. Also under this overhang a narrow groove runs towards the mid-line.

—The posterior septomaxillary foramen (*f.smx.p.*, Fig. 6) is situated within the region usually called the "septomaxillary foramen". This latter, at the junction between the septomaxilla and the maxilla, is really a small depression, the maxillo-septomaxillary fossa (*fo.mx.smx.*, Fig. 5), within which open two foramina (Fig. 6). The anterior of these is the posterior septomaxillary foramen (*f.smx.p.*), the posterior (according to Watson, 1914) is the end of the lacrimal duct (*or.a.c.la.*(?)).

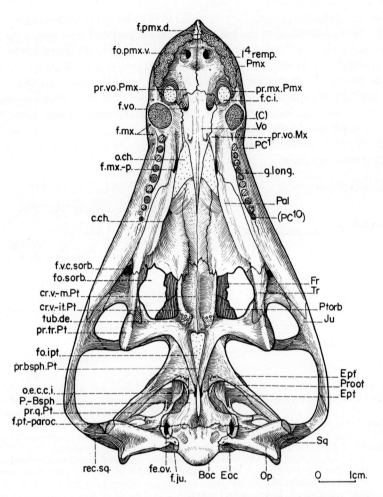

Fig. 2 *Regisaurus jacobi*—Ventral view of the skull ($\times$ 1). *Boc*, basioccipital; *(C)*, canine alveolus; *c.ch.*, crista choanalis; *cr.v.-it.Pt*, ventro-intermediate crest of the pterygoid; *cr.v.-m.Pt*, ventro-medial crest of the pterygoid; *Eoc*, exoccipital; *Ept*, epipterygoid; *f.c.i.*, foramen for the lower canine; *fe.ov.*, fenestra ovalis; *f.ju.*, jugular foramen; *f.mx.*, maxillary foramen; *f.mx.-p.*, maxillo-palatine foramen; *fo.ipt.*, interpterygoid vacuity; *fo.sorb.*, suborbital vacuity; *fo.pmx.v.*, ventral premaxillary fossa; *f.pmx.d.*, dorsal premaxillary foramen; *f.pt.paroc.*, pterygo-paroccipital foramen; *Fr*, frontal: *f.v.c.sorb.*, foramen for the vessels of the suborbital canal; *f.vo.*, vomerine foramen; *g.long.*, longitudinal groove; I^4 *remp.*, fourth replacing incisor; *Ju*, jugal; *Mx*, maxilla; *o.ch.*, internal choanae opening; *o.e.c.c.i.*, external opening of the canal for the internal carotid; *Op*, opisthotic; *Pal*, palatine; *P-Bsph*, para-basisphenoid complex; PC^1, first postcanine; (PC^{10}), tenth postcanine alveolus; *Pmx*, premaxilla; *pr.bsph.Pt*, basiphenoid process of the pterygoid; *pr.mx.Pmx*, maxillary process of the premaxilla; *Proot*, prootic; *pr.q.Pt*, quadrate process of the pterygoid; *pr.tr.Pt*, transverse process of the pterygoid; *pr.vo.Mx*, vomerine process of the maxilla; *pr.vo.Pmx*, vomerine process of the premaxilla; *Ptorb*, postorbital; *rec.sq.*, squamosal recess; *Sq*, squamosal; *Tr*, transversum; *tub.de.*, dentigerous tuberosity; *Vo*, vomer.

The foramina of the snout have been studied at length in *Moschowhaitsia* by Tatarinov[1] (1964, pp. 77-78), who interprets the premaxillary foramina as exits for the branches of the medial ramus of the ethmoidalis nerve. According to the same author, the septomaxillary canal, situated between the anterior and posterior septomaxillary foramina, could give passage to the lateral ramus of the ethmoidalis nerve, by analogy with the situation in lizards. Bahl's (1937) study of *Varanus* led me to the same conclusion. The most recent suggestion about the function of the foramen, described as the "orifice of the lacrimal duct" by Watson, is made by Tatarinov, who calls it the "septomaxillary foramen". Tatarinov considers that the lateral ramus of the ethmoidalis nerve passes through this opening. If so, this ramus might divide after leaving the foramen; one branch of it might go inward through the septomaxillary canal and another branch could possibly come out through the facial opening of the maxillo-septomaxillary fossa.

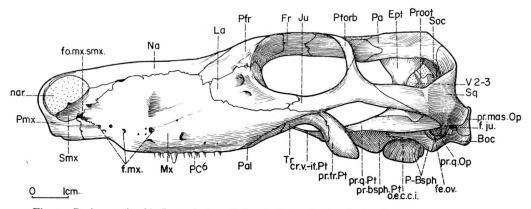

Fig. 3 *Regisaurus jacobi*—Lateral view of the skull (×1). *Boc*, basioccipital; *cr.v.it.Pt*, ventro-intermediate crest of the pterygoid; *Ept*, epipterygoid; *fe.ov.*, fenestra ovalis; *f.ju.*, jugular foramen; *f.mx.*, maxillary foramen; *fo.mx. smx.*, maxillo-septomaxillary fossa; *Fr*, frontal; *Ju*, jugal; *e.o.c.c.i.*, external opening of the canal for the internal carotid; *La*, lacrimal; *Mx*, maxilla; *Na*, nasal; *nar.*, naris; *Pa*, parietal; *Pal*, palatine; *P-Bsph*, para-basisphenoid complex; *PC⁶*, sixth postcanine; *Pfr*, prefrontal; *Pmx*, premaxilla; *pr.bsph.Pt*, basisphenoid process of the pterygoid; *pr.mas.Op*, mastoid process of the opisthotic; *Proot*, prootic; *pr.q.Op*, quadrate process of the opisthotic; *pr.q.Pt*, quadrate process of the pterygoid; *Ptorb*, postorbital; *pr.tr.Pt*, transverse process of the pterygoid; *Smx*, septomaxilla; *Soc*, supraoccipital; *Sq*, squamosal; *Tr*, transversum; *V 2-3*, notch in the epipterygoid for the maxillary and mandibular branches of the trigeminal nerve.

The *maxilla* (*Mx*) is high and very much elongated and has a pitted and grooved surface. Its anterior limit is at the level of the fifth incisor alveolus and of the posterior border of the naris. Posteriorly, it extends as far as the back of the orbit and though the maxilla is much lower at this level, it still forms half of the height of the suborbital arch. Besides the usual contacts in therapsids with the other bones of the snout, the maxilla has the classical contact with the prefrontal (*Pfr*) as in other therocephalians and bauriamorphs and differing from the situation in cynodonts. Above the lower edge of the bone, and more or less parallel to it, are numerous foramina (*f.mx.*), probably for the cutaneous branches of the superior alveolar nerve.

There are no precanines, but a diastema, in front of the single canine, is occupied by a ridge of the maxilla. The maxilla bears ten postcanines.

---

[1] The author uses here Bahl's (1937) terminology, which differs from Tatarinov's: posterior premaxillary foramen, Bahl = dorsal premaxillary foramen, Tatarinov; dorsal premaxillary foramen, Bahl = ventral premaxillary foramen, Tatarinov; the ventral premaxillary foramen, Bahl, is situated on the palate.

The *nasals* (*Na*) are long and although their breadth does not vary much, they are slightly broader just behind the nares and at the level of the maxillary-prefrontal contact, and they narrow at midlength and posteriorly. They have two lateral projections overlying the frontals (*Fr*), but they are much shorter in the mid-line.

The *lacrimal* (*La*) is very much expanded on the face, much more than in *Ictidosuchops intermedius*, a fairly closely related form. In front of the orbit it is nearly as long as the prefrontal, and much higher.

The *prefrontal* (*Pfr*) is more expanded anteriorly than posteriorly and has a fairly extensive contact with the maxilla but none with the postorbital (*Ptorb*).

The *frontals* (*Fr*) are nearly twice as long as broad and they contribute widely to the orbital borders.

The *postorbital* (*Ptorb*) is short anteriorly. Its long, thin lateral process forms a complete postorbital arch with the jugal, as in *Ictidosuchoides longiceps* and *Silphoictidoides ruhuhuensis*. Its posterior flange tapers and reaches about half-way along the junction between the parietals.

There is neither a postfrontal nor a preparietal.

The junction between the *parietals* (*Pa*) occupies a quarter of the length of the skull but the sagittal suture is only visible in front. There is a sagittal crest posteriorly which divides anteriorly into two postorbital-parietal ridges; between these latter the parietals are flat. There is no pineal foramen, but an extremely slight broadening of the sagittal suture, like a scar, may be a vestige of it. The intertemporal breadth is moderate, being greater than in *Ictidosuchoides longiceps* but less than in *Ictidosuchops intermedius*.

2. *Occiput* (Fig. 4 and Plate II, 2)

The parietals contribute more to the occipital surface, than in *Ictidosuchoides longiceps*. In contrast, in *Ictidosuchops intermedius* they do not appear posteriorly at all.

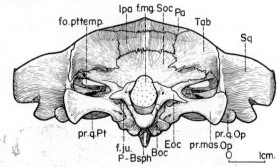

Fig. 4 *Regisaurus jacobi*—Occipital view of the skull (× 1). *Boc*, basioccipital; *Eoc*, exoccipital; *f.ju.*, jugular foramen; *f.mg.*, foramen magnum; *fo.pttemp.*, posttemporal fossa; *Ipa*, interparietal; *Pa*, parietal; *P-Bsph*, para-basisphenoid complex; *pr.mas.Op*, mastoid process of the opisthotic; *pr.q.Op*, quadrate process of the opisthotic; *pr.q.Pt*, quadrate process of the pterygoid; *Soc*, supraoccipital; *Sq*, squamosal; *Tab*, tabular.

The occiput is very broad and low, and the occipital crest is very sharp.

The rectangular-shaped *interparietal* (*Ipa*) is very broad and low.

The *tabular* (*Tab*) is also broader than high, but much higher than in *Ictidosuchops intermedius*, as represented by Crompton (1955, fig. 2A, page 60). It forms only the dorso-lateral corner of the posterior opening of the posttemporal fossa (*fo.pttemp.*).

3. *Cheek* (Fig. 3 and Plate II, 1, 3)

The *jugal* (*Ju*) is slender and especially short anteriorly, extending forwards only to the middle of the inferior border of the orbit. (In *Ictidosuchops intermedius* and *Ictidosuchoides* it extends in front of the orbit.) It meets the postorbital so as to form a complete postorbital bar, as in *Ictidosuchoides*, but this situation differs from that in *Ictidosuchops*, in which the small vertical process of the jugal does not

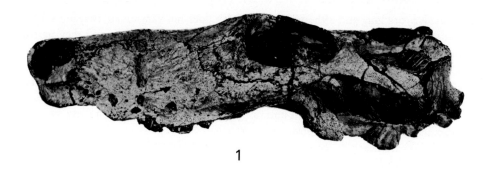

1

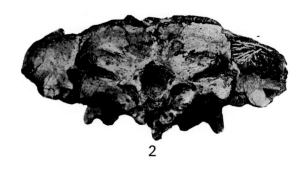

2

3

Plate II *Regisaurus jacobi*—*Lystrosaurus*-zone of Zeekoeigat, Venterstad, Bethulie (South Africa). F. R. Parrington field No. F.R.P./1964/27.

Fig. 1 Left lateral view of the skull (approx. ×1).
Fig. 2 Occipital view of the skull (approx. ×1).
Fig. 3 Right lateral view of the skull (approx. ×1).

meet the postorbital. The jugal forms part of the lateral floor of the orbit (Fig. 1) in contact with the lacrimal and the transversum (Tr = ectopterygoid); this latter bone is largely covered dorsally by the pterygoid.

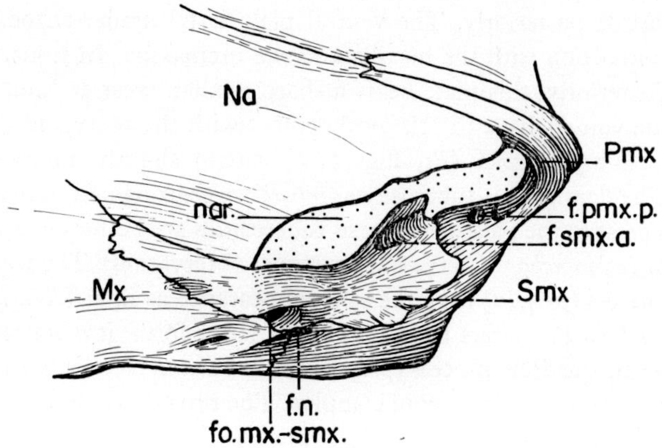

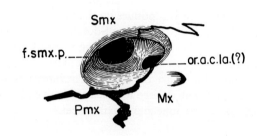

Fig. 5 *Regisaurus jacobi*—Oblique view of the right side of the snout, (no proper scale). *f.n.*, nervous foramen; *fo.mx-smx.*, maxillo-septomaxillary fossa; *f.pmx.p.*, posterior premaxillary foramen; *f.smx.a.*, anterior septomaxillary foramen; *Mx*, maxilla; *Na*, nasal; *nar*, naris; *Pmx*, premaxilla; *Smx*, septomaxilla.

Fig. 6 *Regisaurus jacobi*—Detail of the left maxillo-septomaxillary fossa (no proper scale). *f.smx.p.*, posterior septomaxillary foramen; *Mx*, maxilla; *or.a.c.la.* (?), anterior orifice of the lacrimal canal (?); *Pmx*, premaxilla; *Smx*, septomaxilla.

The *squamosal* (Sq) is important, forming nearly all the posterior wall of the temporal fossa by means of three processes (Fig. 7):

—A dorsal process ($pr.d.Sq$ = process 1, cf. Crompton's nomenclature, 1955, page 154) overlapping the parietal laterally.

—An intermediate process ($pr.it.Sq$ = process 2, *ibid*), in contact with the supraoccipital and the prootic.

—An antero-ventral process ($pr.a.-v.Sq$) fused with the central process of the prootic ($pr.c.Proot$, Fig. 9).

The recess ($rec.sq.$, Figs 2, 7) for the quadrate is deep, but the quadrate and the quadratojugal were free enough to be lost in the process of fossilisation.

As in *Ictidosuchops* and *Ictidosuchoides*, the zygomatic process of the squamosal does not reach the postorbital arcade.

4. *Palate* (Fig. 2 and Plate I, 2)

The palate is characterised by large suborbital vacuities and a rudimentary secondary palate. This secondary palate is formed not only by the contact between maxilla and vomer, but also anteriorly by a contact between premaxilla and maxilla alongside the vomer. The bony choanae open at a level slightly posterior to the canines.

The *premaxillary bones* (Pmx) make a contribution to the secondary palate, with a relatively broad posterior expansion of their median or vomerine process ($pr.vo.Pmx$). Anteriorly, they are marked by a pair of important ventral premaxillary fossae (*fo.pmx.v.*), perhaps for the anterior small branches of the *arteria maxillaris*. By their median and lateral processes extending posteriorly, the premaxillae nearly surround the foramina for the lower canines (*f.c.i.*).

The paired vomers (*Vo*) are completely fused anteriorly, but the suture between them is still visible posteriorly. The ventral plate is triangular-shaped, broadening anteriorly and is in sutural connection with the maxilla and the premaxilla. In front, two large vomerine foramina (*f.vo.*) open. Posteriorly, this plate bears a sharp median crest and, on each side, a strong tuberosity. Posteriorly the vomers have the classical contact with the pterygoids, which is not found in gorgonopsians.

The *pterygoids* (*Pt*, Figs 2, 9) extend slightly anteriorly to the suborbital vacuities (*fo.sorb.*). The dentigerous tuberosities (*tub.de.*) are strong and bear numerous teeth, and they are prolonged by a crest to the palatines. These tuberosities are better developed, and their teeth more numerous, than in *Ictidosuchops intermedius* (personal observation). The transverse process (*pr.tr.Pt*) and the quadrate process (*pr.q.Pt*) have the same appearance as in *Ictidosuchops intermedius*. The basisphenoid process (*pr.bsph.Pt*) differs from that of the latter form: it is reflected and the hollowed lamina connecting it to the quadrate process (*pr.q.Pt*) has a sutural contact with the basipterygoid process (*pr.bspt.P-Bsph*) of the para-basisphenoid complex. The broad interpterygoid fossa (*fo.ipt.*) extends anteriorly between the transverse processes of the pterygoids.

The vomerine process of the maxilla (*pr.vo.Mx*, Fig. 2) is prolonged posteriorly by a sharp *crista choanalis* (*c.ch.*), which continues on the palatine (*Pal*) and stops in front of the suborbital vacuities.

The *palatine* (*Pal*), though well-developed, does not extend posteriorly between the suborbital vacuities as it does in *Bauria*, according to Brink (1963, fig. 7).

The triradiate *transversum* (*Tr* = ectopterygoid) is short posteriorly and, as in *Ictidosuchops* and *Ictidosuchoides*, but unlike the condition in *Bauria*, it does not approach the tip of the transverse process of the pterygoid. A large foramen for the vessels of the suborbital canal (*f.v.c.sorb.*) opens where the transversum, palatine and maxilla meet.

ENDOCRANIUM

The endocranium appears to be less ossified in *Regisaurus* than in *Ictidosuchops intermedius*.

In ventral view, the para-basisphenoid complex (*P-Bsph*, Figs 2, 9) is X-shaped. The anterior part of the basipterygoid process is covered by the pterygoid. The keel (*q.P-Bsph*, Figs 7, 9) is completely free from the pterygoid except at its antero-dorsal part. At half-length the para-basisphenoid complex is constricted but it broadens again posteriorly and offers a wide sutural contact with the basioccipital, between the two *fenestrae ovales* (*fe.ov.*). The complex forms about a quarter of the 'lip' surrounding each fenestra. Laterally the alar process is covered by the prootic. On each side of the keel is the small external foramen of the canal for the internal carotid (*o.e.c.c.i.*). Posteriorly to the keel, the para-basisphenoid complex is deeply hollowed between the two *tubercula spheno-occipitalia* (*tub.sph.-oc.*, Fig. 9).

The *epipterygoid* (*Ept*, Figs 7, 10 and Plate III, 1) is a convex, plate-like bone which does not seem to have any sutural contact with the other bones, except for a very short suture with the supra-occipital: it appears merely to abut against the parietal and to stand on the quadrate process of the pterygoid. The upper part of the *processus ascendens* of the epipterygoid expands anteriorly and posteriorly in a dorsal lamina (*l.d.Ept*, Figs 7, 10) but the breadth of the ascending ramus of this process remains the same as in *Ictidosuchops* and its allies and in the Bauriidae. At mid-height there is a minute posterior apophysis (*a.p.Ept*, Fig. 10) possibly for the separation between the rami 2 and 3 of the trigeminal nerve. The basal part of the epipterygoid extends also anteriorly and posteriorly as two processes. The postero-ventral process (*pr.p.-v. Ept*) broadens posteriorly and reaches the squamosal without being attached to it. As for the antero-ventral process (*pr.a.-v. Ept*), it tapers anteriorly and is folded internally so as to clasp the quadrate process of the pterygoid. The antero-dorsal process of the prootic (*pr.a.-d. Proot*) approaches the upper part of the epipterygoid, although all the rest of the former bone is well separated from the latter, leaving room for a large *cavum epiptericum* (*c.ept.*).

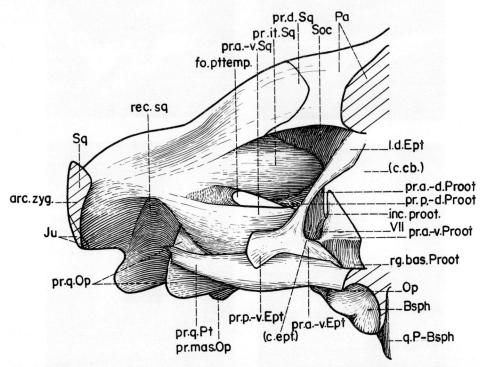

Fig. 7 *Regisaurus jacobi*—Anterior view of the posterior wall of the right temporal fossa (approx. ×2·5). *arc.zyg.*, zygomatic arcade; *Bsph*, basisphenoid; (*c.cb.*), brain cavity; (*c.ept.*), cavum epiptericum; *fo.pttemp.*, posttemporal fossa; *inc.proot.*, incisura prootica; *Ju*, jugal; *l.d.Ept*, dorsal lamina of the epipterygoid; *Op*, opisthotic; *Pa*, parietal; *pr.a.-d.Proot*, antero-dorsal process of the prootic; *pr.a.-v.Ept*, antero-ventral process of the epipterygoid; *pr.a.-v.Proot*, antero-ventral process of the prootic; *pr.a.-v.Sq*, antero-ventral process of the squamosal; *pr.d.Sq*, dorsal process of the squamosal; *pr.it.Sq*, intermediate process of the squamosal; *pr.mas.Op*, mastoid process of the opisthotic; *pr.p.-d.Proot*, postero-dorsal process of the prootic; *pr.p.-v.Ept*, postero-ventral process of the epipterygoid; *pr.q.Op*, quadrate process of the opisthotic; *pr.q.Pt*, quadrate process of the pterygoid; *q.P-Bsph*, keel of the para-basisphenoid complex; *rec.sq.*, squamosal recess; *rg.bas.Proot*, basal area of the prootic; *Soc*, supraoccipital; *Sq*, squamosal; *VII*, foramen for the facial nerve.

The otic bones

Contrary to what has been observed in therocephalian B by Olson (1944, pages 11-15, fig. 6 C) and in *Ictidosuchops intermedius* by Crompton (1955, pages 165-167, fig. 6 A, B, C), there is no single periotic ossification in *Regisaurus*; instead the prootic (*Proot*) and the opisthotic (*Op*) are two well separated bones.[1] Fig. 9). The suture between them runs in the middle of the inner wall of the post-temporal fossa (Fig. 8 and Plate III, 1). It starts on the upper edge of this fossa, where the squamosal, opisthotic and prootic meet. First it goes down and interiorly, and afterwards externally, separating the

---

[1] The author has also observed two separated bones in a whaitsid and a moschorhinid. Hence, this structure is probably more common than is thought.

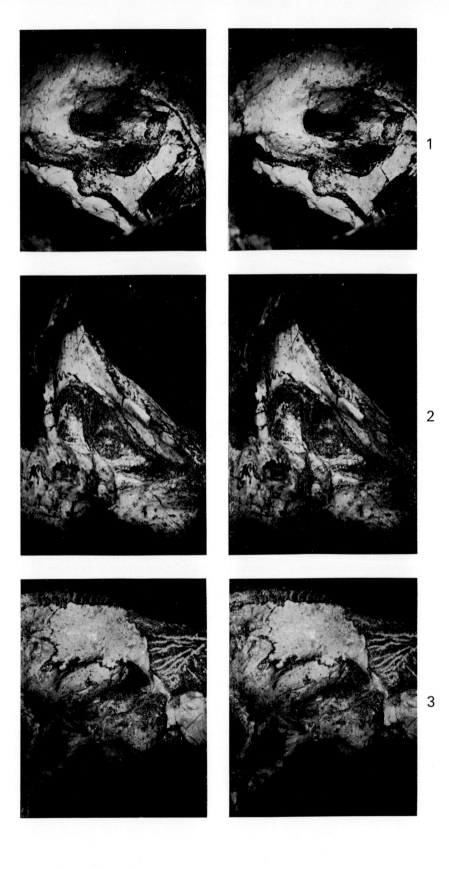

paroccipital process of the opisthotic from the postero-ventral process of the prootic (*pr.p.-v. Proot*).[1] This suture can be followed, in ventral view (Fig. 9), up to the 'lip' of the *fenestra ovalis*.

The *prootic* is a very complex bone with numerous processes (Figs 7, 8, 9, 11 and Plate III, 1, 2, 3). It is in contact with the squamosal, the supraoccipital, the opisthotic and, by its base, with the para-basisphenoid complex. However, its anterior part passes under the epipterygoid without touching it (Fig 7).

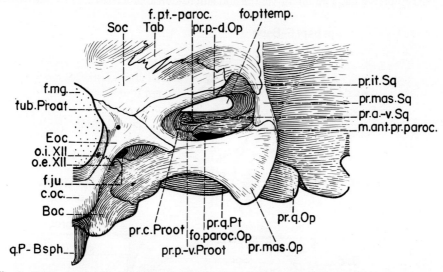

Fig. 8 *Regisaurus jacobi*—Detailed view of the lower part of the right side of the occiput (approx. ×2·5). *Boc*, basioccipital; *c.oc.*, occipital condyle; *Eoc*, exoccipital; *f.ju.*, jugular foramen; *f.mg.*, foramen magnum; *fo.paroc.Op*, paroccipital fossa of the opisthotic; *fo.pttemp.*, posttemporal fossa; *f.pt.-paroc.*, pterygo-paroccipital foramen; *m.ant.pr.paroc.*, anterior wall of the paroccipital process; *o.e.XII*, external foramen for the hypoglossal nerve; *o.i.XII*, internal foramen for the hypoglossal nerve; *pr.a.-v.Sq*, antero-ventral process of the squamosal; *pr.c.Proot*, central process of the prootic; *pt.it.Sq*, intermediate process of the squamosal; *pr.mas.Op*, mastoid process of the opisthotic; *pr.mas.Sq*, mastoid process of the squamosal; *pr.p.-d.Op*, postero-dorsal process of the opisthotic; *pr.p.-v.Proot*, postero-ventral process of the prootic; *pr.q.Op*, quadrate process of the opisthotic; *pr.q.Pt*, quadrate process of the pterygoid; *q.P-Bsph*, keel of the para-basisphenoid complex; *Soc*, supraoccipital; *Tab*, tabular; *tub.Proat*, tuberosity for the articulation with the proatlas.

It is composed, anteriorly and ventrally, of a basal area (*rg.bas.*, Fig. 11), of which the lower limit is in sutural contact with the para-basisphenoid complex from a level posterior to the external opening of the canal for the internal carotid and up to the out-turned 'lip' of the *fenestra ovalis*. A sharp crest separates this basal part from the upper portion of the bone.

Above this basal area, a wing-shaped antero-ventral process (*pr.a.-v.* = ossified *pila antotica*, cf.

---

[1] This lamina is labelled 'mastoid process' in Fig. 6 of Crompton (1955, page 166), apparently by accident as a result of the guide line extending too far. It is in contradiction with all the other figures of the article.

Plate III *Regisaurus jacobi*—*Lystrosaurus*-zone of Zeekoeigat, Venderstad, Bethulie (South Africa). F. R. Parrington field No. F.R.P./1964/27.

Fig. 1 Epipterygoid and anterior view of the posterior wall of the temporal fossa (right side) (×2).
Fig. 2 Ventral view of the posterior part of the skull (left side) (×2).
Fig. 3 Detailed view of the lower part of the occiput. (right side) (×2).

Crompton, 1955, page 180) stretches out broadly. Differing from the condition in *Ictidosuchops intermedius*, this process does not seem to be in contact with the basipterygoid process, but rises well above that level. The antero-ventral process is separated by a broad *incisura prootica* (*inc. proot.*) from the antero-dorsal process (*pr.a.-d.* = part of the *taenia marginalis* cf. Crompton *op. cit.*) which is shorter and more slender. Contrary to the condition in *Ictidosuchops intermedius* (Crompton *op.cit.* page 166), the *incisura prootica* is visible in lateral view (Fig. 10): only its anterior portion is hidden by the epipterygoid. The antero-ventral and antero-dorsal processes are in a nearly parasagittal plane.

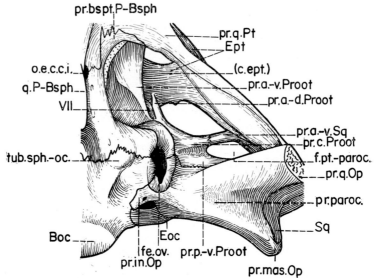

Fig. 9 *Regisaurus jacobi*—Ventral view of the posterior part of the left side of the skull (approx. × 2·5). *Boc.* basioccipital; (*c.ept.*), cavum epiptericum; *Eoc*, exoccipital; *Ept*, epipterygoid; *fe.ov.*, fenestra ovalis; *f.pt.-paroc.*, pterygo-paroccipital foramen; *o.e.c.c.i.*, external opening of the canal for the internal carotid; *pr.a.-d.Proot*, antero-dorsal process of the prootic; *pr.a.-v.Proot*, antero-ventral process of the prootic; *pr.a.-v.Sq*, antero-ventral process of the squamosal; *pr.bspt.P-Bsph*, basipterygoid processes of the para-basisphenoid complex; *pr.c.Proot*, central process of the prootic; *pr.in.Op*, internal process of the opisthotic; *pr.mas.Op*, mastoid process of the opisthotic; *pr.paroc.*, paroccipital process; *pr.p.-v.Proot*, postero-ventral process of the prootic; *pr.q.Op*, quadrate process of the opisthotic; *pr.q.Pt*, quadrate process of the pterygoid; *q.P-Bsph*, keel of the para-basisphenoid complex; *Sq*, squamosal; *tub.sph.-oc.*, tuberculum spheno-occipitale; *VII*, foramen for the facial nerve.

Behind a dorsal notch (*d.n.*) posterior to the *incisura prootica*, the prootic bends medially, forming a dorsal plate (*l.d.*) which has a suture with the supraoccipital.

More posteriorly (Figs 7, 8), two processes extend laterally inside the posttemporal fossa: a postero-dorsal process (*pr.p.d. Proot*) and a postero-ventral process (*pr.p.-v. Proot*). The postero-dorsal process (Fig. 7) is in contact with the intermediate process of the squamosal and with the opisthotic, and forms a part of the anterior border of the roof of the posttemporal fossa. The postero-ventral process (Fig. 8), lying on the paroccipital process of the opisthotic, participates in forming a small part of the internal floor of the posttemporal fossa. This postero-ventral process participates also in the posterior wall of the pterygo-paroccipital foramen (*f.pt.-paroc.*) above the 'lip' of the *fenestra ovalis*. The prootic contributes a quarter of the border of this latter fenestra.

The prootic has one additional lateral process, the central process (*pr.c.*, Figs 9, 11 and Plate III, 2), which fuses with the antero-ventral process of the squamosal anteriorly to the post-temporal fossa.

The probable opening for the facial nerve (*VII*, Fig. 11) is large and is situated under the *incisura prootica* (and so in a slightly higher position than that in *Ictidosuchops intermedius*) at the bottom of a small cupule for the gasserian ganglion.

Above this cupule there is a delicate rising crest. Below, running from the *fenestra ovalis* to the lower point of the antero-ventral process (*pr.a.-v.*), a strong crest emphasizes the angulation between this process and the basal area of the prootic.

In the upper part of the prootic, and anteriorly, the base of the central process is strengthened by two crests (Fig. 11): the lower one extends up to the middle of the base of the antero-dorsal process (*pr.a.-d.*); the upper one, sharper, goes up to the middle of the dorsal notch, after forming a sharp elbow half-way. The postero-dorsal process is slightly strengthened by an anterior blunt crest.

The *opisthotic* (*Op*) is also a complex bone with three main processes: a dorsal process (*pr.p.-d. Op*), an internal process (*pr.in.Op*, Fig. 9 and Plate III, 2) and the paroccipital process (*pr.paroc.*, Fig. 9 and Plate III, Fig. 2, 3).

The dorsal process is connected internally with the prootic and laterally, at the upper edge of the posttemporal fossa, over a short length with the squamosal and over a greater length with the supra-occipital.

The internal process is connected with the basioccipital and the exoccipital and forms part of the floor of the jugular foramen.

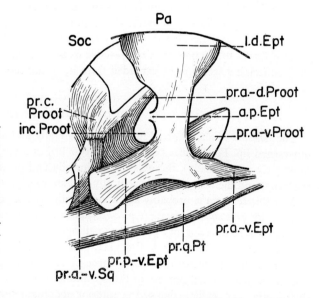

Fig. 10 *Regisaurus jacobi*—Lateral view of the right epipterygoid (approx. ×2·75). *a.p.Ept*, posterior apophysis of the epipterygoid; *inc. proot.*, incisura prootica; *l.d. Ept*, dorsal lamina of the epipterygoid; *Pa*, parietal; *pr.a.-d.Proot*, antero-dorsal process of the prootic; *pr.a.-v. Ept*, antero-ventral process of the epipterygoid; *pr.a.-v.Proot*, antero-ventral process of the prootic; *pr.a.-v.Sq*, antero-ventral process of the squamosal; *pr.c.Proot*, central process of the prootic; *pr.p.-v.Ept*, postero-ventral process of the epipterygoid; *pr.q.Pt*, quadrate process of the pterygoid; *Soc*, supraoccipital.

The paroccipital process forms the floor of the posttemporal fossa; it is connected internally and anteriorly with the prootic (Fig. 8 and Plate III, 3) and is subdivided laterally into two processes (Figs 7, 8, 9 and Plate III, 3): an anterior, or quadrate process (*pr.q.Op*), which makes contact, but no suture, with the quadrate and the squamosal, and a posterior, or mastoid process (cf. Crompton, 1955) (*pr.mas.Op*), which is partly in contact with the squamosal and partly free. On the ventral surface of the bone these two processes are separated by a triangular groove with an apex directed toward the mid-line. This groove, which deepens laterally, is considered as the roof of the middle ear. The upper part of the paroccipital process where it forms the floor of the posttemporal fossa is deeply hollowed. The posterior wall of this hollow, or paroccipital fossa (*fo.paroc.Op*, Fig. 8), is at the level of the mastoid process and its anterior wall at the level of the quadrate process. This structure is unknown in modern reptiles. The anterior wall of the paroccipital process also forms the posterior border of the pterygo-paroccipital fossa (*f.pt.-paroc.*, Fig. 8).

On the occipital face of the opisthotic, a small foramen is visible, perhaps for a vessel.

The *supraoccipital* (*Soc*) of *Regisaurus* appears, in occipital view (Fig. 4), as a wide tripartite bone. Unlike the condition in *Ictidosuchops intermedius*, there is a sutural connection between the supraoccipital and the interparietal, the differentiation of the dorsal and lateral processes of the supraoccipital being well marked because of a ventro-medial extension of the tabulars. Anteriorly, the supraoccipital reaches the epipterygoid and is in sutural connection with it, a situation which is similar to the one suggested for *Ictidosuchops intermedius* by Crompton (1955, page 167 and fig. 4A).

The *exoccipital* (*Eoc*) is a small triangular bone, a characteristic feature in *Regisaurus* is that it contributes neither to the floor of the *foramen magnum* nor to the occipital condyle (Figs. 4, 8), a situation differing from that of all closely related forms. The lateral process is long and pointed, resembling the one of *Ictidosuchoides longiceps* rather than that of *Ictidosuchops*, in which it is much shorter. The tuberosity for the proatlas (*tub. Proat*) is well developed, and near its base is a tiny foramen, probably for a blood vessel.

The *basioccipital* (*Boc*) is well exposed in occipital (Figs 4, 8) and ventral (Figs 2, 9) views, and it alone forms the occipital condyle. Inside the *foramen magnum* the internal opening for the hypoglossal nerve (*o.i. XII*) is situated at the junction between the basioccipital and exoccipital bones. Laterally to the well developed *tuberculum spheno-occipitale* (*tub.sph.-oc.*), an extension of the basioccipital contributes to the border of the *fenestra ovalis* (*fe.ov.*, Fig. 9) though in *Ictidosuchops intermedius*, the basioccipital seems to be excluded from it, according to Crompton (1955, page 155).

THE PRINCIPAL FOSSAE AND OPENINGS OF THE POSTERIOR PART OF THE SKULL

The *fenestra ovalis* (*fe.ov.*, Fig. 9) opens laterally and slightly ventrally. Its border is a rounded out-turned lip. Four bones participate in this lip, in more or less the same proportions: the prootic (anteriorly and dorsally), the opisthotic (dorsally and posteriorly), the basioccipital (posteriorly and ventrally) and the basisphenoid (ventrally and anteriorly). Crompton (1955) has stated that in *Ictidosuchops* this fenestra is probably surrounded only by the periotic bone.

The wide *posttemporal fossa* (*fo.pttemp.*) shows a complicated structure with different bones forming the edges of its anterior and posterior openings.

The anterior opening (Fig. 7; Plate III, 1) is surrounded, dorsally and laterally, by the intermediate process of the squamosal (*pr.it.Sq* = process 2, cf. Crompton 1955), and ventrally by the anterior wall of the paroccipital process of the opisthotic, which is partly covered medially and anteriorly by the prootic. The medial edge of the anterior opening is formed by the prootic.

The posterior opening (Figs 1, 8; Plate III, 3) of the posttemporal fossa faces postero-dorsally. It is mainly surrounded by two bones: dorsally and laterally by the squamosal, (with a short participation by the tabular), and dorsally, medially and ventrally by the opisthotic.

The floor of the posttemporal fossa is broader than its roof and is at two levels: the upper level is internal and formed by the opisthotic covered by the prootic (the suture between the two bones runs half-way from the anterior and posterior openings). The lower level is the bottom of the paroccipital fossa (*fo.paroc.Op*). Crompton (op.cit.) has observed the same feature in *Ictidosuchops intermedius*. As stated above, this condition does not occur in modern reptiles and its significance remains unknown.

The *pterygo-paroccipital foramen* (*f.pt.-paroc.*, Figs 1, 9) is very broad. It is limited anteriorly by the antero-ventral process of the squamosal (= process 3 cf. Crompton) and the central process of the prootic. Posteriorly it is limited by the anterior wall of the paroccipital process and the postero-ventral process of the prootic.

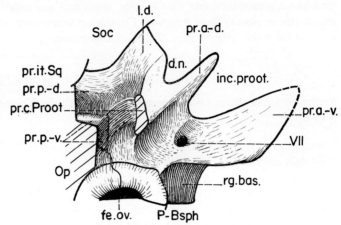

Fig. 11 *Regisaurus jacobi*—Lateral view of the right prootic (no proper scale). *d.n.*, dorsal notch; *fe.ov.*, fenestra ovalis; *inc.proot.*, incisura prootica; *l.d.*, dorsal lamina; *pr.a.-d.*, antero-dorsal process; *pr.a.-v.*, antero-ventral process; *P-Bsph*, para-basisphenoid complex; *pr.c.*, central process; *pr.it.Sq*, intermediate process of the squamosal; *pr.p.-d.*, postero-dorsal process; *pr.p.-v.*, postero-ventral process; *rg.bas.*, basal area; *Op*, opisthotic; *Soc*, supraoccipital; *VII*, foramen for the facial nerve.

The *foramen magnum* (*f.mg.*, Fig. 4) is slightly higher than broad. Its dorsal edge is formed solely by the supraoccipital, its lateral edges by the exoccipitals and its ventral edge by the basioccipital.

The *jugular foramen* (*f.ju.*, Figs 2, 3, 8, 9) is a large opening, mostly in the exoccipital, but part of its floor is formed by the basioccipital and the opisthotic.

II—LOWER JAW

The lower jaw (Fig. 12) is represented only by the posterior half of the dentary. The coronoid process of this bone is long but slender. On its internal side, the meckelian sulcus (*s.m.*) broadens posteriorly.

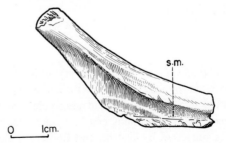

Fig. 12 *Regisaurus jacobi*—Internal view of the posterior part of the left dentary ($\times$ 1). *s.m.*, meckelian sulcus.

III—DENTITION

The dental formula is 6 I, 1 C, 10 PC. The dentition is very damaged and, on the premaxilla, only the alveoli of the incisors remain, although there is a tiny replacing fourth incisor on each side. On the maxilla, the canines are lost. On the right side most of the postcanines are also missing and one can observe, from the front: the first tooth; the alveolus of the second; the third tooth; two alveoli; the sixth tooth; one alveolus; the eighth postcanine; a tiny ninth postcanine and a small alveolus. On the left side, the situation is a little better, and there are: a first postcanine behind a 'scar' posterior to the canine's alveolus; a second tooth not so damaged; a small third one nearly complete; one alveolus; three other postcanines; one alveolus; one tooth and one alveolus. Most of the teeth are broken, so one cannot be sure of their structure. From what remains on the left side they appear to be conical, as far as the more anterior postcanines are concerned.

TAXONOMIC POSITION OF *REGISAURUS*

Regisaurus jacobi shares some common features with the Therocephalia and the Bauriamorpha; that is to say:

presence of suborbital fossae;
frontals contributing widely to the dorsal orbital borders;
no contact between the lacrimals and nasals;
intertemporal region narrow;
single occipital condyle;
processus ascendens of the epipterygoid narrow.

According to the definition of the Bauriamorpha given by Watson and Romer (1956, pages 72-73) and by Brink (Scaloposauria; 1965, page 137)[1], *Regisaurus* can be referred to this group on the following grounds:

temporal fossa relatively small;
zygomatic arcade delicate;
postorbital bar delicate;
nasal with little variation of breadth;
pineal foramen absent;
posttemporal fossa large;
parietal crest low;
suborbital fossa large;
interpterygoid vacuity of large size, of which the anterior end lies between the transverse flanges of the pterygoid.

In their classification of 1956 (page 72), Watson and Romer consider the Bauriamorpha as "theriodonts descended from the Therocephalia and inheriting many features of that group, but characterised by the tendency to develop many advanced characters". From what is actually known of those two related groups, two suppositions can be made: either the Bauriamorpha descended from the Therocephalia, as suggested by Watson and Romer, or Bauriamorpha and Therocephalia descended from the same, unknown ancestor. In the case of the second proposition, the immediate common ancestor of these two groups would have had the following features: numerous precanines; two canines; numerous postcanines; presence of a postfrontal; vomero-pterygoid contact on the primary palate; vomerine teeth; pterygoid teeth on the intermediate and lateral ventral ridges—besides the primitive characters

---

[1] The author does not take into account those characters, noted by Watson and Romer and by Brink, which occur in such other groups as Therocephalia and Cynodontia.

shared by the Therocephalia and the Bauriamorpha (discussed by Romer, 1969)—and also a broad posttemporal fossa, suborbital vacuity and a parietal region in process of constriction.

The subdivision of the Bauriamorpha is quite different according to various authors.

Haughton and Brink (1954) consider the families Lycideopsidae, Ictidosuchidae, Scaloposauridae and Bauriidae as families of the Therocephalia.

Crompton (1955) split the Scaloposauridae into five groups based on different degrees of evolution of the "postorbital arcade, pineal foramen, canine, postcanine teeth and secondary palate". The second group of Scaloposauridae (group B) includes *Ictidosuchops intermedius*, one of the closest forms to *Regisaurus*. Yet *Ictidosuchoides longiceps* remains in the Ictidosuchidae. The Bauriamorpha correspond to the family Bauriidae alone.

Watson and Romer (1956) place in the Bauriamorpha, besides the Bauriidae, the families Lycideopsidae, ?Ictidosuchidae, Nanictidopsidae, Silpholestidae, Scaloposauridae and Ericiolacertidae. Some genera are removed from the Ictidosuchidae and Scaloposauridae and are distributed between the families Nanictidopsidae and Silpholestidae. Some others are considered as *incertae sedis*.

Later Brink (1963, 1965) recognised an infraorder Scaloposauria including Ictidosuchidae, Scaloposauridae and Bauriidae.

Because the descriptions of many of the specimens classified by these authors are incomplete, the author does not intend to discuss them at this stage, and only notes that from its general aspect, *Regisaurus* appears close to such forms as:

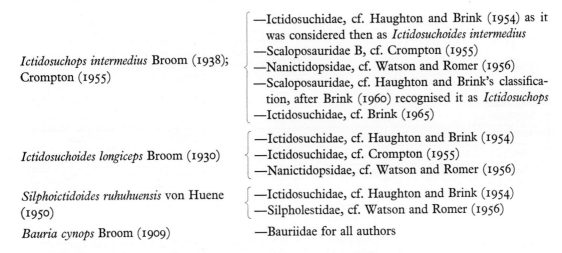

Ictidosuchops intermedius Broom (1938); Crompton (1955)
- Ictidosuchidae, cf. Haughton and Brink (1954) as it was considered then as *Ictidosuchoides intermedius*
- Scaloposauridae B, cf. Crompton (1955)
- Nanictidopsidae, cf. Watson and Romer (1956)
- Scaloposauridae, cf. Haughton and Brink's classification, after Brink (1960) recognised it as *Ictidosuchops*
- Ictidosuchidae, cf. Brink (1965)

Ictidosuchoides longiceps Broom (1930)
- Ictidosuchidae, cf. Haughton and Brink (1954)
- Ictidosuchidae, cf. Crompton (1955)
- Nanictidopsidae, cf. Watson and Romer (1956)

Silphoictidoides ruhuhuensis von Huene (1950)
- Ictidosuchidae, cf. Haughton and Brink (1954)
- Silpholestidae, cf. Watson and Romer (1956)

Bauria cynops Broom (1909)
- Bauriidae for all authors

The general appearance of the *Regisaurus* skull and the pattern of its palate give the impression that it "may be the direct ancestor to *Bauria*", as suggested by Parrington (1967, page 166).

What also strikes one is that, in the palate, *Regisaurus* appears more advanced than *Ictidosuchoides longiceps* and even than *Ictidosuchops intermedius*.

A detailed study has permitted the drawing up of Table 1.

In addition to this table, it has to be noted that in *Regisaurus*, *Ictidosuchops*, *Ictidosuchoides* and *Silphoictidoides* the tabular is completely, or almost completely, excluded from the edge of the posterior opening of the posttemporal fossa. In *Bauria*, the tabular, according to Brink (1963), still participates broadly to this edge.

From this table, it appears that *Regisaurus* is a genus different from the others cited.

Table 1

	Silphoictidoides ruhuhuensis East Africa *Cistecephalus* zone	*Ictidosuchoides longiceps* South Africa *Cistephalus* zone	*Ictidosuchops intermedius* South Africa *Cistecephalus* zone	*Regisaurus jacobi* South Africa *Lystrosaurus* zone	*Bauria cynops* South Africa *Cynognathus* zone
Intertemporal region	rather narrow	narrow (Crompton, 1955)	fairly broad (Crompton, 1955)	of medium breadth	very narrow (Brink, 1963)
Size of orbits	medium	medium	medium	large	medium or large
Postorbital bar	complete	complete (Crompton, 1955)	incomplete (Crompton, 1955)	complete	incomplete
Parietal crest	?	high (Crompton, 1955)	absent (Crompton, 1955)	small	not high but sharp (Brink, 1963)
Pineal foramen	large	rather small	medium	none (scar)	none
Premaxilla	long internasally	?	?	long internasally	usually represented as short
Lacrimal	long and high	long, of medium height	reduced anteriorly, low	long and very high	very small
Jugal	passes anteriorly to the anterior edge of the orbit	passes anteriorly to the anterior edge of the orbit	passes anteriorly to the anterior edge of the orbit	very short anteriorly, limit far behind the anterior edge of the orbit	passes anteriorly to the anterior edge of the orbit
Periotic area	?	?	one periotic bone (Crompton, 1955)	two well separated bones: prootic and opisthotic	?
Contribution of the parietal to the occiput	small	small	none	extensive	none
Interparietal	narrow	broad	medium	broad	medium
Tabular	low	of medium height	low	high	rather high
Occipital condyle	tripartite[1]	tripartite[1]	tripartite[1]	only formed by the basioccipital	tripartite[1]
Secondary palate	?	no primitive secondary palate (Boonstra, 1934; Crompton, 1955)	primitive secondary palate well developed (Crompton, 1955)	primitive secondary palate very well developed, showing not only a vomer-maxilla contact but also a premaxilla-maxilla connection alongside the vomer	secondary palate extended far posteriorly—premaxilla connection covering nearly the whole vomer
Precanines	2	2 (Crompton, 1955)	3 (Crompton, 1955) 2 (Brink, 1960)	none	none
Canines	1?	2	1	1	1
Pterygoid teeth	apparently absent	?	present, but not very numerous	very numerous on a high dentigerous tuberosity	absent

[1] Occipital condyle formed by the two exoccipitals and the basioccipital.

CONCLUSIONS

In its secondary contact between the maxilla and premaxilla alongside the vomer, as well as that between the maxilla and vomer, in its pineal foramen being reduced to a scar, and in the absence of precanines and the presence of a single main canine, *Regisaurus* resembles the Bauriidae. But it differs from them in still possessing a good number of primitive characters, such as:

lacrimal greatly expanded;
pterygoid teeth still abundant;
postorbital bar complete;
interparietal region broad;
extensive contribution of the parietals to the occiput.

In these characters *Regisaurus* is not only more primitive than *Bauria* but also than *Ictidosuchops*, although it is more recent.

Some characters which are peculiar to *Regisaurus*, such as the jugal extremely short anteriorly and the occipital condyle formed only by the basioccipital, seem also to exclude that genus from being the direct ancestor of *Bauria*. It may nevertheless be concluded that *Ictidosuchops* and *Regisaurus* each give a partial indication of the probable nature of the ancestor of *Bauria*.

SUMMARY

The skull of a new genus and species of mammal-like reptile, *Regisaurus jacobi*, from the Lower Trias of Cape Province, has been described with special reference to the foramina of the snout, the secondary palate, the otic bones and the posttemporal fossa.

Regisaurus has many characters in common with *Silphoictidoides ruhuhuensis*, *Ictidosuchoides longiceps* and *Ictidosuchops intermedius* and in certain features seems more advanced, but in others remains more primitive. Some peculiar characters exclude it from being the direct ancestor of *Bauria*.

ACKNOWLEDGMENTS

This paper is dedicated to Dr F. R. Parrington, F.R.S., in appreciation of his advice and encouragement, and of the loan of the beautiful specimen of *Regisaurus* already mostly prepared. I would like to express also my thanks to Prof. J. P. Lehman for his help and advice; to Dr A. d'A. Bellairs for interesting discussions; to Prof. A. S. Romer and Dr A. J. Charig for useful suggestions and to Drs K. A. Joysey and T. S. Kemp for patiently improving the English version of this paper. Material for comparison has been borrowed from the Bernard Price Institute, Johannesburg, where the specimens have been prepared by Mr J. W. Kitching; from the British Museum (Natural History), London, where the type of *Ictidosuchoides longiceps* has been further prepared by Mr R. Croucher; and from the von Huene collection in the Institut für Geologie und Paläontologie der Universität, Tübingen. These excellent preparations are here gratefully acknowledged. My thanks are due also to Mlle J. Crapart who made and corrected the drawings with indefatigable patience and great competence, and to Mr Potiquet for the photographs.

REFERENCES

BAHL, K. N. 1937. Skull of *Varanus monitor* (Linn.). *Rec. Ind. Mus.*, **39**, 133-174.
BOONSTRA, L. D. 1934. A contribution to the morphology of the mammal-like reptiles of the suborder Therocephalia. *Ann. S. Afr. Mus.*, **31**, 215-267.
—— 1953. A new scaloposaurian genus. *Ann. Mag. nat. Hist.*, (ser. 12), **6**, 601-605.

BRINK, A. S. 1960. On some small therocephalians. *Palaeont. afr.*, **7**, 155-182.
—— 1963. On *Bauria cynops* Broom. *Palaeont. afr.* **8**, 39-56.
—— 1965. A new ictidosuchid (Scaloposauria) from the *Lystrosaurus*-zone. *Palaeont. afr.*, **9**, 129-138.
CROMPTON, A. W. 1955. A revision of the Scaloposauridae with special reference to kinetism in this family. *Navors. Nas. Mus., Bloemfontein*, **1**, 149-183.
HAUGHTON, S. H. and BRINK, A. S. 1954. A bibliographical list of Reptilia from the Karroo Beds of Africa. *Palaeont. afr.*, **2**, 1-187.
OLSON, E. C. 1944. Origin of mammals based upon cranial morphology of the therapsid suborders. *Spec. Pap. Geol. Soc. Am.*, No. **55**, 1-136.
PARRINGTON, F. R. 1967. The origins of mammals. *Advanc. Sci.*, **24**, 165-173.
ROMER, A. S. 1969. The Chañares (Argentina) Triassic reptile fauna. I. A new chiniquodontid cynodont, *Probelesodon lewisi*—cynodont ancestry. *Breviora (Mus. comp. Zool. Harvard Univ.)*, **333**, 1-24.
TATARINOV, L. P. 1964. Contribution to the anatomy of the therocephalian skull. *Palaeont. Žh.*, **2**, 72-84 (in Russian).
WATSON, D. M. S. 1914. Notes on some carnivorous therapsids. *Proc. zool. Soc., Lond.* (for 1914), 1021-1038.
WATSON, D. M. S. and ROMER, A. S. 1956. A classification of the therapsid reptiles. *Bull. Mus. comp. Zool. Harv.*, **114**, 1-89.

T. S. KEMP

The jaw articulation and musculature of the whaitsiid Therocephalia

INTRODUCTION

The Whaitsiidae have long been recognised as a somewhat specialised family of the therocephalian theriodonts. They are restricted in time to the Upper Permian and so far have been found only in South Africa (Haughton and Brink, 1954), East Africa (Huene, 1950) and Russia (Tatarinov, 1963). Although they have the majority of typical therocephalian characters the group remains well differentiated from other families by its possession of certain specialised features, particularly the absence of suborbital fenestrae in the palate, loss of the postcanine teeth in all except a few primitive forms where more or less vestigial postcanines persist, the development of palatal processes dividing the internal nares into anterior and posterior parts, a very broadly expanded processus ascendens of the epipterygoid, loss of the interpterygoid vacuity, and a curiously bulbous muzzle (Watson and Romer, 1956).

Previous work on the group has been almost exclusively of a descriptive nature; the account by Brink (1956) of the small form *Aneugomphius ictidoceps* Broom and Robinson is particularly valuable. Tatarinov (1963; 1964) has provided us with a detailed interpretation of the snout region of the primitive Russian form *Moschowhaitsia vjuschkovi* Tatarinov.

In the present account, a detailed description of two of the more problematical regions of the mandibular apparatus, the jaw articulation and the angular bone with its reflected lamina, is given and an attempt to assess their functional significance is made.

Material from Dr F. R. Parrington's collection, now presented to the University Museum of Zoology, Cambridge, has been at my disposal and I have made use of the following specimens:

T 900. *Whaitsia sp.* A large skull complete with the lower jaws, rather weathered in certain regions and distorted laterally. The specimen has been prepared by mechanical means to expose the lower jaws and the right quadrate region completely.
Locality; Doornplaas, Graaff-Reinet, South Africa.
Horizon; *Cistecephalus* Zone, Upper Permian.
Field Number; F.R.P. 1964/30.
T 901. Indeterminate whaitsiid. A perfectly preserved articular bone lacking only the tip of the retroarticular process. No preparation was needed.
Locality; Stockley's site B4 (Stockley, 1932), Katumbi vawili, Ruhuhu Valley, Tanzania.
Horizon; Kawinga Formation (Charig, 1963), Upper Permian.
Field Number; F.R.P. 44. Collected by Dr F. R. Parrington in 1933.

T 902. cf. *Notosollasia lückhoffi* Huene. A fairly large skull with the lower jaws present but badly weathered posteriorly. The skull has been prepared completely in acetic acid.

Locality; Site B19 (Stockley, 1932), between Matamondo and Linyana, Ruhuhu Valley, Tanzania.

Horizon; Kawinga Formation (Charig, 1963), Upper Permian.

Field Number; F.R.P. 92. Collected by Dr F. R. Parrington in 1933.

THE JAW ARTICULATION

In the specimen of *Whaitsia sp.* the quadrate-quadratojugal complex could not be removed from the surrounding bones but, because it is slightly displaced ventrally and anteriorly, it was possible to remove a great deal of the matrix from around it, revealing most of its structure (Fig. 1).

The quadratojugal lies in a narrow incision in the squamosal, alined dorso-medially and running antero-laterally. The incision is open posteriorly (Fig. 1B), ventrally and anteriorly (Fig. 1F and 1H) and the quadratojugal emerges anteriorly from the slit as a free edge. The anterior opening of the slit continues above the upper end of the quadratojugal in a dorso-medial direction across the front face of the squamosal, marking the upper limit of the quadrate recess of the squamosal. The quadratojugal itself is a thin triangular sheet of bone with its medial face slightly concave from front to back (Fig. 1H) and it fits fairly closely into the squamosal incision so that it too is oriented dorso-medially and antero-laterally. The front edge is straight and it curls medially forming a vertical, medial-facing groove into which the lateral face of the quadrate fits. Below the level of the squamosal the quadratojugal is narrower and thicker and it forms a process that rests on the dorsal surface of the swollen lateral articular region of the quadrate. Here the quadratojugal canal runs in an anterior direction between the two bones.

The quadrate is generally quadrilateral with very narrow dorsal and lateral faces and a broader triangular medial face. It is swollen below the level of the squamosal to form the articulating condyles and, more medially, the stapedial process. The pterygoid wing rises from the front of the stapedial process and is a substantial antero-medial extension of the antero-medial edge of the quadrate.

The most important features of the quadrate are the joints that it makes with the adjacent bones. That with the quadratojugal has been mentioned, the narrow lateral edge of the quadrate fitting into the vertical medial-facing groove along the front edge of the quadratojugal (Fig. 1E).

The posterior face of the quadrate (Fig. 1D) has a vertical, rounded ridge running down its medial-most part. Laterally to the ridge the bone is in the form of a narrow vertical trough and then the lateral part of this posterior face is perfectly flat, terminating as the narrow edge that fits into the quadratojugal. The medial ridge fades out about halfway down the quadrate while the trough expands from side to side so that the full width of the lower half of the quadrate is gently concave. The recess of the squamosal, into which the back of the quadrate fits (Fig. 1F), conforms to the quadrate in that the upper medial part is a deep vertical groove lateral to which is a low, broad ridge, also vertically alined. The more lateral part of the recess is flat and featureless and its lateral termination is marked by the anterior opening of the slit for the quadratojugal. The lower part of the recess is only partially exposed in this specimen but appears to be flat, and it ends ventrally as a thin horizontal edge interrupted by a pronounced notch, just medial to the quadratojugal slit, that coincides with the position of the posterior foramen of the quadratojugal canal (Fig. 1B).

The broad lower part of the medial face of the quadrate (Fig. 1A) is in tight contact with the distal end of the paroccipital process. Above the level of the contact the medial face is concave from front to back and as far as can be seen the same is true of the medial face of the quadrate below the level of the contact. Although nothing can be seen of the contacting face of the quadrate itself in this specimen,

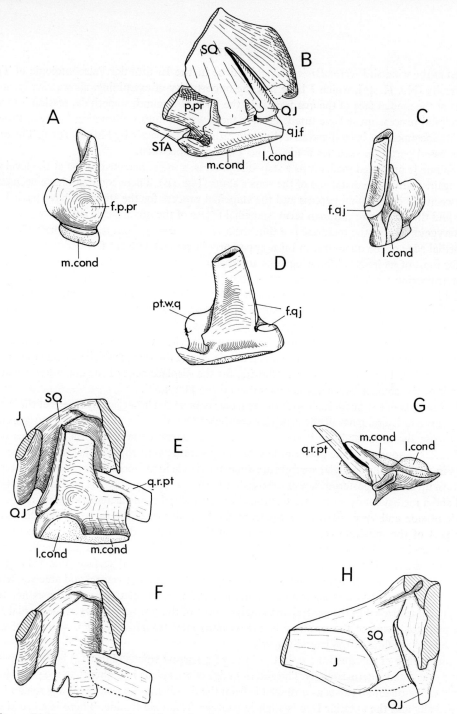

Fig. 1 Right quadrate complex and associated structures of *Whaitsia sp.*, (T 900), as preserved, × ⅔.

A, quadrate in medial view (based partly on another specimen, see text). **B**, quadrate complex *in situ*, posterior view. **C**, quadrate in lateral view. **D**, quadrate in posterior view. **E**, quadrate complex *in situ*, anterior view. **F**, squamosal recess for quadrate complex in anterior view. **G**, quadrate in dorsal view. **H**, quadratojugal *in situ*, medial view.

Abbreviations; *f.qj* facet for the quadratojugal, *f.p.pr* facet for the paroccipital process, *J* jugal, *l.cond* lateral condyle, *m.cond* medial condyle, *p.pr* paroccipital process, *pt.w.q* pterygoid wing of the quadrate, *qj.f* quadratojugal foramen, *QJ* quadratojugal, *q.r.pt* quadrate ramus of the pterygoid, *SQ* squamosal, *STA* stapes.

a specimen of the whaitsiid *Notosollasia lückhoffi* Huene in the Institüt der Palaeontologie of Tübingen University (No. K. 45), which I have had the opportunity of examining, shows clearly that the lower part of the medial face of the quadrate is in the form of a round, smooth depression into which the paroccipital process must have fitted. Clearly there was no suture formation between these two bones. (The specimen has been illustrated (Fig. 45) rather inadequately by Huene, 1950). The present Fig. 1A is based partly on sketches made of the Tübingen specimen.

The quadrate is produced medially as a stapedial process, a thick, narrow process at the level of the condyles against which the distal end of the stapes abuts (Fig. 1B). There could not have been enough space between the paroccipital process and the stapedial process for the stapes to have passed above the latter and thus the more common term 'stapedial recess of the quadrate' is not really appropriate.

The pterygoid wing of the quadrate is a thin sheet of bone arising from the lower half of the sharp antero-medial edge of the quadrate, and it is apparently in contact with the front edge of the stapedial process for most of its length. The wing runs antero-medially and it is markedly convex from top to bottom, terminating medially as a thin, convex, but 'finished' edge. For its full length the wing lies against the front face of the quadrate ramus of the pterygoid but a thin layer of matrix separates these two processes suggesting that there was no suture formation between them, a conclusion supported by the slight ventral displacement of the quadrate which has left the pterygoid quite undamaged. The quadrate ramus of the pterygoid continues postero-laterally right up to the body of the quadrate and there is a slight ridge running horizontally along that length of the ramus which contacts the quadrate, just below the dorsal edge of the ramus. Allowing for the displacement of the quadrate, it marks the dorsal limit of the contact between the quadrate and the pterygoid.

The final contact of the quadrate is with the articular bone of the lower jaw by means of the condyles. The posterior and lateral parts of the condyles are slightly weathered and the ventral surfaces are not visible because the articular bone is still in place. Nevertheless certain important features are displayed. Below the squamosal the posterior face of the quadrate swells slightly backwards to the ventral margin (Fig. 1D), but the condylar surfaces are almost invisible in posterior view, showing no tendency at all to turn up onto the posterior face of the quadrate. Also in this view there is no distinction between a lateral and a medial condyle, the two together forming a continuous articulating surface that is flat from side to side and very slightly convex from the back forwards for as far as can be seen. The anterior part of the quadrate condyle is entirely different, being divided into two distinct parts. Below the main body of the quadrate a lateral condyle protrudes forwards as a massive hemispherical boss with the articulating surface extending right round onto its dorsal surface (Fig. 1C and 1E). A medial condyle lies below the pterygoid wing of the quadrate at a more ventral and anterior level than the lateral condyle. It has a straight anterior margin and the articulating surface does not turn up dorsally, being restricted to the ventral and anterior faces of the condyle. Between the medial condyle and the pterygoid wing of the quadrate is a deep horizontal trough (Fig. 1E), which is terminated laterally by the swelling of the lateral condyle.

The dorsal edge of the quadrate is in the form of a narrow sulcus (Fig. 1G) lacking a periosteal finish and orientated antero-laterally. The anterior edge of the sulcus is higher than the posterior edge. The body of the quadrate twists as it descends from the dorsal edge so that the lower region is alined latero-medially and the anterior face is slightly concave from side to side. There is a broad shallow depression in the front face immediately above and medial to the boss of the lateral condyle (Fig. 1E).

The articular bone of the *Whaitsia sp.* is weathered laterally, the retroarticular process is broken off at the base, and the articulating surfaces are almost completely obscured by the quadrate (Fig. 4). The general form is clear, however, and as far as a comparison can be made it agrees in detail with the isolated hind end of a lower jaw (T 901, see page 213). Since this specimen proved important in elucidating the mechanism of the jaw hinge discussed later, it is necessary to establish its whaitsiid nature beyond doubt. The specific points of resemblance between this isolated specimen and the hind end of the lower jaw of *Whaitsia sp.* are the overall shape and size; the marked anterior lip of the lateral condyle; the more or less flat medial condyle; the absence of any trace of a dorsal process behind either condyle so that the articulating surface is smoothly continuous with the retroarticular process; the continuity of the lateral and medial condyles in posterior view; the ventrally pointing retroarticular process; the orientation of the articular surfaces to face postero-medially; the posterior extent of the surangular bone right to the edges of the condyles; the position and orientation of the surangular rising steeply from just in front of the lateral condyle; the correspondence of the articular surfaces with what could be seen of the quadrate condyles of *Whaitsia sp.*

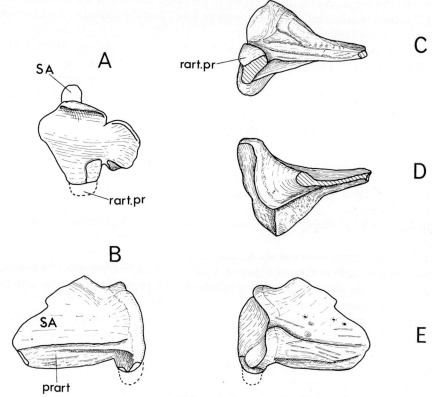

Fig. 2 Isolated articular of an indeterminate whaitsiid, (T 901), as preserved, ×⅔.
A, posterior view. **B**, lateral view. **C**, ventral view. **D**, dorsal view. **E**, medial view.
Abbreviations; *prart* sutural surface for the prearticular, *rart.pr* retroarticular process, *SA* surangular.

The fragment actually consists of the complete articular bone with a part of the surangular still attached. The prearticular has been lost but the sutural surface of the articular for it is preserved in splendid detail. The structure of the specimen is illustrated in Fig. 2, and description is confined to the nature of the articulating surfaces.

When the fragment is orientated in the closed jaw position, the lateral and medial condyles together form an articulating surface that faces posteriorly and about 20° medially. Only to a very slight extent can it be said to face dorsally (Fig. 2D). The lateral condyle is the larger and is more or less roller-shaped. A prominent dorsal lip curves up anteriorly and actually overhangs slightly so that the anterior-most part of the condyle faces ventrally (Fig. 2A and 2E). The posterior part of the lateral condyle curves smoothly downwards onto the back of the retroarticular process, and the middle region of the lateral condyle is produced laterally to increase its width (Fig. 2A).

The medial condyle is continuous with the lateral condyle except in that it lacks the antero-dorsal lip, and so its articulating surface is more restricted anteriorly. The curvature of the medial condyle is saddle-shaped; it has a concave form from front to back that corresponds to the curvature of the lateral condyle but it is slightly convex from side to side with the result that the medial part is turned down a little.

There are two striking anomalies in this region of the skull. The first is the failure of the quadrate to make rigid sutural attachments to the surrounding bones and the second is the orientation of the condylar surfaces in the horizontal plane.

To take the first point first, the evidence that the quadrate-quadratojugal complex was in some degree movable relative to the surrounding bones (i.e. streptostylic) is considerable, and the following suite of facts taken together can be explained by no other reasonable hypothesis.

(a) The quadrate is displaced in the specimen of *Whaitsia sp.* without any damage having been caused either to the quadrate complex or to the adjacent bones.

(b) The quadratojugal is held apparently loosely in a slit in the squamosal and it in turn supports the quadrate.

(c) Evidence cited earlier indicates that the distal end of the paroccipital process fitted into the smooth rounded depression in the medial face of the quadrate, without any suture formation.

(d) The pterygoid wing of the quadrate broadly overlaps the quadrate ramus of the pterygoid but with no suture formation.

(e) The quadrate fits conformably into the quadrate recess of the squamosal but again there is no suture formation.

(f) The notch in the ventral border of the squamosal adjacent to the posterior opening of the quadratojugal canal suggests itself as a device to prevent occlusion of the foramen when the quadrate moved.

(g) The sulcus along the dorsal edge of the quadrate could represent a site for the attachment of a muscle or ligament running postero-dorsally to control quadrate movement.

That a streptostylic quadrate is possible in the therapsids is shown by the condition in the gorgonopsids (Kemp, 1969), where there is a ball-and-socket arrangement between the quadrate and the squamosal allowing the lower jaw to move antero-posteriorly. In the present case of the whaitsiids the mode of quadrate movement must have been quite different. Of the six principal degrees of freedom possible (shift along or rotation about the three principal axes, vertical, lateral and longitudinal), several can be dismissed as involving improbable events. An antero-posterior shift would be prevented by the quadratojugal in its slit in the squamosal, unless there were a simultaneous lateral shift. However, any such lateral shift would result in disarticulation of the quadrate from both the paroccipital process and the pterygoid and further, the ridge-groove system between the back of the quadrate and the squamosal would be inhibitory. Equally, of course, a pure lateral shift of the quadrate can be discounted. A ventral-dorsal shift of the quadrate would be more reasonable although the ball-and-socket joint between the paroccipital process and the quadrate would not be appropriate, and in any case such a movement does not seem to serve any functional requirement.

Of the possible axes of rotation, a vertical axis would cause disarticulation of the quadrate from the squamosal and pterygoid and it is difficult to see how the quadratojugal would allow it. Equally, rotation about a transverse axis would result in loss of contact between the back of the quadrate and the squamosal and again the orientation of the quadratojugal is difficult to reconcile.

The most reasonable interpretation is that the quadrate was capable of rotation about a longitudinal axis providing that the axis ran at the level of the distal end of the paroccipital process (Fig. 3A). The

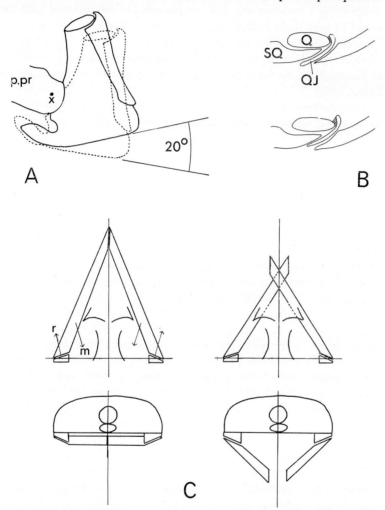

Fig. 3 Mechanism of the whaitsiid jaw hinge.

A, rotation of the quadrate complex shown diagrammatically in posterior view; x is the longitudinal axis of rotation. **B**, diagrammatical horizontal section through the squamosal and quadrate complex illustrating the slight distortion between the quadrate and the quadratojugal during the rotation of the complex. **C**, the effect of angular displacement of the axis of the hinge joint on jaw opening. Dorsal view above and posterior view below. The actual effect is a combination of these two effects.

quadrate would then rotate up and down about the distal end of the paroccipital process, explaining why this joint is in the nature of a ball-and-socket. There would be almost no tendency for the quadrate to disarticulate from the quadrate ramus of the pterygoid because this joint would lie more or less along the axis of rotation. Instead, the quadrate would slide up and down upon the quadrate ramus, thus perhaps accounting for the convex form of the pterygoid wing of the quadrate in a dorso-ventral direction. The principal motion imparted to the quadratojugal would be up and down in its slit in the squamosal, although it would also suffer a slight tendency to move latero-medially. Because of its orientation in the slit, a lateral movement would inevitably be accompanied by an anterior movement, and a medial movement by a posterior one, but the geometry of the region is such that the antero-posterior shift would be very slight indeed and could comfortably have been accommodated by an elastic fibrous connection between the quadrate and the quadratojugal instead of being imparted to the quadrate. Indeed, this slight distortive motion of the quadratojugal is probably the reason why the quadrate and quadratojugal do not form a single virtually fused complex as, for example, in the gorgonopsids (Kemp, 1969).

Rotation about a longitudinal axis, as here proposed, means that the quadrate will remain in close, extensive contact with the squamosal and the ridge-groove system between these two bones is appropriate, and finally, the notch in the ventral edge of the squamosal can be explained as a device allowing the quadrate complex to move up and down without occluding the posterior opening of the quadratojugal canal.

Before considering the functional significance of this type of streptostyly we must turn to the second enigmatic feature referred to above, the orientation of the articulating surfaces relative to the skull axes. As viewed from above, the axis of the hinge joint is alined at about 20° from the transverse axis of the skull so that it runs antero-medially. This will be referred to as the angular displacement in the horizontal plane. Again, as viewed from behind, the joint axis is alined at about 20° from the horizontal so that it runs ventro-medially, an angular displacement in the transverse plane. Thirdly, it was seen that the articulation in the jaw-closed position was more or less vertical, the articulatory surface of the articular facing very much posteriorly.

Because the articular bears against the anterior face of the quadrate, and hardly at all against the ventral face, except at relatively large gapes, there could have been no possibility of a propalinal (antero-posterior) movement of the lower jaw upon the quadrate. Equally the complete absence of any kind of dorsal process of the articular behind the condyles indicates that there could have been no antero-posterior, streptostylic, shift of the quadrate carrying the lower jaw forwards (a conclusion reached on different grounds above). Indeed, these latter two arguments may be taken further for they indicate that the articular must have been held firmly in place on the quadrate by muscles with a large posteriorly directed force, an opinion supported by the occurrence of the large antero-dorsal process in front of the lateral articular condyle and the absence of any direct support for the quadrate dorsally. The identity of the muscle with the large, posteriorly directed force is clearly the temporalis, inserting on the great coronoid eminence and coronoid process of the dentary, and this must have been the major jaw-closing muscle of the skull. (The use of the term 'temporalis' muscle is not intended to indicate necessary homology with the muscle of the same name in mammalian terminology. It is restricted to mean that part of the reptilian capiti mandibularis that runs from the medial and posterior parts of the temporal fenestra.) The temporalis is related to the greatly enlarged temporal fenestra. However, enlargement of the fenestra has been by posterior and medial expansion rather than lateral expansion, leading to the long narrow intertemporal region of the skull with a 'fan' of muscle fibres radiating from the medial and posterior edges of the fenestra to the lower jaw.

However, such a muscle could hardly avoid having a large medially-directed component to its force which would tend to pull the articular bone off the quadrate in a medial direction. Thus the lower jaw would pivot about the lateral pterygoid flange of the palate and the symphysis would be forced open. This tendency has been overcome by the hinge having its axis alined antero-medially; it is more able to resist the posterior component of the temporalis muscle because the articulating face of the quadrate faces more or less posteriorly, and the hinge itself must have been essentially a roller-bearing in order that a firm contact between the articular and the quadrate was maintained at all gapes. But with such a system alone the lower jaw could not possibly open. A moment's reflection indicates that, because of the angular displacement of the hinge axis in the horizontal plane, as the jaw opened about the articulation it would tend to shift medially as well, which would of course be impossible because of the symphysial connection between the two mandibles (Fig. 3C).

Consider now the ventro-medial orientation of the hinge axis in the transverse plane, as seen from behind. A rotation of the lower jaw downwards about the axis would, due to this particular angular displacement of the axis, be accompanied by a lateral shift of the jaw and thus by a tendency for the symphysis to come apart (Fig. 3C). Therefore, the effect of the angular displacement of the hinge axis in the horizontal plane and in the transverse plane respectively tend to cancel the tendency of the jaws to move bodily in the horizontal plane. However, it can be demonstrated geometrically that if the lower jaw is not to deviate from a parasagittal plane as it opens and closes there are no two possible fixed values for the angles of deviation of the hinge axis in the two respective planes, horizontal and transverse. Instead, if the angle of deviation from the normal in one of the planes is fixed, then the angle of deviation from the normal in the other plane is a variable dependent on the angle of gape. (Alternatively, the two deviations could both be variable, as a pair of co-dependents together dependent on the angle of gape, but this is regarded as less likely on functional grounds discussed below.)

Since, as suggested above, the angle of deviation in the horizontal plane is presumably related to the direction of the net force of the temporalis muscle fibres, it would be expected that this angle would be fixed at an appropriate angle and that the functionally less critical angle of deviation of the hinge axis in the transverse plane would be the variable. It is surely no coincidence, therefore, that the morphology of the quadrate complex indicates that it probably rotated about a longitudinal axis thus varying the angle that its condyle axis made in a transverse plane.

With the observed deviation from the normal in the horizontal plane about 20°, the hinge axis can be calculated as requiring an angular excursion of about 20° in the transverse plane in order to allow a gape of 60°, a figure which appears to be well within the potentiality of mobility of the quadrate complex.

THE ANGULAR AND REFLECTED LAMINA

The angular bone consists of two distinct parts, the main body and the reflected lamina. The main body is a thin high sheet applied to the similarly oriented vertical sheet of the surangular. On the left side of the skull of *Whaitsia sp.* the reflected lamina was damaged and weathered and so the remains of it were removed to expose the lateral surface of the main body of the angular (Fig. 4C). The dorsal half is a broad, smooth, slightly concave area bounded above and behind by the swollen dorsal part of the surangular, and limited antero-dorsally by the exceedingly well-developed angular crest which stands almost 1 cm proud of the main body of the angular. The posterior edge of the crest is at right angles to the plane of the angular and is slightly hollowed out between its proximal and distal edges. In contrast, the anterior face of the crest slopes less steeply onto the angular. Ventrally the angular crest merges with the thickened part of the angular below the intramandibular fenestra (see below), and dorsally it peters out just below the surangular-angular suture.

Below the concave region the lateral face of the angular is very slightly convex down to the straight ventral margin (preserved undamaged on the right side of the skull) which runs in close contact with the ventral margin of the prearticular (Fig. 4E).

In front of the angular crest the angular rapidly loses height and parts company with the surangular

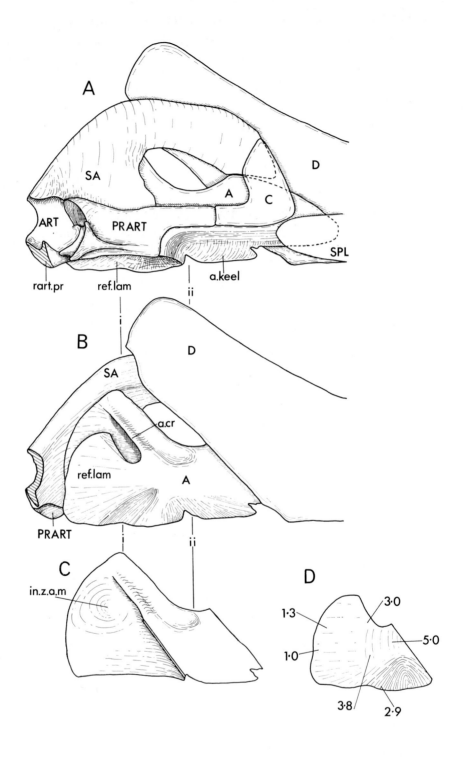

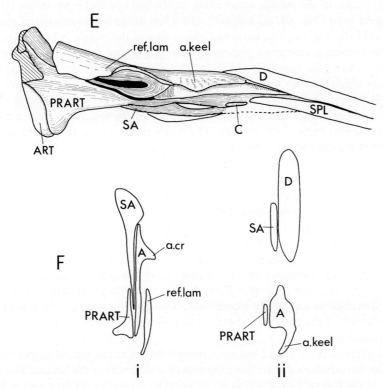

Fig. 4 The hind end of the lower jaw of *Whaitsia sp.* (T 900), slightly restored. × ⅔.

A, medial view. **B**, lateral view. **C**, lateral view of the angular with the reflected lamina removed. **D**, inner surface of the reflected lamina, showing the thickness of the bone in millimetres at certain points. **E**, ventral view. **F**, transverse sections at i and ii as indicated.

Abbreviations; *A* angular, *a.keel* angular keel, *a.cr* angular crest, *ART* articular, *C* coronoid, *D* dentary, *in.z.a.m* insertion of the zygomatico-angularis muscle, *PRART* prearticular, *rart.pr* retroarticular process, *ref.lam* reflected lamina, *SA* surangular, *SPL* splenial.

leaving a perfectly natural, large intramandibular fenestra. Below the fenestra the angular runs medially to the dentary as a broad sheet which is greatly thickened along a horizontal line from the lower end of the angular crest forwards and a well-developed keel descends from the thickening to form a natural anterior continuation of the reflected lamina (Fig. 4B and 4C). There is an almost horizontal ridge supporting the lateral face of the keel, below which the keel is slightly bent medially in an apparently natural fashion, and a series of course corrugations run antero-ventrally down the keel from the strengthening ridge.

The reflected lamina of the angular arises along an oblique line that is an antero-ventral continuation of the angular crest (Fig. 4B, 4C and 4D), and it lies in the same parasagittal plane as the keel of the main part of the angular. The horizontal thickening of the angular which is present below the intramandibular fenestra is continued posteriorly onto the lamina with the result that, although not apparent in lateral view, the lamina is markedly thicker in its central region than in the dorsal and posterior regions which are very thin. The ventral region is intermediate in thickness (Fig. 4D).

The lamina is completely free of the more medial parts of the jaw except along the line of attachment and it is held sufficiently laterally of the main body of the angular to leave a gap, the angular recess, between the two parts of the angular. The width of the recess varies between about 0·8 cm dorsally and about 0·4 cm postero-ventrally.

The greater part of the lateral face of the lamina is slightly convex from top to bottom and there is no indication that this is other than the natural condition. In the antero-ventral region, however, there is a marked trough running postero-ventrally to the lower margin of the lamina and a second depression commences on the antero-ventralmost part of the lamina to be continued extensively on the ventral keel of the angular anteriorly to the lamina, where it represents the medial bending of the keel noted earlier. Most significantly these two depressions separate a broad ridge which, because the lamina is of constant thickness in this region, is an external reflection of a large trough on the internal surface of the lamina. The trough coincides with the region where the ventral edge of the main body of the angular turns postero-medially (Fig. 4E). Thus there is here a large expansion of the angular recess which may be described as a ventrally-facing fossa, bounded anteriorly by the line of attachment of the lamina to the rest of the angular and constricted dorsally and posteriorly by the angular recess regaining its normal width.

The lateral surface of the reflected lamina is smooth except in the ventral region where a series of almost imperceptible striations radiate from the area of attachment of the lamina. The medial surface of the lamina of the left side of *Whaitsia sp.* (Fig. 4D). shows no features except for the central thickening referred to. Although there are no striations, a series of minute markings run across the bone surface, in a posterior direction over the dorsal region and a postero-ventral direction below. The ventral edge of the lamina is probably broken and so it is not possible to judge the original extent to which the lamina extended ventrally below the level of the rest of the jaw.

Owing in great measure to the absence of any corresponding structure in modern vertebrates, the functional significance of the reflected lamina of the angular has remained perplexing. Parrington (1955) reviewed briefly previous theories and offered his own interpretation based on the assumption that the lamina was the functional forerunner of the mammalian angular process and was thus the site of insertion of an antero-dorsally directed superficial masseter muscle, external to the jaw, derived as a slip of the capiti mandibularis block of more primitive reptiles. This theory has been discarded by Crompton and Hotton (1967, p. 23) and is not accepted by Barghusen (1968) in the case of therapsids generally, or by Kemp (1969) specifically in the case of the Gorgonopsia and their ancestors. Tempting as Parrington's explanation is, it does not satisfactorily account for the pattern of strengthening ridges seen on the lamina.

It is essential to appreciate that the detailed form of the lamina differs greatly from group to group among the therapsids and, therefore, whatever its primitive function might have been in the sphenacodont pelycosaurs it subsequently became modified in different ways in order, presumably, to fulfil different functional requirements. Barghusen (1968) concluded that the particular type of reflected lamina seen in the Therocephalia and Bauriamorpha was for the attachment of ventrally-directed musculature, especially the branchio-mandibularis muscle functioning to depress the hyoid apparatus, and also perhaps intermandibularis muscles running between the two mandibles. He doubts that the pterygoideus muscle would have attached to "... the delicate and reduced lamina of *Bauria* ..." (*op. cit.* page 20). The structure of the angular and reflected lamina described here for *Whaitsia* is very similar to the structure in *Bauria* and other therocephalians, but the further details available lead to an interpretation of its significance that is quite different from that of Barghusen, whose hypothesis may be objected to on two general grounds. First it seems incongruous that such a highly complex structure should have evolved in order to provide attachment for musculature that is adequately accommodated by a conventional jaw in other reptiles, and second there is no explanation for the very considerable extent of the lamina and its associated angular recess.

The reflected lamina effectively doubles the posterior region of the lower jaw, leaving the angular recess as a substantial gap between the two parts. Because the inner and outer walls of the recess both consist of the same bone, the angular, it is hard to see the structure as a device for controlling the movements of two parts of the jaw involved in an intramandibular kinetic joint. The only other reasonable explanation is that the structure allows two muscles, running in very different directions, to insert in the same region of the lateral side of the lower jaw. The nature of two such muscles is, I believe, apparent.

The angular crest bounds a shallow depression in the lateral wall of the angular, internal to the reflected lamina, in a manner suggesting that a muscle ran more or less posteriorly from the crest. The raised ridge-plus-depression system is commonly found in vertebrates when a muscle inserts at a low angle to a bone surface, in which cases the ridge is found to provide a site of insertion for the distal extremity of the muscle (compare, for example, the masseteric fossa of mammals). Further support for the hypothesis of a posteriorly directed muscle running from the angular crest is the freedom of the dorsal edge of the reflected lamina which would be necessary for the muscle to escape from the angular recess and for the lamina to ride over the lateral surface of the muscle as the gape of the jaws altered. The precise form of the angular crest too is suggestive as it is not symmetrical in section but has a steeper posterior edge than anterior edge (Fig. 4C and 4F), which implies that the function of the crest is not merely to strengthen the angular. This latter observation too suggests that there was no muscle running anteriorly from the crest because the form of the anterior edge of the crest is unsuitable and, in any case, there is no evidence that the intramandibular fenestra was involved in jaw musculature (Kemp, in press).

A muscle inserted on the angular crest could have run in any direction from postero-dorsally to almost ventrally as far as the crest is concerned. However, the freeing of the dorsal edge of the lamina strongly suggests that it ran approximately posteriorly and in no other direction does there seem to be a suitable site of origin from the skull. As it is, however, the muscle would reach the skull in the region of the squamosal immediately lateral to the quadratojugal. The squamosal continues laterally away from the quadrate complex as a broad, smooth area facing somewhat ventrally (Fig. 1E) and leaving quite a large gap between the lower jaw and the posterior part of the zygomatic arch. There is no reason for such a gap unless it were to allow for muscles to run from the lateral surface of the lower jaw to the squamosal. Thus the evidence for this muscle is strong and, in order to avoid prejudicial implications of homology with other therapsid groups, it may be given the name of 'zygomaticoangularis'.

Turning now to the reflected lamina, we may recall that the pattern of strengthening consists of a horizontal thickening, below which are two broad depressions that widen as they run downwards and separate a broad ridge on the lateral surface of the lamina. This structure clearly indicates that muscles ran in an approximately ventral direction from the lamina. If one regards only the lateral surface of the lamina as having carried muscle fibres, then one is forced to the conclusion of Barghusen (1968) that such fibres could not have turned medially below the ventral edge of the lamina because the lamina is not adequately supported against a medially-directed force. Thus the pterygoideus musculature must be discounted as having had a large insertion on the lamina. On the other hand, the manner in which the angular recess opens ventrally as a large ventrally-facing fossa (Fig. 4E) strongly suggests that a muscle entered the recess from below and inserted onto the inner wall of the reflected lamina and probably also on the lateral wall of the body of the angular. Such a muscle could quite well have turned medially below the ventral edge of the jaw because the body of the muscle filling the angular recess would itself prevent the reflected lamina from being forced medially. Yet as far as the lamina itself was concerned, it would be required to resist a muscle force applied in a ventral direction. There is of course no reason why some of these muscle fibres should not have attached higher up on the inner surface of the lamina, alongside the zygomatico-angularis muscle, thus accounting for the great dorsal extent of the lamina alongside that muscle. Equally, given this basic system of muscle fibres attached to the inner surface of the lamina there is no reason why certain of the fibres should not attach to the lateral surface of the lamina as well and turn medially below the ventral edge of the lamina, accounting for the fine striations on the lateral lamina surface. Such musculature as proposed would be pterygoideus, at least in a general topological sense and would presumably have originated from the back of the well-developed lateral pterygoid flange of the palate, the quadrate ramus of the pterygoid and possibly the expanded epipterygoid and ventral keel of the parasphenoid-pterygoid as well.

Although we can go far towards understanding the structure of the angular in terms of the pterygoideus muscles, it is quite likely that the reflected lamina had other muscles attached to it as well. Topologically the lamina is a posterior continuation of the ventral keel of the angular and to judge from the condition of the sphenacodont pelycosaurs (Romer and Price, 1940), this may be true in a morphological sense too. Thus it is probable that those muscles like the branchiomandibularis and intermandibularis which attach to the angular keel in other reptiles, did in fact insert on the reflected lamina in whaitsiids, but, as suggested earlier, it is hard to imagine that they played an important part in determining the structure of this region of the jaw.

The ventral keel of the angular in front of the reflected lamina shows clear indications of muscle attachment to the lateral surface, in the form of the antero-ventrally directed striations. The medial bending of this keel suggests that the muscle fibres in question turned medially below the ventral margin, but whether they were a further part of the pterygoideus muscle system or perhaps an intermandibularis cannot be determined.

DISCUSSION

The nature of the hinge mechanism has already been accounted for as a necessary modification consequent upon the enlargement of the temporalis musculature. It was a device that overcame the increased net medially directed force of the jaw closing muscles. The organisation of the angular and its reflected lamina can be explained also as a result of modifications which were necessary in consequence of this increase in the temporalis muscle. The pterygoideus musculature was the second important muscle mass and it too had increased greatly in size to judge from its extensive area of origin from the skull.

Most of the fibres were directed antero-dorsally and so it must have had the important function of providing a reaction to the posteriorly directed component of the temporalis, so relieving the hinge of this function to some degree. Nonetheless, the pterygoideus must be regarded primarily as a jaw closing muscle and thus its insertion had to be as low down as possible on the jaw. When the jaw hinge was depressed in order to increase the moment produced by the temporalis muscle, the moment of the pterygoideus muscle must inevitably have tended to be reduced. The obvious solution to this

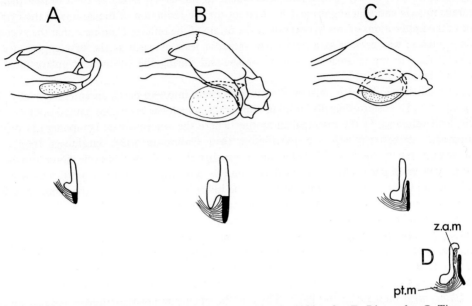

Fig. 5 Evolution of the reflected lamina, as illustrated by **A**, *Ophiacodon*. **B**, *Dimetrodon*. **C**, *Titanophoneus*. In each case the inner surface of the hind end of the jaw is shown with the extent of the pterygoideus insertion. The extent of the reflected lamina is indicated by a dashed line in **B** and **C**. A cross section through the angular is shown below each and the angular keel or its presumed homologue the reflected lamina is in black. **D** is a similar section of a whaitsiid.

Abbreviations; *pt.m* pterygoideus muscle, *z.a.m* zygomatico-angularis muscle.
(A and B after Romer and Price, C after Orlov.)

would have been to develop a deep ventral keel in the angular region of the jaw, as suggested by Romer and Price (1940) in the case of the sphenacodont pelycosaurs. As the pterygoideus increased in size however, such a keel would have had to become unwieldy in size. The alternative solution to apply then would have been to wrap the muscle around the ventral margin of the jaw and insert it laterally, effectively causing all the muscle fibres to act at the level of the ventral edge of the jaw, a level low enough to give the muscle an adequate lever arm to the hinge. If, however, the lateral surface of the jaw in this region were already occupied by a zygomatico-angularis muscle insertion, it would have been unavailable for pterygoideus insertion. Instead, therefore, the attachment of the angular keel moved dorsally up the lateral surface of the angular, creating the angular recess between the body of the angular and the keel, which then becomes known as the reflected lamina. The pterygoideus muscle continued to insert on the morphologically internal surface of the angular, but in mechanical effect it had shifted to the lateral mandibular surface. Evidence that this is a feasible view of the reflected lamina is indicated in Fig. 5, where the series *Ophiacodon*, *Dimetrodon* and *Titanophoneus* can be interpreted as three phases in the evolutionary process.

It might reasonably be objected that the usual way of inserting two muscles in the same region of a bone is to attach one of them to the fascia overlying the other (Crompton, pers. comm.). The answer in this case is perhaps that historically the pterygoideus has remained associated with the morphologically internal surface of the angular, never having really invaded the lateral surface of the jaw at all, and thus it could only have achieved its present functionally lateral position because the internal surface of the angular itself migrated as the inner surface of the reflected lamina. Subsequently, the pterygoideus could possibly gain a new attachment to the lateral fascia of the external part of the temporalis muscle and this might partially account for the reduction of the lamina in the cynodonts.

One of the implications of this hypothesis of the origin of the reflected lamina is that the zygomatico-angularis muscle was present at a very early stage, at least as early as the first appearance of the reflected lamina in the sphenacodont pelycosaurs, and during the subsequent enlargement of the lamina in later therapsids.

Thus we see the tendency to polarise the muscle system into two parts, an antero-dorsally running pterygoideus and a postero-dorsally running temporalis, which has been commented upon frequently, usually with reference to the mammalian system where the antero-dorsal component is principally the superficial masseter muscle (e.g. Parrington, 1934; Crompton, 1963; Barghusen, 1968).

The advantages of this kind of arrangement are that the individual muscle fibres can be longer than if they inserted perpendicularly to the jaw, and by inserting the postero-dorsal muscle above the hinge level and the antero-dorsal muscle below the hinge level, the length of the moment arm to the hinge of both muscles is increased. A third advantage is that there will be a reduction of the net component of the muscle force along the horizontal line of the jaw.

Another possible advantage not commented upon before concerns the change in the geometrical arrangements of the two muscles with a change in gape. Fig. 6 compares the disposition of the fibres of the two muscles with the jaws closed and with the jaws open. First it may be noted that the length of the temporalis fibres ($t.m$) has increased much more rapidly than the length of the pterygoideus fibres ($pt.m$) with increase in the gape. Thus at the open-jaw position the temporalis fibres must have been approaching their maximum extension and thus the force they can produce had fallen greatly (see Kemp, 1969, page 46, and Gans and Bock, 1965, for discussion of this phenomenon). On the other hand, the small extension of the pterygoideus fibres suggests that they could be producing more or less their maximum force. The second point to notice is that the length of the moment arm to the hinge of the temporalis muscle has decreased with the increase in gape, while that of the pterygoideus has increased. These two points together indicate that the antero-dorsal muscle (pterygoideus) was the more important for producing a high closing moment at large gapes and that the postero-dorsal muscle (temporalis) was the more important at low gapes, and that one result of this type of muscle organisation was to increase the gape over which an effective closing force was available.

There is, however, an important implication of this interpretation. Because the jaw hinge is adapted to resist a largely posteriorly directed force the articular contacts the anterior face of the quadrate when the jaws are closed (Fig. 6A). A line, X, can be drawn along which disarticulation of the articular from the quadrate could occur in an almost dorsal direction and it must have been necessary at all times that the total muscle force lay in such a direction that it prevented such disarticulation. When the jaws were closed, assuming that the temporalis was producing much the larger force, the net force would have been close to a right angle to line X and the friction between the articular and the quadrate would comfortably prevent disarticulation (Fig. 6A). With the jaws open, however, assuming that the pterygoideus was producing the larger force, the resultant muscle force would lie at a much smaller angle to line X and the friction between the articular and the quadrate might well have been too small to prevent disarticulation along line X (Fig. 6B). If part of the temporalis were

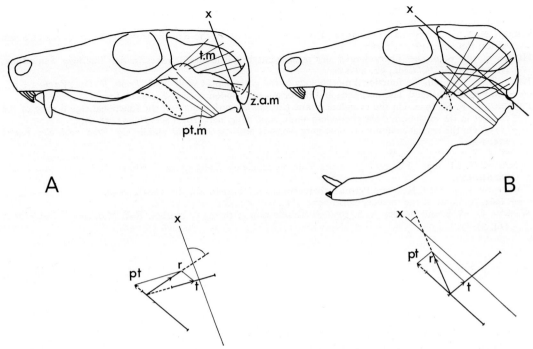

Fig. 6 Effect of gape upon the muscle forces. **A**, jaws closed. **B**, jaws opened. Line X is the line of potential disarticulation (see text). A vector diagram of the forces produced by the muscles temporalis (*t.m*) and pterygoideus (*pt.m*) is shown below in each case to demonstrate the probable change in angle that the resultant force (*r*) makes with line X, if the effect of the zygomatico-angularis muscle (*z.a.m*) is ignored. (Outline of the skull after Brink's reconstruction of *Aneugomphius ictidoceps* Broom and Robinson.)

to have gained an origin from the skull near to the level of the axis of the jaw hinge, it would have undergone a very small increase in length as the jaws opened and could thus have been arranged to maintain an almost isometric, maximum contraction throughout the range of the jaw gape. And if it inserted onto the dorsal half of the lower jaw it would produce a force with a significant component resisting disarticulation of the lower jaw along line X at all gapes. Thus there must have been a high evolutionary premium to develop such a muscle for it would allow the full utilisation of the potential of the pterygoideus muscle for providing a large closing force at wide angles of gape. The suggested zygomatico-angularis has just such properties, as too does that part of the temporalis muscle that originated from the skull on the ventral downgrowth of the squamosal medial to the quadrate.

REFERENCES

BARGHUSEN, H. R. 1968. The lower jaw of cynodonts (Reptilia, Therapsida) and the evolutionary origin of mammal-like adductor jaw musculature. *Postilla*, **116**, 1-49.

BRINK, A. S. 1956. On *Aneugomphius ictidoceps* Broom and Robinson. *Palaeont. afr.*, **4**, 97-115.

CHARIG, A. J. 1963. Stratigraphical nomenclature in the Songea series of Tanganyika. *Rec. Geol. Surv. Tanganyika*, **10**, 47-53.

CROMPTON, A. W. 1963. On the lower jaw of *Diarthrognathus* and the origin of the mammalian lower jaw. *Proc. zool. Soc. Lond.*, **140**, 697-750.

CROMPTON, A. W. and HOTTON, N. III. 1967. Functional morphology of the masticatory apparatus of two dicynodonts (Reptilia, Therapsida). *Postilla*, **109**, 1-51.

GANS, C. and BOCK, W. 1963. The functional significance of muscle architecture. A theoretical analysis. *Ergebn. Anat. EntwGesch.*, **38**, 115-142.

HAUGHTON, S. H. and BRINK, A. S. 1954. A bibliographical list of the Reptilia from the Karroo beds of South Africa. *Palaeont. afr.*, **2**, 1-187.

HUENE, F. VON 1950. Die Theriodontier des ostafrikanischer Ruhuhu-Gebietes in der Tübinger Sammlung. *Neues Jb. Geol. Palaeont.*, **92**, 47-136.

KEMP, T. S. 1969. On the functional morphology of the gorgonopsid skull. *Phil. Trans. R. Soc.*, B**256**, 1-83.

—— (In press.) Whaitsiid Therocephalia and the origin of cynodonts. *Phil. Trans. R. Soc.*

PARRINGTON, F. R. 1934. On the cynodont genus *Galesaurus*, with a note on the functional significance of the changes in the evolution of the theriodont skull. *Ann. Mag. nat. Hist.* (Ser. 10), **13**, 38-67.

—— 1955. On the cranial anatomy of some gorgonopsids and the synapsid middle ear. *Proc. zool. Soc. Lond.*, **125**, 1-40.

ROMER, A. S. and PRICE, L. W. 1940. Review of the Pelycosauria. *Spec. Pap. Geol. Soc. Am.*, **28**, 1-538.

STOCKLEY, G. M. 1932. The geology of the Ruhuhu coalfields, Tanganyika Territory. *Q.Jl. geol. Soc. Lond.*, **88**, 610-622.

TATARINOV, L. P. 1963. New late Permian therocephalian. *Palaeont. Zh.* (for 1963), 76-94.

—— 1964. Anatomy of the therocephalian head. *Palaeont. Zh.* (for 1964), 72-84.

WATSON, D. M. S. and ROMER, A. S. 1956. A classification of therapsid reptiles. *Bull. Mus. comp. Zool. Harv.*, **114**, 37-89.

A. W. CROMPTON

The evolution of the jaw articulation of cynodonts

INTRODUCTION

The progressive increase in the size of the dentary in the mammal-like reptiles, and especially in the cynodonts, and the accompanying decrease in the size of the bones forming the reptilian jaw joint (quadrate and articular) have been well documented and much discussed since these forms were first discovered during the last century. These changes preceded the establishment of a mammalian jaw joint between the squamosal and the dentary and the development of a three-boned mammalian middle ear. It has generally been assumed that the jaw articulation of cynodonts was formed solely by a condyle of the quadrate and a glenoid formed by the articular plus surangular (Crompton, 1963a and b). This assumption implies that in the stages immediately preceding the establishment of a contact between the dentary and the squamosal the bones forming the jaw articulation must have been very small in comparison with the remainder of the jaw (dentary). It is generally accepted that the dentary established contact with the squamosal lateral to the reptilian jaw joint and that, once established, the point of contact rapidly enlarged to form a condyle and allowed the reduced quadrate and articular to assume a sound conducting role. I (Crompton, 1963a and b) argued that there must have been some advantage to inserting all the jaw closing muscles on the dentary, but the enlargement of the dentary was accompanied by a reduction in the size of the bones forming the jaw joint and, therefore, the forces generated by the jaw closing muscles had to be alined so as to reduce the forces to which the weakened jaw joint was subjected during mastication. This is a fundamental feature of the jaw mechanics of modern mammals as well (Crompton and Hiiemäe, 1969).

Barghusen and Hopson (1970) have attempted to give a more complete explanation for the increase in the size of the dentary and the parallel and independent increase in the size of this bone in at least five cynodont lineages. They have claimed that once the adductor musculature of cynodonts was differentiated to form a mammalian pattern, e.g. in early Triassic cynodonts such as *Thrinaxodon*, contact between the dentary and the squamosal would appear to be a probable event. The size of the dentary relative to the remainder of the lower jaw in early cynodonts (procynosuchids) is approximately the same as that of earlier pristerognathid therocephalians, late Permian scaloposaurids, and early Triassic bauriamorphs. A remarkable, but seldom stressed, fact (Barghusen, 1968) is that in the Therocephalia and Bauriamorpha there was no progressive increase in the size of the dentary over and beyond that attained by the earliest known representatives; the dentary of the last of the known bauriamorphs, *Bauria*, is relatively no larger than that of a pristerognathid from the *Tapinocephalus*-zone. The reason for the conservative lower jaw in these forms was that they never developed a fully mammalian arrangement for the jaw closing muscles; a distinct masseter muscle was never evolved

(Barghusen, 1968). In cynodonts more advanced than *Thrinaxodon* there was a rapid decrease in the size of the postdentary bones. Barghusen and Hopson (1970) relate this in part to the transfer of the pterygoideus musculature from the postdentary bones to the dentary. They also argue that the great mass of jaw musculature attached to the dentary exerted a posterodorsally-directed thrust on the lower jaw and that this thrust created a tendency for the dentary to be forced upwards and backwards on the postdentary bones. This thrust also created a downwards and forwards reaction force at the jaw joint. In order to overcome this thrust and to compensate for the decrease in the size of the postdentary bones, they argued that the base of the coronoid process grew posteriorly over the diminishing postdentary bones and formed a brace. As the dentary part of the brace increased in size it established contact with the squamosal to establish a new articulation which "... would have added a buttress to aid the postdentary bones in resisting such a reaction force...." Although Barghusen and Hopson do not discuss it, their explanation assumes that before the contact between the squamosal and the dentary took place a strong primary reptilian articulation must have been retained. This would appear to be a necessary assumption because there is little point in strengthening the dentary-postdentary bone contact if at the same time the area of actual contact between the bones forming the jaw joint was diminishing and therefore weakening the jaw joint. However, it is generally assumed that the primary reptilian jaw joint was minute in the transitional forms. Romer (1970) has briefly discussed additional connections between the surangular and the squamosal in forms such as *Trirachodon* and *Massetognathus*. These connections, he has pointed out, help to strengthen the weak articular-quadrate articulation.

In advanced forms such as the tritylodontids the postero-ventral edge of the coronoid process supports two horizontal ridges. These were presumably originally established to strengthen the contact between the reduced postdentary bones and the coronoid process. However, in the tritylodontids only the medial ridge supported the postdentary bones. I (Crompton, 1963) suggested that these ridges helped to strengthen the coronoid process to resist the forces to which it was subjected by the increased mass of jaw musculature inserting on this region of the dentary. In earlier forms (e.g. *Thrinaxodon*) the base of the coronoid process was strengthened by the deep postdentary bones but in advanced forms these bones were too small to give the support needed and new horizontal ridges provided the necessary support. Consequently, the articular process of the dentary developed not only to strengthen the dentary-postdentary bone contact, but also to strengthen the coronoid process itself.

In our review of the origin of mammals (Hopson and Crompton, 1969) we concluded that the two principal groups of early mammals (i.e. therians and nontherians) were probably fairly closely related and were derived from forms closely allied to the early Galesauridae (= Thrinaxodontidae). Romer (1969a and 1970) has recently described a Middle Triassic cynodont, *Probainognathus jenseni*, in which a subsidiary jaw joint is present between a glenoid in the squamosal and the surangular plus the posterior end of the dentary. Despite this advanced feature, the remainder of the skull is not much advanced beyond typical members of the Galesauridae such as *Thrinaxodon*. Romer has claimed that there is nothing in the structural pattern of a form such as *Probainognathus* to debar it from a position directly ancestral to a primitive mammal. Barghusen and Hopson (1970) originally claimed that the absence of cingulum cusps in *Probainognathus* tended to remove it from this position, but recently Hopson (personal communication) has discovered cingulum cusps in *Probainognathus sp.*; their arrangement is similar to those of *Thrinaxodon*.

Previous work on the jaws of mammal-like reptiles (Crompton, 1963a and b; Barghusen, 1968; Barghusen and Hopson, 1970) only consider biting across the postcanines and do not consider either biting across the incisors or the effects of a single part of the jaw muscle mass contracting forcibly.

Both would have subjected the jaw joint to significant forces and it is difficult to envisage how a fragile jaw joint could have withstood large forces generated in this way.

The main purpose of this paper is to provide evidence which resolves the apparent paradox involved in the evolution of the mammalian jaw; a progressive increase in the strength of the lower jaw and an increase in the relative mass of jaw musculature coupled with an apparent decrease in the relative size and strength of the jaw joint. Detailed re-examination of this region has brought to light details which have not been reported previously and which are pertinent to the above discussion. In the following section the jaw articulations of several cynodonts will be described.

DESCRIPTION OF THE JAW JOINT OF SEVERAL CYNODONTS

Thrinaxodon (Plate I and Fig. 1)

It was decided to start this section with a description of *Thrinaxodon* for two reasons: (1) the Galesauridae (= Thrinaxodontidae) appear to be ancestral to several of the lines of advanced cynodonts, and (2) abundant, well-preserved, and beautifully prepared skulls of this genus are available for study. Details of the jaw articulation of *Thrinaxodon* are shown in Fig. 1. In posterior view (Fig. 1A) the quadrate (Q) is roughly 'L' shaped with the horizontal limb extending laterally. The vertical part of the 'L' is broad and its external edge is drawn out to form a distinctive flange (*pf*) which fits into a notch (*qn*) in the external edge of the squamosal (SQ). This notch is a feature unique to cynodonts and is not present in the therocephalians and their descendants, the bauriamorphs.[1] The remainder of the posterior surface of the quadrate (vertical part of the 'L') lies in a shallow pocket in the anterior surface of the squamosal. The horizontal part of the 'L' forms the articular surface or condyle and part of the ventral surface of the quadratojugal (Q-J) rests on the dorsal surface of the lateral extension of the quadrate. The quadratojugal has a short spur which extends up the external edge of the vertical limb of the quadrate.

External to the notch in the squamosal for the quadrate is a second notch (*q-jn*), which partially supports the dorsal tip of the quadratojugal. The internal border of this notch is formed by a narrow ventrally directed spur of the squamosal (*sq sp*). This spur separates the two notches and is in intimate contact with the posterior surface of the high ascending ramus of the quadratojugal. Only the dorsal third of the external border of the quadratojugal notch is actually in contact with the quadratojugal. Ventral and external to the point of contact with the quadratojugal the edge of the squamosal curves away from the quadratojugal. This feature has been illustrated before (Hopson, 1966, Fig. 4B) but its significance has not been commented on. An important feature of the quadratojugal is that its lower end extends laterally beyond the lateral termination of the quadrate (see Fig. 1A). This will be discussed below. Fig. 1B is an anterior view of the quadrate-quadratojugal complex. It illustrates the tongue and groove join between the external surface of the quadratojugal and the edge of the squamosal, and the wide gap or notch separating the squamosal and external surface of the quadratojugal below this point of contact. Between the quadratojugal and the posterior termination of the jugal (*J*) the ventral surface of the squamosal extends downward as a wide flange (*art fl*) which will be referred to as the articular flange of the squamosal. The postero-medial edge of this flange forms the external free border of the quadratojugal notch. The postero-ventral region of the articular flange approaches a distinct swelling or boss on the surangular which will be referred to as the articular boss of the surangular (*art b*). This is best seen in an external view (Fig. 1C) of the articulation. A groove in the ventral surface of the quadratojugal presumably accommodated a cartilaginous continuation of the

---

[1] Dr T. S. Kemp has drawn my attention to the presence of a similar quadrate notch in whaitsiids, which is described in his contribution to this volume.

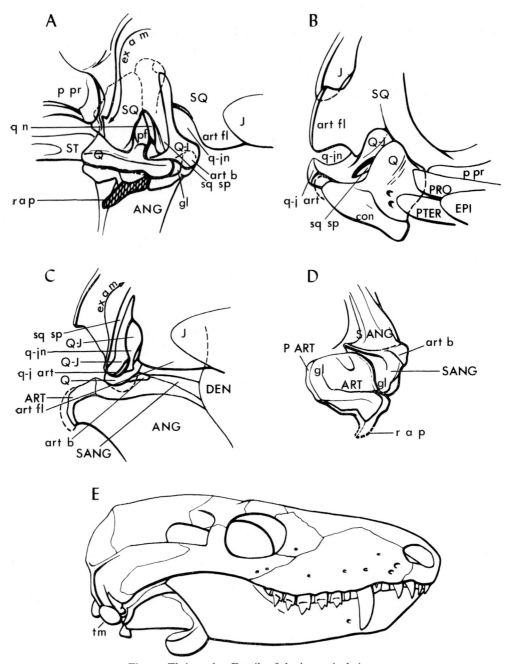

Fig. 1 *Thrinaxodon*. Details of the jaw articulation.
A, posterior, **B**, oblique ventral, and **C**, lateral views; **D**, posterior view of the postdentary bones; **E**, lateral view of the skull of *Thrinaxodon*. See page 251 for abbreviations used in this and all following figures.

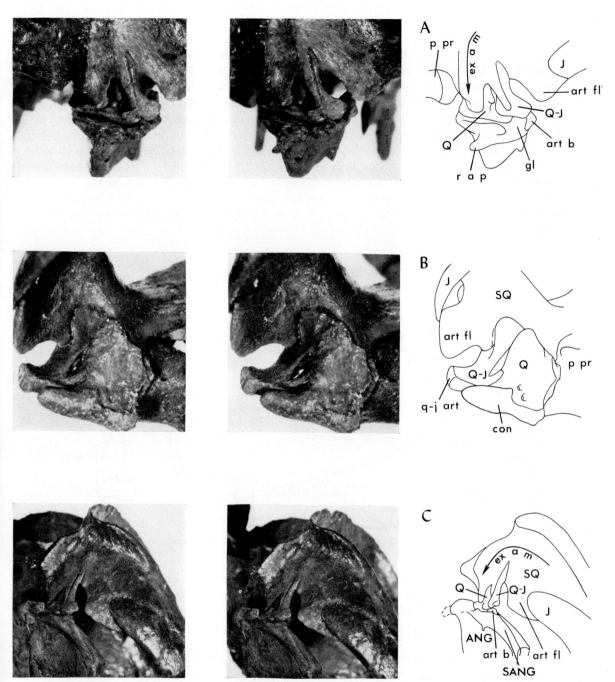

Plate I *Thrinaxodon*. **A**, posterior, **B**, antero-ventral, and **C**, lateral views of the jaw-joint.

quadrate but external to this groove, a small triangular area (*q-j art*). is present. This area closely approaches the ventral tip of the articular flange of the squamosal and lies directly behind the articular boss of the surangular (Fig. 1A and 1C) and will be referred to as the articular facet of the quadratojugal. The articular boss of the surangular is a small but distinct feature on the postero-external surface of the surangular (*SANG.*). It lies directly above the glenoid (Fig. 1A, *gl*). In posterior view (Fig. 1D) it is roughly triangular in shape and has a roughened surface indicating that in life either a tendon inserted there or that it was covered by cartilage. Both the articular (*ART*) and the surangular form the glenoid cavity. The articular boss is easily damaged and has been lost during preparation in most of the existing specimens of *Thrinaxodon*.

A consideration of more advanced cynodonts, and especially *Probainognathus* (Romer, 1969b and 1970), suggests that in *Thrinaxodon* cartilage and/or connective tissue joined the articular boss of the surangular, the articular facet of the quadratojugal, and the postero-ventral edge of the articular flange of the squamosal. These bony structures and interconnections appear to have strengthened the primitive reptilian joint between the articular plus surangular and the quadrate plus quadratojugal.

Therocephalians and Procynosuchids

(Figs 2 and 3)

An articular boss of the surangular and an articular flange were not present in therocephalians or bauriamorphs, but in other respects the structure of the jaw joint is similar to that described for *Thrinaxodon*. The external and posterior view of the lower jaw of a pristerognathid therocephalian is shown in Fig. 2.

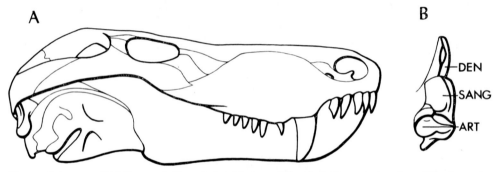

Fig. 2 Pristerognathid therocephalian. **A**, lateral view of the skull; **B**, posterior view of the lower jaw.

The dorsal surface of the glenoid and dentary contact is a smooth curve and is not interrupted by a small boss. This curve terminates in the distinct lip which forms the dorsal edge of the part of the glenoid formed by the surangular; a similar lip is present in *Thrinaxodon*. A wide gap separates the ventral edge of the squamosal from the posterior region of the surangular. In *Thrinaxodon* this gap is filled by the articular flange of the squamosal.

Details of the quadratojugal in therocephalians are unfortunately not available, but in *Bauria* (Crompton, 1955) the quadratojugal has a distinct flange extending laterally beyond the lateral termination of the quadrate. This flange may have had a ligamentous connection with the surangular.

Several specimens of South African Permian procynosuchids are present in the collections of South African museums and especially in the Bernard Price Institute for Palaeontological Research in

Johannesburg. However, in no one of these specimens is the jaw articulation either exposed or preserved to show the detail seen in the acid prepared specimens of *Thrinaxodon*. In Fig. 3 an attempt has been made to reconstruct the jaw articulation of *Leavachia*. This is based upon two specimens in the Bernard Price Institute (Nos. BPI 234 and 357) and a specimen in the South African Museum (No. K338).

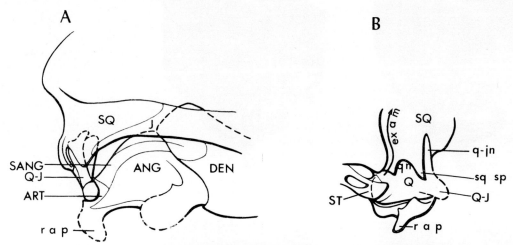

Fig. 3 *Leavachia sp.* **A**, lateral, and **B**, posterior views of the jaw articulation.

An articular boss of the surangular and articular flange of the squamosal also appear to be absent in *Leavachia*. In Fig. 3 the posterior and lateral views of the articular region of the procynosuchid *Leavachia* are illustrated. Shallow notches are present in the ventral edge of the squamosal for the quadrate and quadratojugal. However, less of the posterior surface of the quadrate is covered by the squamosal than in *Thrinaxodon*. The external surface of the dorsal tip of the quadratojugal is grasped by the squamosal but an articular flange of the squamosal is not present.

The external termination of a bulky stapes (*ST*) rests in a shallow pocket on the postero-medial side of the quadrate. The tympanic membrane was presumably supported by both the squamosal and the posterior surface of the quadrate. Tatarinov (1968) has described a lateral mandibular articulation between the surangular and the quadratojugal in the Permian cynodont *Dvinia*. It appears to be an articulation between a surangular boss and quadratojugal but, unfortunately, the details of the articulation are not clear. It does, however, indicate that strengthening of the jaw joint was initiated in Russian procynosuchids and ligamentous strengthening may have been present in South African procynosuchids although the osteological features associated with this strengthening were not developed in the latter.

Trirachodon (Plate II and Fig. 4)

Gomphodont cynodonts of the *Cynognathus*-zone fall into two distinct families: the Diademodontidae and the Trirachodontidae. Both appear to have been derived from the Galesauridae. The jaw articulation in *Trirachodon* is beautifully preserved in a specimen belonging to the South African Museum (No. K12168). The articular boss of the surangular is enlarged (Fig. 4) in comparison with *Thrinaxodon*. It can be seen in a posterior and lateral view that the boss has migrated ventrally so that its centre is more closely alined with the transverse axis of the glenoid than in the earlier cynodont *Thrinaxodon*.

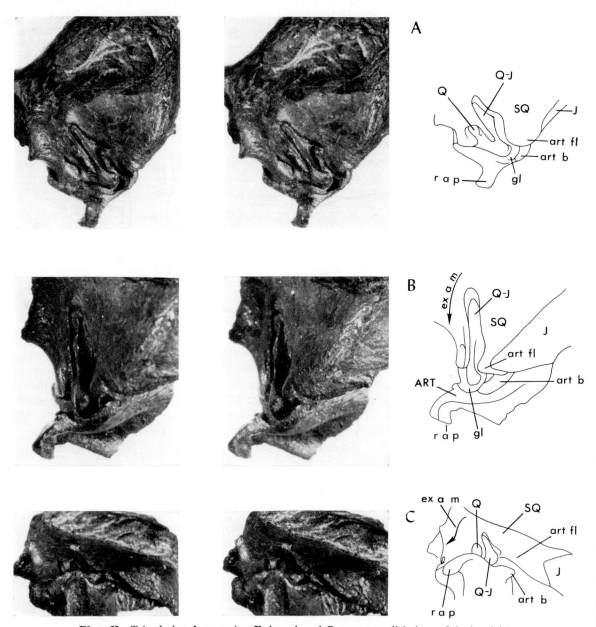

Plate II *Trirachodon*. **A**, posterior, **B**, lateral, and **C**, ventro-medial views of the jaw-joint.

The boss has a slightly curved surface; the medial part faces backwards and upwards whereas the lateral part faces backwards, upwards, and slightly outwards. In comparison with *Thrinaxodon*, the articular flange of the squamosal in *Trirachodon* has enlarged and is firmly buttressed by the jugal. It has grown downwards, backwards, and inwards so that it lies directly behind as well as lateral to the surangular boss (Fig. 4C and 4D). There is no doubt that in *Trirachodon* a subsidiary mandibular articulation was present between the squamosal and the surangular. Because the surangular boss lies closer to the articular glenoid than was the case in *Thrinaxodon*, and because the quadrate appears to have been relatively loosely held in the squamosal, contact between the surangular boss and the squamosal could be maintained when the jaw was opened. The quadratojugal contact with the surangular seen in *Thrinaxodon* has been replaced in *Trirachodon* by the articular flange of the squamosal which has extended medially behind the surangular to make contact with the surangular boss. Consequently, the lateral extension of the quadratojugal is absent in *Trirachodon* and other gomphodont cynodonts.

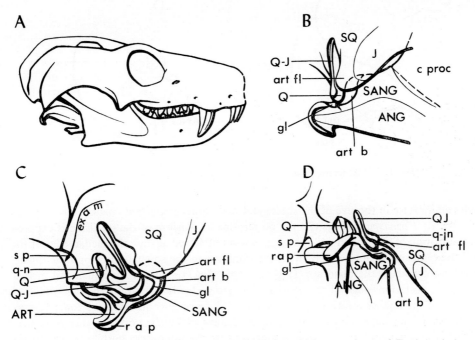

Fig. 4 *Trirachodon sp.* **A**, lateral view of the skull; **B**, lateral, **C**, posterior, and **D**, ventral views of the jaw joint.

As in *Thrinaxodon*, only a small fraction of the postero-internal edge of the squamosal grasps the antero-external region of the dorsal tip of the quadratojugal. The remainder of the edge is separated by a wide space from the quadratojugal.

Relative to the size of the dentary, the quadrate, quadratojugal, articular, and surangular are smaller in *Trirachodon* than in *Thrinaxodon*, but a well-developed squamosal-surangular articulation is present. Therefore, although the primitive reptilian joint is relatively weaker this was compensated for by strengthening the contact between the articular flange of the squamosal and the surangular.

The dorsal edge of the surangular when seen in lateral view has the shape of a wide inverted 'V'.

The dorsal surface of the anterior arm of the 'V' is broad and is closely applied to the ventral surface of the coronoid process of the dentary (*c proc*). Only a short distance, therefore, separates the posterior point of the dentary from the surangular boss.

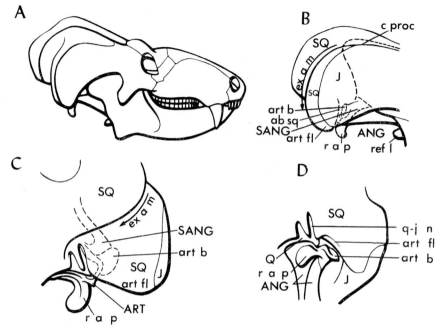

Fig. 5 *Diademodon sp.* **A**, lateral view of the skull; **B**, lateral, **C**, posterior, and **D**, ventral views of the jaw joint.

Besides an increase in the size of the surangular boss, other important changes have taken place in *Trirachodon*. In *Thrinaxodon* (Fig. 1A) the postero-medial surface of the quadrate is exposed below the ventral margin of the squamosal. The stapes was buttressed against the medial edge of the quadrate and the tympanic membrane lay in an emargination in the squamosal behind the quadrate and may not have contacted the quadrate (Hopson, 1966). In *Trirachodon* the squamosal medial to the groove for the posterior flange of the quadrate has grown further ventrally to reach the level of the ventral edge of the medial condyle of the quadrate. This prevents an extensive contact between the medial surface of the stapes and the quadrate. A shallow pit in the posterior surface of the ventral extension of the squamosal in *Trirachodon* suggests that the stapes was buttressed against the squamosal at this point. Unfortunately, the stapes was lost in this specimen. A contact between the stapes and the quadrate was present in earlier cynodonts and apparently for this reason it has been tacitly assumed that this contact was retained in more advanced forms and is homologous with the stapes-incus (= stapes-quadrate) contact of mammals. It has already been shown (Crompton, 1972) that in the Middle Triassic traversodonts and late Triassic tritylodontids the lower jaw was drawn backwards during the final stages of the power stroke of mastication. In tritylodontids and possibly in some traversodontids the movement involved was made possible not necessarily by a loose joint between the articular and the quadrate (as in dicynodonts) but by a loose connection between the quadrate plus quadratojugal and the squamosal. As the tympanic membrane appears to have been wholly supported by the squamosal in these forms, rather than partially by the quadrate as in earlier forms such as *Thrinaxodon*, the quadrate could move freely without directly affecting the tympanic membrane.

However, if the stapes contacted both the tympanic membrane and the quadrate, movement of the quadrate would involve the stapes and seriously affect sound conduction. A loss of contact between the stapes and the quadrate would be one way of solving this problem. *Trirachodon* and some of the Middle Triassic cynodonts chose this route.

Diademodon (Plate III and Fig. 5)

The structure of the jaw joint in *Diademodon* is essentially the same as that of *Trirachodon* (Fig. 5). I have described the articular complex of *Diademodon* (Crompton, 1963), but in the material upon which that description was based, the surangular was damaged. In a new specimen (U.S. Nat. Mus. No. 5622), in which the lower jaw is preserved *in situ*, a large articular boss is present. Fig. 5B-D are reconstructions of the articulation. A prominent feature of *Diademodon* is the massive articular flange of the squamosal, i.e., the ventral part of this bone lying between the notch for the quadratojugal and the postero-ventral part of the jugal. The antero-medial face of this flange (*ab sq*, Fig. 5B and 5D) articulates directly with the large articular boss of the surangular. This flange and the adjacent portion of the jugal are so massive in *Diademodon* that the articulation is almost completely hidden in lateral view (Fig. 5A). The primitive jaw articulation of *Diademoden* is smaller relative to the skull size than in either *Trirachodon* or *Thrinaxodon*, but this was more than compensated for by the large articular boss-squamosal contact.

Cynognathus (Fig. 6)

The articular complex of *Cynognathus* (Fig. 6A) is essentially the same as that of *Diademodon*. A large articular boss is present on the surangular and the articular flange of the squamosal is powerfully developed and solidly buttressed by the jugal. Fig. 6 is partially based upon a cast of the lower jaw of

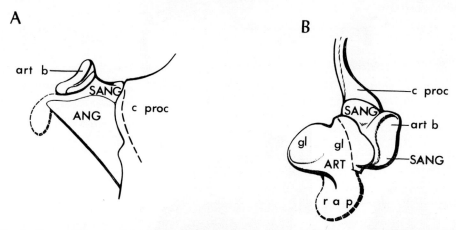

Fig. 6 *Cynognathus sp.* **A** and **B**, lateral and posterior views of the lower jaw component of the jaw joint.

Cynognathus belonging to the Department of Paleontology, University of California at Berkeley (N. 4906). In this specimen the articular complex has been freed from the dentary. A distinct promontory or buttress is present on the dorsal surface of the surangular where this bone contacts the dentary. This feature marks the posterior extent of the dentary. In *Thrinaxodon* a shallow groove in the dorsal edge of the surangular marks the posterior point of the coronoid process of the dentary

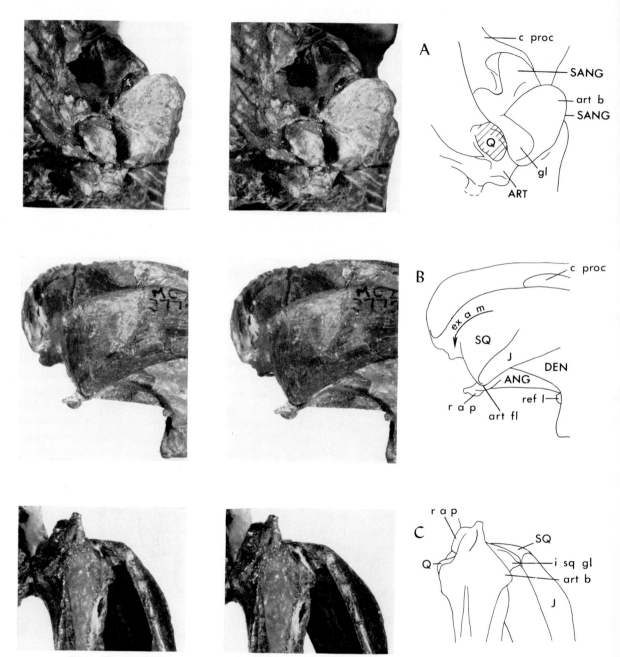

Plate III **A**, posterior view of the postdentary bones of *Diademodon*; **B**, lateral, and **C**, ventromedial view of the jaw-joint of *Probelesodon*.

but in *Trirachodon*, *Diademodon*, and *Cynognathus* a distinct buttress was developed to contact the wide ventral edge of the coronoid process. In crushed skulls the dentary and the surangular are often separated and the dentary forced backwards relative to the surangular. The buttress is important because it indicates the true posterior extent of the dentary contact with the surangular.

Massetognathus and Scalenodon (Fig. 7)

The best preserved and most plentiful material of Middle Triassic gomphodont cynodonts is the South American Chañares genus *Massetognathus* described by Romer (1967). In this genus (Fig. 7) a well

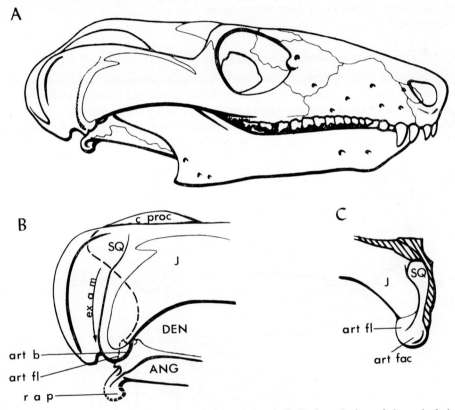

Fig. 7 *Massetognathus pascuali*. **A**, lateral view of the skull; **B**, lateral view of the articulation; **C**, internal view of the zygomatic arch.

developed articular boss on the surangular is present. The distance between the articular boss and the buttress to support the back end of the dentary was reduced in comparison with *Diademodon* and *Trirachodon* so that the posterior point of the dentary closely approached the subsidiary articulation between the surangular and the articular flange. In contrast to *Diademodon* and *Trirachodon* a distinct facet or incipient glenoid is present on the anterior and internal surface of the articular flange of the squamosal (Fig. 7C) indicating the point of articulation with the surangular. In *Diademodon* the articular flange of the squamosal is large and the articular boss of the surangular faced almost directly backwards so that the articular surface did not spread on to the external surface of the boss as it does in the more lightly built forms such as *Massetognathus* and *Trirachodon*.

The surangular-articular complex is not preserved in any of the available specimens of East African traversodonts but in a new skull of *Scalenodon sp.* the ventral edge of the articular flange of the squamosal is bulbous and appears to have formed part of a subsidiary jaw articulation. When the jaw was drawn forwards in traversodonts at the beginning of the masticatory cycle (Crompton, 1972) contact between the back of the articular boss and the squamosal was presumably lost, but maintained between the squamosal and the external surface of the articular boss. The quadrate and quadratojugal were presumably drawn forward out of their loose housing in the squamosal. During the power stroke at the end of the masticatory cycle the lower jaw was drawn backwards; the quadratojugal-quadrate complex rammed home and the surangular boss was forced against the squamosal. Details of the jaw articulation of the large traversodonts such as *Exaeretodon* cannot be obtained from existing literature.

Probelesodon and Probainognathus (Plates III and IV and Figs 8 and 9)

Romer (1969a and b, 1970) has placed the South American carnivorous cynodonts in the family Chiniquodontidae and has recognized four genera: *Chiniquodon, Belesodon, Probelesodon,* and *Probainognathus*. The latter two genera are based upon several well-preserved skulls.

In *Probelesodon* a small articular boss of the surangular is present (Fig. 8). It is directed posterolaterally towards an incipient glenoid in the articular flange of the squamosal. Only a short distance

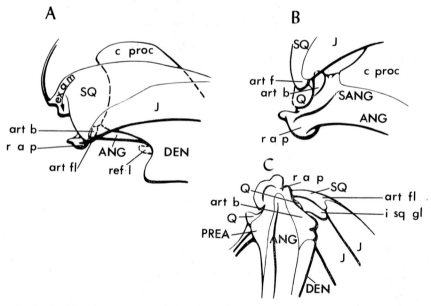

Fig. 8 *Probelesodon lewisi.* **A**, lateral, **B**, oblique, and **C**, ventral views of the articulation.

separates the posterior tip of the dentary from the articular boss of the surangular. The jaw articulation is not unlike that of *Massetognathus* although there are minor differences such as in the shape of the jugal-squamosal suture. The articular surface of the quadrate faces downwards and slightly forwards when the jaw is firmly closed and in this aspect *Probelesodon* resembles *Thrinaxodon*.

The jaw articulation of *Probainognathus* is considerably advanced beyond that of any of the cynodonts considered so far. Romer (1969b and 1970) has pointed out that an incipient condyle or glenoid

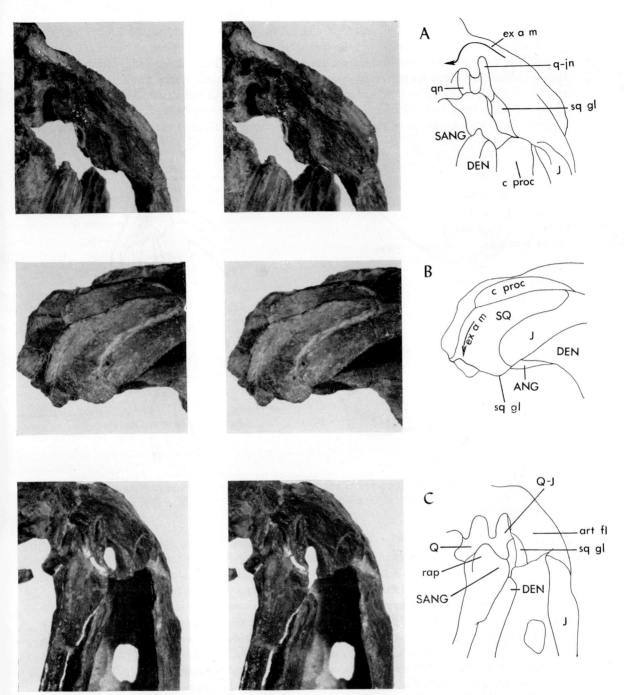

Plate IV *Probainognathus*. **A**, ventral view of articular region (the postdentary bones and quadrate-quadratojugal complex all missing); **B**, lateral view of the same specimen; **C**, ventral view of skull in which the accessory jaw bones and quadrate-quadratojugal complex are in place.

fossa is present in the squamosal of this form and has suggested that it articulated directly with the posterior end of the dentary and part of the surangular.

Details of the jaw articulation in this form are given in Fig. 9. These differ slightly from Romer's because, although the jaws are preserved *in situ* in many of the specimens, they are all slightly crushed and it is difficult to reconstruct the jaw articulation with too much confidence. The glenoid (*sq gl*) in this form is formed on the internal and ventral surface of a massive buttress or boss on the inner surface of the articular flange of the squamosal immediately in front of the shallow notch for the

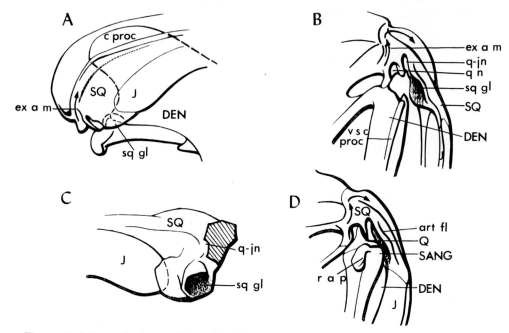

Fig. 9 *Probainognathus jenseni*. **A**, lateral, and **D**, ventral reconstructions of the jaw joint; **B**, ventral view of specimen No. 1964-XI-14-7 from which the postdentary bones, quadratojugal, and quadrate have been lost; **C**, medial view of the zygomatic arch to show the buttress supporting the glenoid in the squamosal.

quadratojugal (Fig. 9B and 9C). The glenoid lies laterally to the articulation between the articular and the quadrate (Fig. 9D). The articular flange, when seen in lateral view (Fig. 9A), has a long exposure separating the jugal from the quadratojugal notch. This is in sharp distinction from all other cynodonts above the level of *Thrinaxodon*. In these forms only a small amount of the articular flange is visible in lateral view but it is more extensive in posterior view. In *Probainognathus* the quadrate-quadratojugal complex is directed forwards and downwards and the height of the squamosal immediately above these bones is considerably reduced in comparison with all other cynodonts considered in this paper. This is a mammalian feature. The notches for the quadrate and quadratojugal are shallow. A clear articular boss of the surangular appears to be absent and if present, could not have been large. A buttress is present on the dorsal surface of the surangular where this bone contacts the postero-ventral tip of the dentary. An attempt has been made to reconstruct the articulation in Fig. 9A. It supports Romer's view that both the dorsal surface of the surangular and the posterior tip of the dentary articulated with a glenoid in the articular flange of the squamosal. From a form such as this the dentary contact could increase to develop a dentary condyle whereas the elements forming the primitive reptilian jaw articulation could become reduced in size and be incorporated into the middle ear.

Although *Probainognathus* has been placed in the Chiniquodontidae there are differences other than the nature of the jaw articulation between this form and the other three genera which Romer has placed in this family. For example, the secondary palate is shorter and the teeth do not have the prominently recurved cusps which characterize *Probelesodon*. The postcanines of *Probainognathus* are almost identical to those of *Thrinaxodon* not only in outline but also in that cingulum cusps of the typical *Thrinaxodon* pattern are present, though much less prominently developed (Hopson, personal communication). *Probainognathus* appears to be closely related to *Thrinaxodon*, on the one hand, and perhaps to *Eozostrodon* (= *Morganucodon*), on the other.

Occlusal planes between the postcanines are absent in *Thrinaxodon*, but present in *Eozostrodon* (Crompton and Jenkins, 1968). These planes are present in *Probainognathus* and, as in *Eozostrodon*, the cingulum cusps are obliterated during an early stage of wear, for which reason they were not present in the specimens described by Romer (1969b and 1970). *Probelesodon* also appears to be related to the Galesauridae (= Thrinaxodontidae) but formed the subsidiary jaw articulation by establishing a connection between the articular boss on the surangular and a heavily buttressed articular flange of the squamosal, rather than by forming a large glenoid facet in the ventro-medial surface of the articular flange.

Ictidosaurs (Fig. 10)

This discussion of the jaw articulation of cynodonts would not be complete without mentioning the ictidosaurs *Diarthrognathus* and *Pachygenelus* (Crompton, 1963a and b; Hopson and Crompton, 1969).

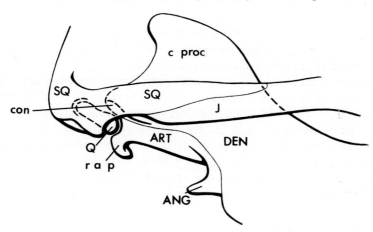

Fig. 10 *Diarthrognathus broomi*. Reconstruction of the jaw articulation.

These forms appear to have arisen from a primitive cynodont, perhaps even more primitive than *Thrinaxodon*. The squamosal of these forms is markedly reduced in height and distinct notches for the quadrate and quadratojugal on the postero-ventral margin of the squamosal are missing. The articular flange of the squamosal (Fig. 10) has a long lateral exposure and the ventral surface of the squamosal in this region is concave. The lateral ridge of the dentary articulates with the concave ventral surface of the squamosal lateral to the primitive reptilian articulation between the quadrate and the articular, but a distinct squamosal glenoid was apparently not present. A small, but distinct, condyle (*con*) is present on the dentary of *Diarthrognathus* but this may not have been present in *Pachygenelus*, although the lateral ridge was well developed in this genus. The structure of the articulation and the nature of the postcanine teeth of the ictidosaurs, the reduction in the number of the incisors, and the retention of an interpterygoid vacuity set the ictidosaurs apart from other cynodont lineages and although they have developed a mammalian jaw articulation, they do not appear to be ancestral to any known Jurassic mammals.

SUMMARY AND DISCUSSION

The jaw articulation of several cynodonts has been described in this paper. It has been concluded that in the earliest cynodonts, the procynosuchids, the articulation was strengthened by ligamentous connections between the surangular and the quadratojugal. In the more advanced Galesauridae, e.g. *Thrinaxodon*, osteological features unique to the cynodonts were developed to strengthen the jaw joint. These were: an articular flange of the squamosal, an articular boss on the surangular, and an articular facet on the quadratojugal. In *Thrinaxodon* these three structures did not come into direct contact but were apparently joined by ligaments. However, in more advanced forms the articular flange of the squamosal and the surangular boss increased in size and a clear subsidiary articulation was established between them lateral to, and in addition to, the primitive cynodont articulation between a condyle formed by both the quadratojugal and the quadrate and a glenoid formed by both the surangular and articular.

An increase in the size of the dentary resulted in a relative decrease in the size of the postdentary bones and it is usually assumed that this also implied a weakening of the primitive reptilian jaw articulation. Changes in the orientation of the jaw muscles could have reduced the magnitude of the forces acting through the jaw joint but a threshold must have been reached beyond which the postdentary bones could be reduced no further without endangering the normal functioning of the jaw joint. Cynodonts were apparently unique among the therapsids in developing subsidiary strengthening of the jaw joint (initially ligamentous and later by a subsidiary jaw joint). This compensated for the reduction in size of the elements forming the primitive reptilian jaw joint. It is generally assumed that as the jaw joint decreased in size in advanced cynodonts, the posterior tip of the dentary gradually grew backwards until it established contact with the squamosal. At that point in time, it has been assumed, the jaw articulation was fragile; i.e., the quadrate and articular were extremely small and the squamosal-dentary contact minute. The squamosal-dentary articulation was therefore looked at as a new articulation which only developed in very advanced cynodonts in which the postdentary bones and quadrate were greatly reduced in size. Once established it was thought that the dentary-squamosal contact would rapidly increase in size to form a dentary condyle and a distinct glenoid in the squamosal. Therefore, in the shift from a 'reptile' to a 'mammal' the nature of the jaw joint changed substantially and involved a rapid strengthening of the jaw joint in the early history of 'mammals'.

The new work on the cynodont jaw reported in this paper has shown that this view is essentially incorrect and that the shift from one type of jaw articulation to the other was gradual and did not involve a weakening and subsequent strengthening of the jaw joint. In advanced cynodonts the primitive jaw articulation (quadrate-articular) and the subsidiary articulation (surangular-squamosal) together formed a substantial articulation. In functional terms the surangular-squamosal articulation was the predecessor of the mammalian dentary-squamosal articulation and its formation was initiated early in cynodont evolution. As the dentary grew backwards, it invaded an existing joint between the surangular and the squamosal; the characteristic feature of mammals, namely the dentary-squamosal articulation, was therefore more a modification of an existing articulation than the establishment of a new articulation. As the dentary invaded the articulation to form a condyle, the surangular part of the articulation, together with the other postdentary bones, could gradually continue to decrease in size.

Barghusen and Hopson (1970) have argued that the articular process of the dentary developed to form a brace to withstand forces acting through the jaw joint. It is now clear that the subsidiary jaw articulation also developed to withstand these forces.

The gradual acquisition of the mammalian jaw articulation and the realization that its functional counterpart was present in early cynodonts supports the view (Hopson and Crompton, 1969; Romer, 1970; and Barghusen and Hopson, 1970) that the reptile-mammal boundary should be based not upon a single character, but upon a suite of characters.

The attainment of a mammalian-type arrangement of jaw muscles in early cynodonts was not the only factor involved in the parallel enlargement of the dentary in diverse cynodont lineages; strengthening of the jaw joint and subsequent establishment of a subsidiary jaw articulation was also an important consideration.

In mammals the muscle forces can be arranged so that the jaw articulation was not load-bearing when the molars were involved in active occlusion (Crompton and Hiiemäe, 1969), but Hiiemäe (1971) has shown that in the rat the jaw joint is load-bearing when the incisors are used. This is also probably true for other mammals and advanced mammal-like reptiles. This fact would have prevented a decrease in the size of the jaw joint below a certain minimum and partially accounts for the development of a subsidiary articulation in cynodonts.

The pattern of development of the subsidiary jaw articulation is not identical in all the cynodont lineages. Two patterns can be recognized. The most important is that found in the line leading from *Thrinaxodon* to *Probainognathus*. In *Thrinaxodon* the jaw joint is apparently strengthened by a connection (not necessarily a direct articulation) between an articular boss on the surangular and the quadratojugal, and one between the surangular and the articular flange of the squamosal. The latter lies above and slightly external to the surangular boss. All that appears to have happened in the transition between *Thrinaxodon* and *Probainognathus* is that the contact between the quadratojugal and the surangular was lost and that the articular flange grew downwards to establish a clear articulation with the surangular. The ventral edge of the articular flange was thickened to form a glenoid fossa. As the dentary increased in size it migrated backwards along the dorsal border of the surangular and invaded the existing articulation between the squamosal and the surangular. Once the new dentary squamosal articulation was established, the quadratojugal-quadrate complex and accessory jaw bones could decrease in size. Although *Probainognathus* is more advanced than *Thrinaxodon* in terms of jaw articulation, the remainder of the skull is not much advanced beyond *Thrinaxodon*. This includes the structure of the postcanine teeth which have a pattern of cingular cusps very similar to that of *Thrinaxodon*. A close relationship between *Thrinaxodon* and *Eozostrodon* has been suggested (Crompton and Jenkins, 1968; Hopson and Crompton, 1969) and *Probainognathus* therefore seems to make an ideal intermediate between *Thrinaxodon* and *Eozostrodon*. This is essentially in agreement with Romer's view (1969b and 1970). Kermack (1967), however, has argued that the structure of the lateral wall of the braincase in *Morganucodon* (= *Eozostrodon*) precludes the cynodonts from being ancestral to this group of early mammals.

The other cynodont lineages; galesaurids→cynognathids, galesaurids→chiniquodontids, galesaurids→diademodontids→traversodontids→tritylodontids, galesaurids→trirachodontids, and a primitive cynodont→diarthrognathids have developed subsidiary jaw articulations in a different way to that of *Probainognathus*. *Cynognathus, Probelesodon, Diademodon*, the early traversodonts (*Scalenodon* and *Massetognathus*) and *Trirachodon* have lost the surangular-quadratojugal contact, strengthened the articular boss of the surangular, and migrated the articular flange of the squamosal posteriorly and inwards so that it lies externally to and behind the articular boss of the surangular, rather than only lateral to it as in *Probainognathus*. In these lines a decrease in the size of the quadrate-quadratojugal-articular complex was compensated for by the development of this strong subsidiary articulation. In advanced members of these lines the distance separating the articular boss from the buttress for the posterior end of the dentary decreased progressively in size so that the posterior tip of the dentary progressively became closer to the subsidiary jaw articulation. Theoretically it is possible that, as in the case of *Probainognathus*, the dentary could have invaded the articulation between the surangular and articular flange of the squamosal. This may have happened in the terminal members of the traversodonts, the tritylodontids. It has already been shown (Crompton, 1963a and b; Barghusen and Hopson, 1970) that the decrease in the size of the postdentary bones in gomphodont cynodonts was accompanied by the development of a broad base to the coronoid process where the latter contacts the dorsal surface of the surangular. In cross-section the postero-ventral tip of the coronoid process has the form of an inverted T-shaped girder. The horizontal arms of the 'T' have been referred to as the lateral and medial ridges. In *Tritylodon* these ridges are extensive but only the medial ridge supports the accessory jaw bones. In *Diarthrognathus* the lateral ridge actually contacts the hollow ventral surface of the squamosal laterally to the primitive reptilian articulation. It is possible that the same may have happened in the advanced tritylodontid, *Tritylodontoideus* (Fourie, 1968). In this case the dentary-squamosal articulation would have developed laterally to the articulation between the surangular boss and the squamosal and might therefore be thought of as a new articulation rather than a modification of an existing articulation. Additional material is required to resolve this point.

ACKNOWLEDGEMENTS

This work was made possible by a grant from the National Science Foundation (GB-4435). I am deeply indebted to Professor A. S. Romer for his permission to study the new South American cynodonts and for reading the manuscript. I am extremely grateful to Drs J. A. Hopson and H. R. Barghusen for their careful reading of the original manuscript and for their useful comments. Most of the specimens illustrated were prepared by Mr C. Schaff of Yale and Mr A. Lewis of Harvard. The illustrations were prepared by Miss M. Estey, the stereo photographs were prepared by Mr A. Coleman of Yale, and several drafts of the manuscript were patiently typed by Mrs G. Dundon.

REFERENCES

BARGHUSEN, H. R. 1968. The lower jaw of cynodonts (Reptilia, Therapsida) and the evolutionary origin of mammalian adductor jaw musculature. *Postilla*, **116**, 1-49.

BARGHUSEN, H. R. and HOPSON, J. A. 1970. Dentary-squamosal joint and the origin of mammals. *Science*, **168**, 573-575.

CROMPTON, A. W. 1955. A revision of the Scaloposauridae with special reference to kinetism in this family. *Navors. nas. Mus., Bloemfontein*, **1**, 149-183.

—— 1963a. On the lower jaw of *Diarthrognathus* and the origin of the mammalian lower jaw. *Proc. zool. Soc. Lond.*, **140**, 697-750.

—— 1963b. The evolution of the mammalian jaw. *Evolution*, **17**, 431-439.

—— 1972. Postcanine occlusion in cynodonts and tritylodontids. *Bull. Br. Mus. (Geol.).* In press.

CROMPTON, A. W. and HIIEMÄE, K. 1969. How mammalian molar teeth work. *Discovery (Magazine of the Peabody Museum of Natural History, Yale University)*, **5**, 23-34.

CROMPTON, A. W. and JENKINS, F. A. JR. 1968. Molar occlusion in late Triassic mammals. *Biol. Rev.*, **43**, 427-458.

HIIEMÄE K. 1971. The structure and function of the jaw muscles in the rat (*Rattus norvegicus* L.). III. The mechanics of the muscles. *J. Linn. Soc. (Zool.)*, **50**, 111-132.

FOURIE, S. 1968. The jaw articulation of *Tritylodontoideus maximus*. *S. Afr. J. Sci.*, **64**, 255-265.

Hopson, J. A. 1966. The origin of the mammalian middle ear. *Amer. Zool.*, **6**, 437-450.
Hopson, J. A. and Crompton, A. W. 1969. Origin of mammals. In Dobzhansky, Th. *et al.* (Eds), *Evolutionary biology*, Appleton-Century-Crofts, New York, **3**, 15-72.
Kermack, K. A. 1967. The interrelations of early mammals. *J. Linn. Soc. (Zool.)*, **47**, 241-249.
Parrington, F. R. 1946. On the cranial anatomy of cynodonts. *Proc. zool. Soc. Lond.*, **116**, 181-197.
Romer, A. S. 1967. The Chañares (Argentina) Triassic reptile fauna. III. Two new gomphodonts, *Massetognathus pascuali* and *M. teruggii*. *Breviora (Mus. Comp. Zool. Harvard Univ.)*, **264**, 1-25.
—— 1969a. The Chañares (Argentina) Triassic reptile fauna V. The new Chiniquodontid cynodont, *Probelesodon lewisi*—cynodont ancestry. *Breviora (Mus. Comp. Zool. Harvard Univ.)*, **333**, 1-24.
—— 1969b. Cynodont reptile with incipient mammalian jaw articulation. *Science*, **166**, 881-882.
—— 1970. The Chañares (Argentina) Triassic reptile fauna. VI. A Chiniquodontid cynodont with an incipient squamosal-dentary jaw articulation. *Breviora (Mus. Comp. Zool. Harvard Univ.)*, **344**, 1-18.
Tatarinov, L. P. 1968. Morphology and systematics of the northern Dvina cynodonts (Reptilia; Therapsida; Upper Permian). *Postilla*, **126**, 1-51.

ABBREVIATIONS USED IN FIGS 1-10

ab sq	anterior border of the squamosal	*PTER*	pterygoid
ANG	angular	*Q*	quadrate
ART	articular	*Q-J*	quadratojugal
art b	articular boss of the surangular	*q-j art*	q-j articular facet on the quadratojugal
art fac	articular facet on the squamosal which articulates with the surangular	*q n*	quadrate notch
		q-jn	quadratojugal notch
art fl	articular flange of the squamosal	*r a p*	retroarticular process
con	condyle	*ref l*	reflected lamina of the angular
c proc	coronoid process	*SANG*	surangular
DEN	dentary	*s p*	shallow pocket for the distal end of the stapes
EPI	epipterygoid		
ex a m	groove for the external auditory meatus	*SQ*	squamosal
gl	glenoid	*sq gl*	glenoid in the squamosal for articulation with the surangular plus dentary
i sq gl	incipient glenoid in the squamosal		
J	jugal	*sq sp*	spur of the squamosal separating the notch for the quadratojugal and quadrate
pf	posterior flange of the quadrate		
p pr	paroccipital process	*ST*	stapes
PREA	prearticular (also *PART*)	*tm*	tympanic membrane
PRO	prootic		

P. M. BUTLER

The problem of insectivore classification

How to classify the insectivores is perhaps the most controversial problem of mammalian taxonomy. The earlier history of the problem has been thoroughly reviewed by Gregory (1910), in his classic work on the Orders of Mammals, and it is only necessary to give a brief summary here by way of introduction.

Hedgehogs, moles and shrews were recognised as three genera by Linnaeus (1758), who included them in his order Bestiae, together with pigs, armadilloes and opossums. This unnatural group, characterised mainly by the elongation of the snout, was broken up by Linnaeus' successors, but the three genera of insectivores were kept together. However, because of their plantigrade feet they were put with the bears and other carnivores into an order Plantigrades by Geoffroy St Hilaire and Cuvier (1795).

It was not until 1811 that the insectivores were separated off in a group on their own. This was done by Illiger, who created for them a family Subterranea, in his order Faculata. The term 'insectivore' was first used in a classification by de Blainville (1816), and Cuvier (1817) recognised "Les Insectivores" as a group, latinised as Insectivora by Bowdich (1821). By this time several new genera had been added: *Desmana*, the American moles *Scalopus* and *Condylura*, the Madagascan *Tenrec* and *Setifer* and the African golden mole *Chrysochloris*.

Between 1822 and 1848 there were described two genera of tree-shrews (*Tupaia*, *Ptilocercus*) from S.E. Asia and three genera of elephant-shrews (*Macroscelides*, *Petrodromus*, *Rhynchocyon*) from Africa. Wagner (1855) included these in the Insectivora, and also *Cynocephalus* (= *Galeopithecus*), a peculiar gliding mammal that had previously been placed near the lemurs or the bats. The inclusion of these forms considerably extended the concept of an insectivore, and it was soon realised that they were not very much like the hedgehogs, moles and shrews that constituted the order in 1821. Peters (1864), for instance, noticed that they had a caecum on the gut, lacking in what might be called the traditional insectivores, and Haeckel (1866) created the suborders Menotyphla, for the forms with a caecum, and Lipotyphla, for those without a caecum. Gill (1872) divided off *Cynocephalus* into a separate suborder, Dermoptera. Following Leche (1885) the Dermoptera have generally been regarded as a separate order, but the remaining Menotyphla have usually been left in the Insectivora.

In 1880 Thomas Huxley applied the idea of evolution to mammalian classification. He regarded the Insectivora as occupying a "central position", because they had retained the ancestral characters of the other orders of eutherian mammals. The idea that all the Eutheria have been derived from the order Insectivora has become widely accepted, but it has had some consequences which have not been altogether good. One of these is the presumption that a character found in the Insectivora is primitive, unless there is good evidence to the contrary. Many undoubtedly primitive characters do survive in various insectivores, such as abdominal testes, small unconvoluted cerebral hemispheres, ring-shaped tympanic bone, the clavicle, plantigrade pentadactyl feet, paired anterior vena cava, and traces of a cloaca in the female. None of these is universal in the Insectivora, and they all occur in other orders, some of which retain primitive characters not found in the Insectivora, e.g. the septomaxilla of edentates. Characters which are clearly not primitive are widespread in the Insectivora, particularly in the Lipotyphla. The often repeated statement that the palatal perforations of the hedgehog are a primitive character shared with marsupials is quite erroneous.

The Insectivora are in fact more diverse anatomically than the members of any other order, so much so that it is impossible to define the order except by exclusion of the specialisations which distinguish the other orders: they do not have rodent incisors, or hoofs, or prehensile feet, or wings, and so on. All we can say is that insectivores are eutherians which do not belong to any of the more clearly defined orders.

Another consequence of Huxley's dictum is that, as the ancestral eutherians were insectivores, any early fossil eutherian not clearly related to one of the other orders is classifiable in the order Insectivora. Palaeontological exploration during the latter part of the last century brought to light a great variety of early Tertiary forms. New orders were created for the larger and more striking of these—Amblypoda, Taeniodonta, Tillodonta—and out of the numerous smaller and less distinctive types there gradually crystallised the concepts of the creodonts, or primitive carnivores, and the Condylarthra, or primitive ungulates. Other forms were recognised as belonging to existing orders, such as primates and rodents. The rest were left in the Insectivora. When Cretaceous eutherians became recognised they were put into the Insectivora almost without question. Thus the fossil Insectivora form an even more diversified assemblage than the living Insectivora.

Simpson (1937, p. 106), in a discussion of the insectivore concept from a palaeontological standpoint, considered that the order contained four different sorts of lesser groups: (1) very primitive placentals whose relationship to later groups is not now recognisable; (2) animals in or near the ancestry of later more specialised insectivores; (3) animals less closely related to later groups, but sharing specialisations with them that seem to indicate that they arose from a common stock with the later groups; (4) derivatives of the basal placental stock that "had begun to diverge markedly from any other groups, without, however, having a sufficiently long history, being sufficiently important faunal elements, or acquiring sufficiently striking special characters to warrant the erection for them of a special order". The effect of this policy has been to use the Insectivora as a waste-basket for tidying up the classification of mammals.

Tidiness is an essential quality of a taxonomist, and there is something to be said for waste-baskets. Groups like the Ganoidei among fishes, the Anapsida among reptiles and the Orthoptera and Neuroptera among insects have served a useful purpose in the past, though with the advance of knowledge they have been superseded by more complex arrangements. Simple classifications are easier to remember and so are more popular with students, and therefore with their textbooks, but where phylogeny has been complex, taxonomy has also to be complex if it is to be realistic. Waste-basket groups like the Insectivora, convenient though they may be at a certain level of knowledge, can be no more than temporary expedients: sooner or later they will have to be sorted out.

The process of sorting has in fact been going on for a very long time, but no generally acceptable solution has yet been reached. Apart from the fact that only a small number of people have interested themselves in the problem, the main reason for the slowness of progress has been the lack of information about the small mammals of the late Cretaceous and early Tertiary, when the eutherian radiation was going on. Most of the genera are known only from teeth and jaws, in only a few cases has the skull been adequately described, and identifiable remains of the postcranial skeleton are very rare indeed.

Recent years have seen a revival of interest in the early placentals, particularly in the United States, and the continuing collection of new material, together with a closer examination and restudy of older museum collections, gives reason to believe that the sequence of phyletic branching of the early placentals, and therefore their classification, will be cleared up in the foreseeable future. It is impossible in the space available to review adequately the work that has been done on this problem: I will content myself with commenting on some of the suggestions that have been made.

Gregory (1910) divided the Insectivora into two orders, Lipotyphla and Menotyphla, corresponding to the suborders distinguished by Haeckel. Gregory thought the Lipotyphla were related to the Carnivora and the Menotyphla to the Primates. There is much to be said for raising the Lipotyphla to ordinal status. They have many common features, including some that are not primitive: for example the absence of the caecum, the reduction of the jugal, the expansion of the maxilla in the orbital wall displacing the palatine, the possession of a mobile proboscis moved by a series of muscles which influence the form of the skull, the reduced pubic symphysis and the haemochorial placenta. They have every appearance of being a natural group descended from a common ancestor, probably in the Palaeocene. The group contains nearly all the living insectivores, comprising 61 genera, making them comparable with the Primates (59 genera), and exceeded only by the Rodentia, Carnivora, Artiodactyla and Chiroptera. The Lipotyphla corresponds to the order Insectivora as understood by Bowdich in 1821, and a case might be made for restricting the term Insectivora to the Lipotyphla alone; however, as Insectivora has been used in a wider sense for over a century this would probably be unacceptable.

Not everyone has admitted the unity of the Lipotyphla. Broom (1916) proposed an order Chrysochlorida for the golden moles on the basis of a number of distinctive features of the skull and of Jacobson's organ. They are highly specialised for burrowing in a manner different from that of the Talpidae, and the modifications of the middle part of the skull are probably due to this; other specialised skull characters are shared with Soricidae and Tenrecidae (Butler, 1956). There is no evidence in favour of a special relationship between the Chysochloridae and the Tenrecidae: these families can be traced back only to the Miocene, but at that time they seem to have been as different as they are today (Butler, 1969). The main feature that these two African families have in common is zalambdodont cheek teeth, a character they share with *Solenodon*, from the West Indies.

The significance of the zalambdodont molar is one of the major outstanding problems of insectivore phylogeny. Gill (1885) divided the Insectivora into two suborders: Zalambdodonta and Dilambdodonta. Recently Van Valen (1967) transferred the suborder Zalambdodonta from the Insectivora to his new order Deltatheridia, placing them near the Palaeoryctidae. Vandebroek (1961) raised the Zalambdodonta to the status of an order, in which he included the Palaeoryctidae. Thenius (1969) regards the Zalambdodonta as evolving independently of the remaining Insectivora, but he excludes the Palaeoryctidae from the group. The rarity of intermediates between dilambdodont and zalambdodont molars and the lack of probable phyletic series in which one type has evolved from the other considerably add to the difficulties. Gregory (1920), Matthew (1937) and at one time myself (1939) have tried to trace an independent line of zalambdodont insectivores back into the Mesozoic, and *Deltatheridium* and *Palaeoryctes* have often been cited as intermediate stages. I cannot see any significant resemblance between the Palaeoryctidae and the Tenrecidae. The skulls of the Miocene tenrecids are quite unlike *Palaeoryctes*, but they do show an approach to erinaceids (Butler, 1969). The teeth of *Palaeoryctes* function in a quite different way: there is a nearly vertical shearing stroke, as shown by wear-scratches, as opposed to the much more transverse movement in tenrecids, which resemble shrews in this respect (Mills, 1965). *Solenodon* also has many shrew-like characters, and McDowell's (1958) comparison of *Solenodon* with *Nesophontes*, a Pleistocene dilambdodont from the West Indies, is very convincing, even though his method of homologising the tooth patterns is not. One is compelled to believe that somehow or other, dilambdodont teeth have been converted into zalambdodont teeth. The independent evolution of zalambdodonty in two different marsupials, *Notoryctes* in Australia and *Necrolestes* in South America (Patterson, 1958), suggests the possibility that zalambdodont teeth may have evolved more than once in the Insectivora.

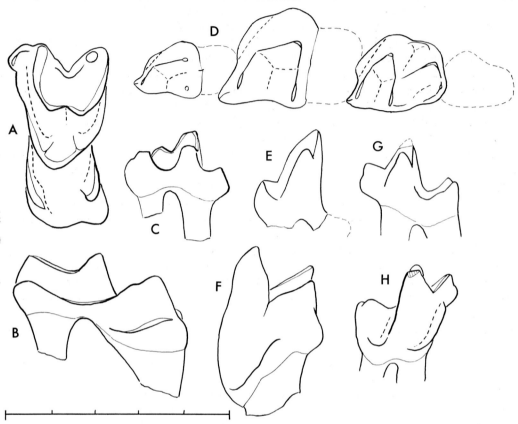

Fig. 1 Isolated teeth of *Butselia biveri* from the Isle of Wight
(From the private collection of Mr R. L. E. Ford).
A, B, C, left M^2 in crown, anterior and buccal views.
D, right P_4 trigonid, M_1 trigonid and M_2, crown view.
E, P_4 trigonid, lingual view. F, M_1, anterior view.
G, H, M_2 lingual and buccal views. The line below represents 5 mm.

Of some interest in this connection is the reported discovery of isolated zalambdodont molars, named *Butselia biveri*, from the L. Oligocene of Hoogbutsel, Belgium, (Quinet and Misonne, 1965). More material of the same species has been collected in the Isle of Wight by Mr Richard Ford, from the lower Hamstead Beds near Yarmouth, and provides additional information. Quinet and Misonne based their new species on the upper molars only, believing that the lower molars represented a different type of zalambdodont; however, the occurrence of upper and lower molars together again in the Isle of Wight, combined with the fact that they are of suitable size and pattern to occlude, strongly indicates that only one species is involved. Mr Ford's collection contains a mandible with a well preserved M_2 and the alveoli of all the teeth as far forward as I_2.

Butselia resembles the zalambdodonts in that the upper molars are transversely broadened; the paracone is well removed from the buccal margin and about midway across the crown when seen in ventral view; the metacone is rather close to the paracone and smaller in size; the buccal styles are developed in the same way as in Tenrecidae and *Solenodon*, the stylocone being about as high as the metacone; in the lower molars the paraconid is well developed, trenchant, and reaches the lingual side of the tooth; the talonid is narrower, shorter and much lower than the trigonid, the entoconid is rudimentary and the crista obliqua meets the trigonid near the mid-line. However, *Butselia* is not a zalambdodont: it has a distinct metacone, placed directly behind the paracone; its protocone is well differentiated, with rudimentary conules and a strong hypocone cingulum; its talonid is larger and more definitely basined than in any zalambdodont, even *Potamogale*. It also has four lower premolars, the first two with single roots, whereas in *Solenodon* and the tenrecids there are never more than three premolars and P_2 is usually two-rooted. Nevertheless, *Butselia* has some additional resemblances to *Solenodon* which may be significant: I_2 is enlarged (whether a small I_1 was present is not known); P_4 is rather similar, with a low anterior paraconid and a high metaconid placed directly lingually to the protoconid; *Solenodon* is the only zalambdodont to retain a hypocone cingulum; there is a long, rugose symphysis which extends back to below P_3; although the mandible is incomplete posteriorly, it can be seen that the coronoid process rises nearly vertically, the mandibular foramen (and therefore probably the condyle) is low, and the ventral edge of the temporal muscle insertion on the medial side is developed into a very prominent ridge. It would be easier to derive the dentition of *Solenodon* from that of something like *Butselia* than from *Nesophontes*.

Such a derivation would involve a further lingual movement of the paracone, a further reduction of the metacone, and a reduction and simplification of the protocone region of the upper molar and of the talonid basin of the lower molar. Functionally, such a change would result in increasing the length of the shearing crests of the paracone and protoconid. The direction of jaw movement does not change: wear-scratches on the lower molars of *Butselia* and *Solenodon* both make an angle of about 40° to the long axis of the roots.

Among fossil Insectivora, a genus that appears to show much resemblance to *Butselia* is *Geolabis* (= *Metacodon*), from the Oligocene of N. America with an ancestry that can be traced back to the Palaeocene. *Geolabis* is uniquely specialised in that its upper molars, despite being transversely developed, have two lingual roots. In other respects it is less advanced in a zalambdodont direction: the paracone and metacone are nearer the buccal edge, though still removed from it more than in erinaceids, for example; the buccal styles, which have the same arrangement as in *Butselia*, are however small and soon worn off; the metacone, though somewhat reduced and close to the paracone, is better developed than in *Butselia;* the talonids of the lower molars, though low and somewhat reduced in length, are broader than in *Butselia*, and the oblique crest meets the trigonid to the buccal side of the mid-line; the entoconid is reduced, but not so much as in *Butselia*; the paraconid is less lingual in position; I_2, though the largest of the incisors (alveoli in Carnegie Museum, 9262), is still comparatively small, and the canine is larger; P_1 and P_2 have two roots. The coronoid process, though it rises steeply, is not quite as upright as in *Butselia*. In *Butselia* there is a strong vertical keel on the anterior surface of the paraconid of the lower molars, only rudimentary in *Geolabis*. It probably replaces the reduced hypoconulid in keeping adjacent teeth together and preventing the impaction of food between them; a trace of a similar structure occurs in *Solenodon*.

McKenna (1960) noticed that *Geolabis* resembles *Solenodon* in a number of features of the skull, for example the incomplete zygomatic arch and the short infraorbital canal. With the help of *Butselia*, it seems feasible to derive *Solenodon* from a geolabid stock.

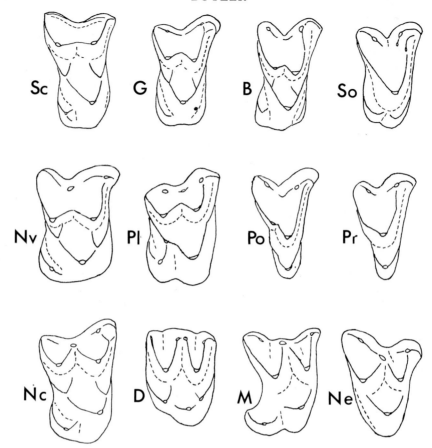

Fig. 2 Right M^2 of various Lipotyphla. Sc, *Scenopagus*; G, *Geolabis*; B, *Butselia*; So, *Solenodon*; Nv, *Nyctitherium velox*; Pl, *Plesiosorex* (after Wilson, 1960); Po, *Potamogale*; Pr, *Protenrec*; Nc, "*Nyctitherium curtidens*"; D, *Desmana*; M, *Myosorex*; Ne, *Nesophontes*. Not to scale.

Geolabis has generally been regarded as a specialised erinaceid. In 1948 I removed it from the Erinaceidae and made it the type of a new family. I still believe there are no good grounds for placing it in the Erinaceidae, despite McKenna's (1960) criticism. I associated it with *Plesiosorex*, which has subsequently become much better known (Wilson, 1960). This M. Oligocene—M. Miocene genus is very similar to *Butselia* in many respects, although it is specialising in a somewhat different direction. The upper molars have well developed buccal styles arranged in the same way as in *Butselia*; the lower molars have lingually placed, trenchant paraconids; the proportions of P_4-M_3 are similar; an incisor, presumably I_2, is enlarged, the canine is small and P_1 and P_2 have single roots; the coronoid process is vertical and far removed from the condyle. *Plesiosorex* differs from *Butselia* in its high, erinaceid-like entoconid, its more strongly developed hypocone, the less lingual position of its paracone and metacone and the greater independence of its metacone. Its infraorbital canal is short and shrew-like. It is clearly not erinaceid, and is probably best left in a separate family Plesiosoricidae, together with *Meterix* (its Pliocene descendant). *Butselia* may be placed in the same family.

Micropternodus is another puzzling form. A specimen in the Field Museum (T1662) shows that it has an incomplete zygomatic arch, and a very short bridge-like infraorbital canal with the lachrymal

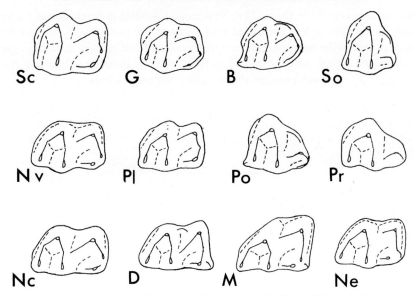

Fig. 3 Right M_2 of the same Lipotyphla as in Fig. 2. Not to scale.

foramen opening in the top of the bridge. The lachrymal foramen is above M^2, indicating that the teeth are placed far forward in relation to the orbit, as in *Talpa*. This, combined with the vertical direction of the coronoid process, implies a small eye and is consistent with the theory of fossorial habits suggested by Reed (1956). *Micropternodus* was put into the Deltatheridia by Van Valen (1967), but Robinson's (1968) view that it is related to *Geolabis* and *Nyctitherium* seems more convincing. *Apternodus* may also belong to the same general group. It was removed from the Lipotyphla by McDowell (1958), but his interpretation of the ear is open to question. In view of the fact that *Apternodus* has zalambdodont molars the results of McKenna's restudy of this animal are awaited with great interest.

In the Eocene of N. America, besides the geolabids, two groups of Lipotyphla can be distinguished and both occur in Europe. One of these, usually referred to as the Adapisoricidae, contains *Entomolestes*, *Scenopagus* and other forms which are almost certainly primitive erinaceids. The last survivor of this group seems to be the Oligocene *Ankylodon*. The second group is the Nyctitheriidae, which is believed to be broadly ancestral to the moles and shrews. According to Robinson (1968), *Nyctitherium velox* is close to *Saturninia*, which Stehlin (1940) interpreted as a primitive shrew, but Robinson is inclined to believe that *Nyctitherium* has talpid rather than soricid affinities. There is very little evidence for or against this view. *Nyctitherium* resembles both shrews and moles in having a mesostyle on the upper molars, and in *N. celatum* (Matthew, 1909) it is incorporated in a W-shaped ectoloph as in shrews and moles (according to Robinson, 1968, this species is probably the same as *N. serotinum* but in my opinion it is not congeneric with *N. velox*). The infraorbital canal of *N. velox* is very short, but it is not known whether the zygomatic arch was complete, as in moles, or incomplete as in shrews.

The primitive erinaceoids, *Entomolestes* and *Scenopagus*, differ from *Nyctitherium* and the geolabids in a number of characters of the teeth. The paracone and metacone are more buccally situated; there is a gently undulating buccal cingulum between the parastyle and the metastyle, but no other stylar cusps; the paraconid, at least on M_2 and M_3, is reduced to a low transverse ridge. The entoconid is higher than the hypoconid, and tends to line up with it to form a ridge along the posterior margin of the tooth with the hypoconulid in the middle. This ridge is similar in height to the paraconid of the following tooth, and a more or less horizontal wear surface (due to the hypocone) involves the edge of the paraconid as well as the hypoconulid. In nyctitheres, and still more in geolabids, the paraconid is trenchant and much higher than the talonid in front of it; its anterobuccal surface forms a semi-vertical shearing plane functioning against the metacone-metastyle crest.

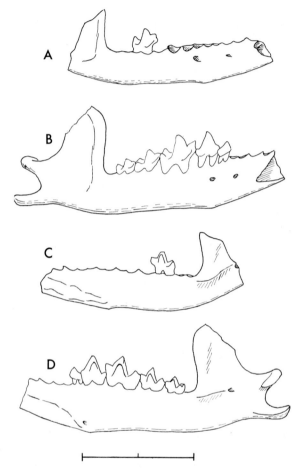

Fig. 4 **A, C**, outer and inner views of mandible of *Butselia biveri*, (specimen in the private collection of Mr R. L. E. Ford). **B, D**, the same, *Plesiosorex cf. soricinoides* (after Viret, 1940). The line below represents 10 mm.

How far the two groups had differentiated in the Palaeocene is uncertain, but some Palaeocene forms like *Leptacodon tener* (McKenna, 1968) and *L. packi* seem to resemble *Nyctitherium*, while others such as *Leipsanolestes siegfriedti* and "*Leptacodon*" *ladae* are more like *Entomolestes*.

In 1956 I divided the Lipotyphla into two sections, for which I used Saban's (1954) terms Erinaceomorpha and Soricomorpha, differing in characters of the skull. This was based mainly on living forms. Unfortunately there are no reasonably good lipotyphlan skulls before the Oligocene, but at that time *Geolabis* was definitely soricomorph in its skull (McKenna, 1960). In the Eocene, *Nyctitherium velox* had a short infraorbital canal as in later soricomorphs but in *Scenopagus* the infraorbital canal was long. There is thus an indication that the erinaceoid Adapisoricidae differed from the nyctithere-geolabid group in their skulls as well as their teeth, and that the erinaceomorph-soricomorph cleavage of the Lipotyphla can be traced back into the Palaeocene. The Tenrecidae, however, do not fall into

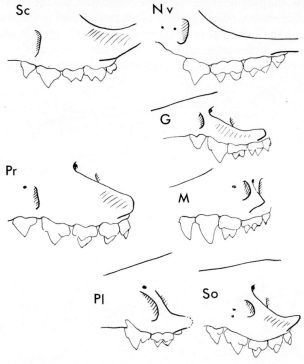

Fig. 5 Maxillae of various Lipotyphla, seen from the left side, to illustrate differences in the infraorbital canal. Not to scale.
Sc, *Scenopagus*; Pr, *Protenrec*; Nv, *Nyctitherium velox*; G, *Geolabis*; M, *Micropternodus*; Pl, *Plesiosorex* (after Wilson, 1960); So, *Solenodon*.

the pattern, for even as late as the Miocene the infraorbital canal was long and erinaceoid-like (Butler, 1969). For this reason it seems advisable to separate them from the Soricomorpha, and I propose a suborder Tenrecomorpha to house them. To emphasise the isolated position of the Chrysochloridae I propose to put them also in a separate suborder, using Broom's name Chrysochlorida. The Soricomorpha would then coincide with the Soricoidea as used by Simpson (1945), except that he put *Geolabis* ("*Metacodon*") in the Erinaceidae and *Solenodon* in the Tenrecoidea. The classification of Lipotyphla would then be as follows:

Order Lipotyphla.
  Suborder Erinaceomorpha.   **Families** Adapisoricidae (including *Ankylodon*), Erinaceidae, ? Dimylidae.
  Suborder Soricomorpha.   **Families** Geolabididae, Plesiosoricidae (including *Butselia*), Solenodontidae, Micropternodontidae, Nyctitheriidae, Talpidae, Soricidae, Nesophontidae, ? Apternodontidae.
  Suborder Tenrecomorpha.   **Family** Tenrecidae (including *Potamogale*).
  Suborder Chrysochlorida.   **Family** Chrysochloridae.

After elimination of the Lipotyphla two living insectivore families are left, the tree-shrews (Tupaiidae) and the elephant-shrews (Macroscelididae). Gregory (1910) put them in an order Menotyphla. They retain primitive characters lost in the Lipotyphla: the caecum (lost in one genus), the normal pubic symphysis, the palatine expansion in the orbital wall, and the large jugal. At the same time they are advanced in other respects: the large brain which extends forward between the enlarged orbits; the entotympanic bone in the bulla. The two families are widely divergent in adaptive trends: tupaiids are squirrel-like and more or less arboreal; macroscelids are terrestrial forms with elongated hind limbs, progressing like rabbits. In tupaiids the testes descend into a scrotum, in macroscelids they are

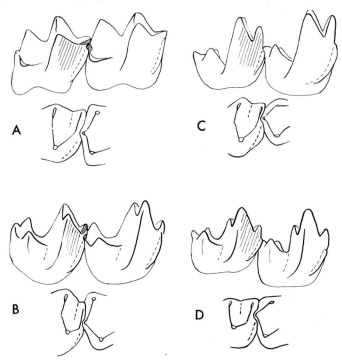

Fig. 6 Semilateral views of right M_1 and M_2 and crown views of the area of contact, to illustrate differences in paraconid function. **A**, *Entomolestes nitens*; **B**, "*Leptacodon*" *ladae*; **C**, *Geolabis mellingeri*; **D**, *Leptacodon packi*. Not to scale.

abdominal. The brain is markedly different in structure in the two families. Macroscelids parallel Lipotyphla in the possession of a mobile proboscis. The fossil history of the macroscelids goes back to the Oligocene (Patterson, 1965), all in Africa, but it throws no light on the origin of the group. No certain fossil tupaiids are known: *Anagale* is now believed to have nothing to do with them, nor indeed with the Primates (McKenna, 1963). *Adapisoriculus*, an extremely poorly known genus from the Palaeocene of Europe, has been claimed as a tupaiid by Van Valen (1965), but his argument is based on too many assumptions to be compelling. *Adapisoriculus* may be a mixodectid: according to Szalay (1969), there are resemblances in the teeth between Mixodectidae and Tupaiidae which might indicate a relationship, but the time interval is so great that little weight can be attached to such resemblances. Practically nothing is known about these Palaeocene mammals except their teeth and jaws, and the general shape of the skull. *Microsyops*, hitherto regarded as a mixodectid, is a probable primate. The Oligocene *Leptictis* (= *Ictops*) has a ring-shaped tympanic lying inside an entotympanic bulla, like

Tupaia (McDowell, 1958; McKenna, 1966) but its teeth are quite different: they show some resemblances to those of macroscelids.

Current opinion is turning against the inclusion of the tupaiids in the Primates, but if their place is not there it is equally not with the lipotyphlous insectivores. Two courses seem to be open: to unite them with the macroscelids and perhaps some fossil families in an order which would correspond to Gregory's Menotyphla, or to put them in an order by themselves. I think the latter is the best course in our present state of knowledge, and suggest the name Scandentia (Wagner, 1855). The macroscelids would make another order, Macroscelidea, as I suggested in 1956.

We are left with a wide range of early fossil placentals, mostly poorly known and of uncertain relationships. Several family groupings can be recognised, but their taxonomy above the family level is in a chaotic state. The Palaeoryctidae (Cretaceous-Eocene) are considered by Van Valen (1966) to be related to some groups of creodonts, and he has proposed a new order, Deltatheridia, to include them. The name is unfortunate, because *Deltatheridium* is one of the most aberrant members of the group and may not belong to it. The Plagiomenidae (Palaeocene and Eocene) are usually put into the Dermoptera because of dental resemblances, but the skull is unknown. The peculiarly specialised Apatemyidae (Palaeocene-Oligocene), reviewed by McKenna (1963), are perhaps worthy of a separate order, for which the name Apatotheria was suggested by Scott and Jepsen (1936). The Pantolestidae (Palaeocene-Oligocene) includes *Chadronia*, the largest of all insectivores, with a skull 15 cm long (Cook, 1952). The Leptictidae were for a long time regarded as relatives of the erinaceoid Lipotyphla, but characters of the orbital wall and of the ear of *Leptictis* show that this is not so; the characters that leptictids share with erinaceids are probably primitive eutherian characters only. There is a considerable amount of leptictid material in American museums which badly needs examination (Szalay, 1966). There was a tendency in the past to use the Leptictidae as a dumping ground for small mammals with unspecialised tribosphenic dentitions; most of the faulty allocations of Tertiary forms have now probably been detected, but whether any Cretaceous forms should be placed in the family is still uncertain. *Gypsonictops* does not fit into the Leptictidae very well; some specimens have five lower premolars, for instance. *Procerberus* has been interpreted as a palaeoryctid by Lillegraven (1969) on grounds that seem more convincing than its reference to the Leptictidae by Sloan and Van Valen (1965).

I have mentioned only a few of these problematic groups. Altogether, more than 50 genera of insectivores, other than Lipotyphla, have been described from the Cretaceous to the Oligocene. It is not possible in our present state of knowledge to give this miscellaneous group more than a waste-basket classification. They are really all Eutheria *incertae sedis*, and it might be more honest, though less tidy, to classify them thus. The alternative is to put them together in an indefinable group. Romer (1966) has done this in creating a suborder Proteutheria, in which he includes the living menotyphlan families as well as the fossil non-lipotyphlan insectivores. Van Valen (1967) adopted Romer's name, but he excluded the macroscelids as a separate suborder though retaining the tupaiids. The name seems to imply a group containing the ancestral placentals, together presumably with those forms which have diverged comparatively little from the basic placental stock. It might be legitimate to include also short-lived more specialised lines, produced in the late Cretaceous and earliest Tertiary, which did not diversify further beyond the family level. To assume that the two living menotyphlan families are much like the earliest placentals is to beg a very big question, and they were certainly not short-lived; I would therefore exclude them both from the Proteutheria and put them into separate orders. The Proteutheria would, on this scheme, rank as an order, taking the place of the Insectivora as the waste-basket which is still unfortunately necessary.

Thus, in my opinion, the order Insectivora should disappear, unless the name is restricted to the

Lipotyphla. The waste-basket still remains, in the form of the Proteutheria. Further research should reduce the size of the Proteutheria, by discovering relationships of some of its groups to other orders, and possibly by making new orders out of other groups. As this process of refinement goes on the nature of the basic eutherian stock will gradually become clear. Perhaps we already know some members of it: *Kennalestes* (Kielan-Jaworowska, 1968) is a likely candidate.

REFERENCES

BOWDICH, T. E. 1821. *An analysis of the natural classification of mammalia.* J. Smith, Paris.
BROOM, R. 1916. On the structure of the skull in *Chrysochloris*. *Proc. zool. Soc. Lond.*, **118**, 446-500.
BUTLER, P. M. 1939. The teeth of the Jurassic mammals. *Proc. zool. Soc. Lond.*, B **109**, 329-356.
—— 1948. On the evolution of the skull and teeth in the Erinaceidae, with special reference to fossil material in the British Museum. *Proc. zool. Soc. Lond.*, **118**, 446-500.
—— 1956. The skull of *Ictops* and the classification of the Insectivora. *Proc. zool. Soc. Lond.*, **126**, 453-481.
—— 1969. Insectivores and bats from the Miocene of East Africa: new material. *Fossil vertebrates of Africa*, ed. L.S.B. Leakey, Vol. **1**, 1-37.
COOK, H. J. 1952. A remarkable new mammal from the Lower Chadron of Nebraska. *Am. Midl. Nat.*, **52**, 388-391.
CUVIER, G. L. C. F. D. 1817. *Le Règne Animal . . .* Déterville, Paris, **1**.
DE BLAINVILLE, H. M. D. 1816. Prodrome d'une nouvelle distribution systématique du règne animal. *J. Phys., Chimie, Hist. nat.*, **83**, 244-267.
GEOFFROY SAINT-HILAIRE, É. and CUVIER, G. 1795. Mémoire sur une nouvelle division des mammifères, et sur les principes qui doivent servir de base dans cette sorte de travail. *Magasin Encyclopédique*, **1** (2), 164-190.
GILL, T. 1872. Arrangement of the families of mammals with analytical tables. *Smithson. misc. Collns*, **230**, 1-98.
—— 1885. Insectivora. *The standard natural history*, S. E. Cassino, Boston, **5**, 134-158.
GREGORY, W. K. 1910. The orders of mammals. *Bull. Am. Mus. nat. Hist.*, **27**, 1-254.
—— 1920. The origin and evolution of the human dentition. Part II. *J. dent. Res.*, **2**, 215-283.
HAECKEL, E. 1866. Systematische Einleitung in die allgemeine Entwicklungsgeschichte. *Generelle Morphologie der Organismen*, Georg Reimer, Berlin, Bd. **2**, pp. XVII-CLX.
HUXLEY, T. H. 1880. On the application of the laws of evolution to the arrangement of the Vertebrata and more particularly of the Mammalia. *Proc. zool. Soc. Lond.*, **1880**, 649-662.
ILLIGER, J. C. W. 1811. *Prodromus Systematis Mammalium et Avium . . .* C. Salfeld, Berlin.
KIELAN-JAWOROWSKA, Z. 1968. Results of the Polish-Mongolian palaeontological expeditions. Part I. Preliminary data on the Upper Cretaceous eutherian mammals from Bayn Dzak, Gobi Desert. *Palaeont. pol.*, **19**, 171-191.
LECHE, W. 1885. Über die Säugethiergattung *Galeopithecus*. *K. svenska VetenskAkad. Handl.*, **21**, No. 11, 1-92.
LILLEGRAVEN, J. A. 1969. Latest Cretaceous mammals of upper part of Edmonton Formation of Alberta, Canada, and review of marsupial-placental dichotomy in mammalian evolution. *Paleont. Contr. Univ. Kans.*, **50**, 1-122.
LINNAEUS, C. 1758. *Systema Naturae per Regna Tria Naturae* 10th ed., Laurentii Salvii, Stockholm. Tom **1**, 52-53.
MATTHEW, W. D. 1909. The Carnivora and Insectivora of the Bridger Basin, Middle Eocene. *Mem. Am. Mus. nat. Hist.*, **9**, 289-567.
—— 1937. Paleocene faunas of the San Juan Basin, New Mexico. *Trans. Am. phil. Soc.* (new ser.), **30**, 1-150.
McDOWELL, S. B. 1958. The Greater Antillean insectivores. *Bull. Am. Mus. nat. Hist.*, **115**, 115-214.
McKENNA, M. C. 1960. The Geolabidinae, a new subfamily of early Cenozoic erinaceoid insectivores. *Univ. Calif. Publs geol. Sci.*, **37**, 131-164.
—— 1963a. New evidence against tupaioid affinities of the mammalian family Anagalidae. *Am. Mus. Novit.*, **2158**, 1-16.
—— 1963b. Primitive Paleocene and Eocene Apatemyidae (Mammalia, Insectivora) and the primate-insectivore boundary. *Am. Mus. Novit.*, **2160**, 1-39.
—— 1966. Paleontology and the origin of the Primates. *Folia primat.*, **4**, 1-25.
MILLS, J. R. E. 1965. The functional occlusion of the teeth of Insectivora. *J. Linn. Soc. (Zool.)*, **47**, 1-25.
PATTERSON, B. 1958. Affinities of the Patagonian fossil mammal, *Necrolestes*. *Breviora*, **94**, 1-14.
—— 1965. The fossil elephant shrews (family Macroscelididae). *Bull. Mus. comp. Zool., Harv.*, **133**, 295-335.
PETERS, W. C. H. 1864. Über die Säugethiere Gattung *Solenodon*. *Abh. Akad. Wiss. Berlin*, (for 1863), 1-22.
REED, C. A. 1956. A new species of the fossorial mammal *Arctoryctes* from the Oligocene of Colorado. *Fieldiana, Geol.*, **10**, 305-311.
ROBINSON, P. 1968. Nyctitheriidae (Mammalia, Insectivora) from the Bridger Formation of Wyoming. *Univ. Wyo. Contr. Geol.*, **7**, 129-138.

ROMER, A. S. 1966. *Vertebrate paleontology*, 3rd edit. Chicago Univ., Chicago and London.

SABAN, R. 1954. Phylogénie des insectivores. *Bull. Mus. natn. Hist. nat.*, Paris (ser. 2), **26**, 419-432.

SCOTT, W. B. and JEPSEN, G. L. 1936. The mammalian fauna of the White River Oligocene. Part I. Insectivora and Carnivora. *Trans. Am. phil. Soc.* (new ser.), **28**, 1-153.

SIMPSON, G. G. 1937. The Fort Union of the Crazy Mountain Field, Montana, and its mammalian faunas. *Bull. U.S. natn. Mus.*, **169**, 1-287.

—— 1945. The principles of classification and a classification of mammals. *Bull. Am. Mus. nat. Hist.*, **85**, 1-350.

SLOAN, R. E. and VAN VALEN, L. 1965. Cretaceous mammals from Montana. *Science, N.Y.*, **148**, 220-227.

SZALAY, F. S. 1966. The tarsus of the Paleocene leptictid *Prodiacodon* (Insectivora, Mammalia). *Am. Mus. Novit.*, **2267**, 1-13.

—— 1969. Mixodectidae, Microsyopidae and the insectivore-primate transition. *Bull. Am. Mus. nat. Hist.*, **140**, 193-330.

THENIUS, E. 1969. Stammesgeschichte der Säugetiere (einschliesslich der Hominiden). Kükenthal's *Handbuch der Zoologie*, **8**, Lieferung 47, 1-368. Walter de Gruyter, Berlin.

VANDEBROEK, G. 1961. The comparative anatomy of the teeth of lower and non-specialized mammals. *International Colloquium on the Evolution of Lower and Non-specialised Mammals*. Brussels, Koninklijke Vlaamse Academie voor Wetenschappen, Letteren en Schone Kunsten van Belgie, pp. 215-320.

VAN VALEN, L. 1965. Treeshrews, primates and fossils. *Evolution*, **19**, 137-151.

—— 1966. Deltatheridia, a new order of mammals. *Bull. Am. Mus. nat. Hist.*, **132**, 1-126.

—— 1967. New paleocene insectivores and insectivore classification. *Bull. Am. Mus. nat. Hist.*, **135**, 217-284.

WAGNER, J. A. 1855. *Die Säugethiere in Abbildungen nach der Natur*. Weiger, Leipzig. Supplementband, Abt. 5, 1-810.

WILSON, R. W. 1960. Early Miocene rodents and insectivores from northeastern Colorado. *Paleont. Contr. Univ. Kans., Vertebrata*, art. **7**, 1-92.

K. A. JOYSEY

The fossil species in space and time: some problems of evolutionary interpretation among Pleistocene mammals

INTRODUCTION

While working on fossil echinoderms some twenty years ago I first became involved with problems regarding the treatment of fossil populations, especially in relation to the study of variation within populations, and methods of making comparison between populations. Under the influence of Dr F. R. Parrington I have become involved increasingly with fossil vertebrates and the present essay emerges from the integration of these two fields of experience.

This essay is based partly upon a lecture entitled 'Evolutionary patterns among fossil vertebrates' presented in 1967, at the same session as Dr Parrington's Presidential Address to the Zoology Section of the British Association for the Advancement of Science. A revised version was delivered, under the present title, to the NATO Advanced Study Institute on Vertebrate Evolution held at Istanbul in 1969.

When these two lectures were prepared it was intended to present a personal view-point and so the resulting essay is somewhat egotistical. I have drawn extensively upon my own previous publications and those of a few others working on Pleistocene mammals, and in consequence the reference list is unbalanced. I am fully aware that others have made some of the same points elsewhere and yet receive no credit here, and I am equally aware that others have presented alternative views which receive no discussion here; I offer my apologies to anyone offended by this treatment.

WHAT IS A FOSSIL SPECIES?

I am not concerned here with those legalistic implications of species definition which are required by formal taxonomy, although many of the points raised in such debates are pertinent to the present discussion. The definition of a fossil species presents a number of peculiar problems which have been debated at length elsewhere under the general title of 'The species concept in palaeontology' (Sylvester-Bradley, 1956). All members of a fossil species have been, at some time in the past, a part of a then living species, and so in my view the definition of a fossil species should take into account some cognisance of the definition of living species.

Among living animals the species was originally used to categorise different kinds of animals. These different kinds were based on observable differences in morphology. It was later recognised that morphological discontinuities usually reflect genetical discontinuities between populations, and that these in turn reflect the breeding boundaries recognised by the animals themselves. Since the advent of "The New Systematics" (Huxley, 1945), the classical breeding test definition of a species has been frustrated to some extent by the recognition of polytypic species. Mayr (1942) discussed several instances in which the opposite ends of a polytypic chain have come into contact with one another, either around the circumpolar route or around some other geographical barrier, and so presented the possibility for interbreeding between the opposite extremes of the same species. Under these circumstances, individuals belonging to two different populations of the same species may not interbreed, but if they do interbreed when the genetical difference between the opposite ends of the cline is considerable then such interbreeding may produce either nonviable or partly sterile hybrids. If the animals themselves are uncertain whether or not they belong to the same species, surely we cannot expect to be able to provide an infallible species definition!

The realistic and practical approach to this sort of problem is to subdivide polytypic species into geographical subspecies, and this is done on the basis of arbitrary geographical criteria, when it is judged that the degree of morphological difference between populations living in different regions is sufficient to make some distinction worthwhile. The majority of individuals in a given geographical subspecies may be morphologically distinct from those belonging to another subspecies, but some individuals are likely to be morphologically indistinguishable from those belonging to another subspecies. Indeed the characteristics of a given geographical subspecies can be described only on a population basis, and in order to attribute a given individual to its correct geographical subspecies it is essential to know the locality from which it came. The subdivision of a polytypic species into geographical subspecies frequently coincides with some geographical barrier which causes a partial break in the distribution of the species, but in other cases the arbitrary subdivision may even follow national boundaries. The point I wish to stress is that geographical subspecies may be described on morphological criteria, but they are defined on geographical criteria.

When considering fossils from a single time horizon I can see no reason why they should be treated any differently from animals living on the present time plane. In practice, zoologists do not use the breeding test in the definition of their species. Apart from the difficulties of such experimentation and culture, it has been found that morphological criteria usually provide sufficient basis to distinguish one species from another. It is true that the neontologist has the additional advantage of the soft anatomy to provide a greater number of morphological criteria, and it is right and proper that the palaeontologist should be conscious of the possible short-comings of an assessment based upon the hard parts alone. Nevertheless, it is relatively easy to perform a control test on the neontological and palaeontological concept of a species. On more than one occasion I have divided a freshly collected sample from a marine dredge into two parts, and allowed one half to rot until only the hard parts remained. The fresh sample has been assessed by a neontologist and the pseudo-palaeontological sample by a palaeontologist. Of course, the latter sample contained no trace of many soft-bodied animals, but the palaeontologist's assessment of the molluscs was essentially the same as that made by the neontologist. It is noteworthy that some of the arthropods were lumped by the palaeontologist, having lost some of the minor spines which are used to distinguish between related living species. In contrast, the echinoids were usually over-split because the palaeontological mind tended to attach too much importance to variation in shape and to the differences between young and adult forms. It is surely worthwhile for the mammalian palaeontologist, and indeed any sort of palaeontologist, to consider whether his concept of the species category is comparable with that employed by students of related living forms. Although it is demonstrable that in some groups the palaeontologist tends to be a splitter and in other groups he tends to lump, I have come to the conclusion that, on the whole, the morphological assessment of a fossil species is comparable with that of the living species.

Turning now to the problem of fossils in a geological succession the situation is, of course, complicated by the addition of the time dimension. In the unusual circumstance of finding a continuous geological succession containing a continuous sequence of evolving morphological types it becomes unavoidable that the boundaries between species must be arbitrarily defined. The size of the taxonomic unit does not affect the issue, so it is irrelevant whether the problem is discussed at the level of the species or the subspecies. The problem is to decide whether the units should be defined on chronological or morphological criteria.

Referring to Fig. 1 (Joysey, 1956), the numbers 1-4 represent a series of geological horizons in ascending order. At horizon 1 the symbols a, b, and c represent varieties within a single population, b being the typical form at that horizon while a and c represent less common varieties. Ascending a geological succession, we sometimes find that evolutionary change results in a progressive change in the relative abundance of the different varieties such that some disappear and others become more common, while new varieties appear. Thus at each horizon in Fig. 1, the middle letter represents the most abundant typical form and the other symbols represent less abundant varieties present at that horizon. If the evolutionary sequence were to be subdivided on a morphological basis then each form would be found to range through several geological horizons, whereas if the boundaries were defined on a chronological basis then several different morphological forms (varieties) would be found to be present in the population at each geological horizon.

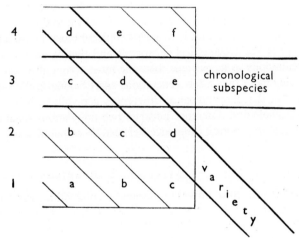

Fig. 1 Diagram to illustrate the two alternative directions in which a continuous evolutionary sequence may be subdivided. Numbers 1-4 represent a succession of geological horizons in ascending order. At each horizon the middle letter represents the typical form in a population at one locality and the other letters represent less common varieties in the same population. (Joysey, 1956.)

It is relatively easy to describe fossil species in morphological terms and at first sight the use of morphological criteria to define the boundaries between species appears to be most in keeping with zoological practice, but in fact this is not so. I advocate that a continuous evolutionary sequence should be arbitrarily subdivided on chronological grounds, and this course is recommended for several distinct reasons.

Firstly, the direction of such boundaries would be parallel to those which are usually provided either by stratigraphical breaks or by gaps in collecting from the succession. Referring again to Fig. 1, if fossils were known only from horizons 1 and 4, the gap between these populations would be of sufficient magnitude that they had no varieties in common, and it is likely that they would be regarded as two species, namely species b at horizon 1 and species e at horizon 4. If the gap were to be filled subsequently by collection from the intervening horizons 2 and 3, then it would be reasonable to refer the intermediate populations to chronological subspecies of the already described species. In this case the population at horizon 2 could be treated as a chronological subspecies of species b, and the population at horizon 3 as a chronological subspecies of species e. It is recognised that the difference between the populations at horizons 2 and 3 would be no greater than the difference between either 1 and 2, or 3 and 4, but it is a matter of convenience to have species boundaries at intervals. In a continuous evolutionary sequence it is inevitable that the species boundaries will be arbitrary in position, and it is likely that they will reflect either present or previous gaps in our knowledge.

A second reason for advocating that an evolutionary sequence should be subdivided on chronological rather than morphological criteria is that the chronological subspecies can be related in concept to the geographical subspecies. It seems reasonable to define chronological subspecies when the degree of morphological difference between two populations has an order of magnitude comparable with that which is regarded as sufficient grounds for the recognition of a geographical subspecies but, of course, in the case of a chronological subspecies the two populations occur in the same region and are separated by an interval of time in the geological succession. Under no circumstances would I suggest that a population be regarded as belonging to a different chronological subspecies just because it occurs at a different geological horizon. This would be just as ridiculous as creating a new geographical subspecies for material from a different region in the absence of any significant morphological difference. In the case of the geographical subspecies it is often possible to confirm that diverse populations belong to a single species by application of the interbreeding test. It is, of course, impossible to apply such a test to fossil material, which makes it more difficult to assess the degree of genetical difference between populations which belong to different chronological subspecies. I know that some of my colleagues find here a philosophical block in attempting to draw an analogy between the concept of the geographical subspecies and that of the chronological subspecies, but if one carries this philosophy to its extreme then one must accept also that it is impossible to confirm that individuals living more than a few generations apart belong to the same species. For my part, I am prepared to accept that morphological differences generally reflect genetical differences and that the degree of morphological difference between two populations separated in time provides some measure of the degree of genetical difference between those populations.

STATISTICAL COMPARISON OF FOSSIL POPULATIONS

Perhaps the most important reason for subdividing an evolutionary sequence on chronological criteria is that the fossils collected from each horizon, being a sample from a population of animals which lived at about the same time, often constitute a sample amenable to statistical comparison. Handpicked specimens selected from the populations present at each of several horizons do not constitute a random sample, and are unsuitable material for statistical analysis. Even at one horizon it is possible for subjective opinions to influence wrongly decisions regarding population boundaries. For example, it might be conjectured that a 'fat' form and a 'thin' form belonged to different 'populations'. If a handpicked sample of each form were to be collected, these would not be random samples. In the event of statistical comparison it is almost inevitable that a significant difference would be found between the samples, and this spurious result would apparently support the conjecture that the two forms were distinct; the importance of randomness in a sample can hardly be overemphasized.

The most basic of all statistical methods is the study of variation of a single character. For example, one may measure egg length in a sample of hen's eggs and then plot a frequency distribution, and calculate the mean length together with the standard deviation. These parameters could then be used as a basis for comparison with a sample of eggs laid by another flock. Similarly the height of different groups of men may be compared, but in order to obtain meaningful information it is essential that the men should all have stopped growing. If one were to mix a number of boys and men the results would be confusing because some of the variation present in the sample would be due to differences in age and from the measurements alone it would be impossible to judge which were boys and which were small men.

By analogy, in the majority of cases the frequency distribution of size of a character is unsuitable as a basis for comparison between samples of fossils, because most fossil populations consist of a mixture of young and old individuals constituting a growth series. Differences in size due to age are generally of a far greater magnitude than the differences in size between individuals of the same age, and so the latter are swamped by the former. Hence, most of the variation present is due to differences in age between the individuals and the value of the mean is meaningless, being dependent upon the relative number of young and adults in the sample (Joysey, 1953).

The frequency distribution of size within a sample of a fossil population of mixed age can, of course, be influenced by a large number of factors. Perhaps the most important of all, the frequency distribution of age in the dead population is usually quite different from that in the living population from which it was derived. This is because the chance of dying usually varies with the age of the individual; in other words, it is unusual for the force of mortality to be constant throughout life, although exceptions to this generalisation are known (e.g. post-fledgling robins). In human terms it is obvious that the age composition in the population of corpses in a cemetery is quite different from that in a living population, although the cemetery population will contain a few individuals representative of each age group. In the event of some catastrophe overwhelming a living population then the age structure of a living population could exceptionally be preserved.

The situation may be further complicated if, in life, the young individuals segregate from the adults, or if the non-breeding males are obliged to live apart from the breeding herd, or if different sections of the population migrate at different times. Any of these behavioural or social phenomena, common among vertebrates, could lead to incomplete representation of the growth series, or even total absence of one sex, in what appeared to be a fossil population. Even after death the treatment of corpses by scavengers is likely to be age dependent, and the distance of transport of corpses in a river may be age-dependent, either of which could lead to a biased accumulation of remains. The chance of fossilisation is likely to vary according to the age of the individual and the robustness of the skeleton. Small and thin bones often tend to be more easily destroyed, and so the older individuals may be preserved preferentially.

In those cases where collection of the fossils depends upon a rock-breaking technique, in contrast to a sieving technique, the chance of finding a small individual is considerably less than that of collecting a larger specimen. It is evident that if a large and a small specimen are similarly placed within a block of rock, then any given break is more likely to expose some part of the larger specimen. The general failure to detect small individuals can be avoided if the rock is broken down to a size smaller than the diameter of the object being sought, and this usually involves an amount of work beyond the patience of the collector. It is for this same reason that the larger species present in a fauna are almost invariably described some years before the smaller species are discovered!

It is indeed a formidable task to attempt to unscramble variation in a growth series unless some independent criterion of individual age is available. Such criteria include the familiar growth-rings in scales, growth-lines in dentine and, more recently, periostial-lines within bones (Morris, 1970). In the absence of such criteria to divide the sample into age groups, the study of variation in size by means of frequency distribution is of very limited value, and quite unsuitable as a basis for comparison between populations. A few structures which do not grow, such as the crowns of brachydont mammalian teeth, are notable exceptions and, unless they have become too worn by use, provide ideal material for the study of variation in size.

Being aware of the usual relationship between size and age, and of the mixture of age groups present in most fossil samples, many workers have attempted to overcome the difficulty by resorting to ratios. This involves the simultaneous study of two variables and, in dividing one variate by the other, it appears at first sight that dependency on the overall size of the individual has been eliminated. Based on this assumption, one can find in the literature many studies of variation based on the value of the ratio of two dimensions. Unfortunately, there is no such easy solution because most animals change their shape as they grow and so the value of the ratio itself changes during individual growth. Hence, the value of the mean of a ratio also depends upon the relative abundance of young and adult individuals in the sample. This feature of ratios is fairly obvious when the dimensions of the two characters, x and y, are plotted against one another on an arithmetic scale graph, and the resulting scatter diagram of the growth series is curved. On the other hand, it has sometimes been assumed, particularly in the field of fish systematics, that when such a scatter diagram follows a straight path it is safe to use the ratio as a basis for comparison between populations. But this is not necessarily so. Unless the straight line also passes through the origin, the value of the ratio at different points along that line cannot be constant. Because of the usual absence in a sample of individuals smaller than a particular size and a proper reluctance to extend the plotted line beyond the smallest individual (into the region of no evidence), many workers fail to observe that their fitted line would have a substantial intercept on one of the axes of the graph. Under these circumstances the value of the ratio does change during growth and so the mean value of the ratio is unreliable as a basis for comparison. Indeed, this is one of the ways in which animals achieve a change in shape during growth without resort to a curvilinear growth relationship between the two characters.

Many of the foregoing difficulties can be overcome by using the parameters of the relative growth curve itself as a basis for comparison between samples. When the scatter diagram approximates to a straight line it is convenient to calculate either the reduced major axis (Kermack and Haldane, 1950), or fit a regression line to the arithmetical data such that $y = a+bx$, where a is the intercept on the y axis and b is the slope of the line. The parameters a and b provide information about the growth series present in each sample and can be used to investigate possible differences between samples (Joysey, 1953, 1959a).

Many scatter diagrams are found to be curvilinear when plotted on an arithmetical scale but, for reasons which have been discussed at length elsewhere (Huxley, 1932), when the data are converted to logarithms the scatter diagram almost invariably follows a straight path. This property of the allometric growth relationship is of great convenience because it can be expressed in terms of the equation $\log y = \log \beta + \alpha \log x$, where α and β are the slope and intercept of the fitted line when x and y are plotted on a double logarithmic scale. The parameters of the allometric growth relationship, α and β can be calculated by a variety of different methods (Gould, 1966), and they provide a valuable tool in the comparison between samples.

A further important practical point is that certain characters which do not grow nevertheless possess a relationship between their dimensions which is analogous to an allometric growth relationship. Hence, it is possible to express the relationship between length and width of brachydont mammalian teeth in terms of an allometric growth equation, even though the sample does not constitute a growth series.

NATURAL SELECTION

Many investigations of variation and relative growth in fossil populations are pursued with the sole intention of using the parameters to test whether or not significant differences can be found between populations from different localities and horizons. But the study of variation in a population can also provide information regarding the effects of natural selection on it. This can most easily be investigated in those animals in which some part of the juvenile skeleton is still recognisable in the adult, as for example in the shells of some molluscs. Then it is possible to make a comparison between the variation present among juvenile members of the population and that present in those parts of the adults which represent their juvenile stages. By this method it has been possible to demonstrate that particular varieties present in the juvenile population either do not reach adulthood, or are under-represented in the adult population, and the inference is that this phenomenon is an expression of natural selection in action.

It is less easy to conduct such studies among vertebrates because many features of the skeleton are resorbed and replaced during growth, but the potential for such studies does exist, particularly among fish and reptiles where, for example, variation may be present in the pattern of dermal bones of the skull. Furthermore, the juvenile fish scale is usually recognisable within the structure of the adult scale. Where it can be established that the number of scale rows, or fin rays, or vertebrae is variable between individuals but is constant during individual growth, then such features are amenable to the investigation of natural selection.

By such methods, Hecht (1952) has shown that in the living gekkonid lizard *Aristelliger*, the juveniles are more variable than the adults in the number of lamellae under their toes. The number of lamellae is of functional importance in maintaining a grip while climbing, and those individuals with a smaller number of lamellae are less likely to survive to reach a large size.

In this context of seeking differences between the variation present among juveniles and that present among adults, it is worth reiterating that even when the relative growth relationship is a straight line the value of the ratio is not growth independent unless that straight line passes through the origin.

Conversely, it is theoretically possible that the effects of natural selection could distort a relative growth line and give it a spurious slope and intercept (Joysey, 1959b). Under the condition of directional selection, as in the case of the gekkonid lizard *Aristelliger*, the differential selection acts against one end of the range of variation present in the juveniles, and these individuals become the juvenile corpses in the cemetery population. Individuals from the other extreme of the range of juvenile variation are more likely to reach a ripe old age, and these would become the specimens representing the adult portion of the population. A line fitted to a scatter diagram would therefore be displaced towards one side of the range of variation which had actually been present in the juvenile population, and it would run obliquely to a line representing the actual growth of those individuals which survived to adulthood (Fig. 2). So far as I am aware this selection anomaly has not been measured, but any such factor which could affect the slope and intercept of a relative growth curve, and perhaps produce a spurious difference between two samples, is worthy of further investigation.

Despite the foregoing problems I am convinced that the parameters of the relative growth relationship are the most convenient tool for making comparisons between fossil populations. I appreciate that multivariate analysis may be more sensitive in detecting differences between populations, but in my experience the tool of bivariate analysis is sufficiently fine for the purpose. It can detect phenotypic differences between samples of the echinoid *Echinus esculentus* dredged only a mile apart on the open sea floor (personal observation), and it has also been used to demonstrate differences, which have been attributed to differential selection, between populations of the echinoid *Echinocardium cordatum* collected from different localities (Nichols, 1962). Similarly, differences can be detected between demes of mammalian populations which undoubtedly belong to the same geographical subspecies, and so these simple statistical methods are already proved capable of picking up genetical 'noise', which we must expect to be superimposed upon the trends of long-term evolution.

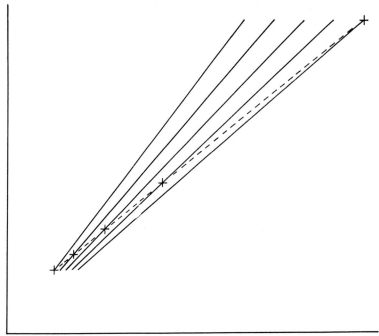

Fig. 2 Diagram to illustrate the effect of directional selection on the slope and intercept of a trend line (broken) fitted to a sample of a dead population. The unbroken lines indicate the expected growth relationship, in several individuals, between any two characters represented on the ordinate and abscissa. A selection anomaly would be produced if throughout the growth series those individuals at the lower end of the range of variation on the abscissa were more likely to die. Arithmetical scale on both axes. See text for further explanation. (After Joysey, 1959.)

PROBLEMS OF EVOLUTIONARY INTERPRETATION

Turning now to another point, there is considerable observational evidence that when environmental conditions change, the flora and fauna usually respond by changes in distribution. Environmental changes are usually too rapid for animals to evolve their way out of trouble, and if conditions become unsuitable then the fauna becomes locally extinct, and if conditions become suitable elsewhere then the species is likely to move into new regions. There is abundant direct evidence of changes in the geographical distribution of particular species within historical times (e.g. Forbes et al., 1958; Joysey, 1963; Hatting, 1963) and even more dramatic evidence of quite different geographical distributions of still extant species during the Pleistocene (Degerbøl, 1964; West, 1968).

If environmental change should cause a shift in the geographical distribution of a species already differentiated into a number of geographical subspecies, then it is reasonable to expect that a given locality previously occupied by one subspecies may come to be occupied by another. Referring to Fig. 3, if one assumes that no detectable evolutionary change occurs in the time span involved, then the succession at any one locality would contain a sequence of forms which would mimic an evolutionary series, and the geographical subspecies B at the lowest horizon would become a chronological subspecies at the middle horizon (Joysey, 1952). Assuming now that some evolutionary change also occurred, so that the succession of forms at one locality could be described as an evolutionary sequence A1-B2-C3, the effect of the changes in distribution would nevertheless be superimposed on the actual evolutionary change, and the apparent rate of evolution would be different from the actual rate of evolution. Unless one could determine the distribution of all the related forms in space and time it would be very difficult to unscramble the relative importance of these two phenomena.

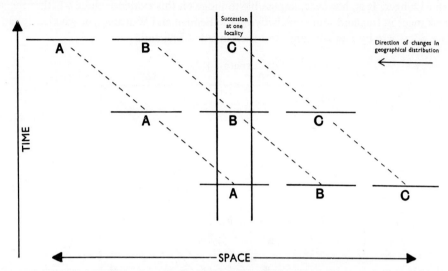

Fig. 3 Diagram to illustrate the effect produced in the geological succession at one locality by changes in the geographical distribution of a polytypic species. At each time horizon the letters represent geographical subspecies. See text for further explanation. (After Joysey, 1952)

It is sufficient here to indicate that apparent fluctuations in the rate of evolution could result from changes in animal distribution, and in particular that a rate of evolution may appear to spurt if a form which has already evolved elsewhere should move into the previous range of a closely related form.

In seeking liaison between the types of evolutionary patterns observed by palaeontologists and the sort of evolutionary mechanisms studied by ecological geneticists, the Pleistocene period is the obvious part of the fossil record to investigate because time is measured in thousands of years rather than millions of years. Particularly in the higher latitudes the unusual fluctuations in the Pleistocene climate and the consequent changes in animal distribution complicate the picture, but this is the only opportunity we have to make detailed comparisons between living populations and ancestral forms which often belong to the same species. Degerbøl of Copenhagen and Kurtén of Helsinki have been pioneers in this field and a few examples drawn from their studies are described here in order to illustrate the contention that geographical and chronological subspecies can be regarded as analogous entities.

Kurtén and Rausch (1959) made biometric comparisons between samples of the glutton living in Alaska and Fennoscandia, and concluded that the two populations are geographical subspecies of the same species (*Gulo gulo*). They found significant differences in several characters of the skull and dentition, one of which is illustrated in Fig. 4. The scatter diagrams show that there is an overlap in the range of variation present in the two regions, and that the allometric relationship between width and length of the upper carnassial is significantly different. Furthermore, a sample of fossil gluttons from the Pleistocene of Europe was found to have more in common with the living American population than with those living in Europe at the present time.

A similar degree of difference is shown in Fig. 5, illustrating covariation between the lengths of the second and third lower premolars in two samples of cave hyaena (*Crocuta crocuta*) from the Pleistocene of England (Kurtén, 1967), but here the difference between the two samples is chronological, one being from the Eemian (interglacial) and the other from the Wurmian (last glaciation). The shift in the allometric axis has been attributed to a genetical change, the second premolar becoming shorter and the third longer. It is, however, impossible to judge on this evidence alone whether the genetical change took place in England sometime between the Eemian and Wurmian, or whether it had already taken place elsewhere by Eemian time, and then invaded England.

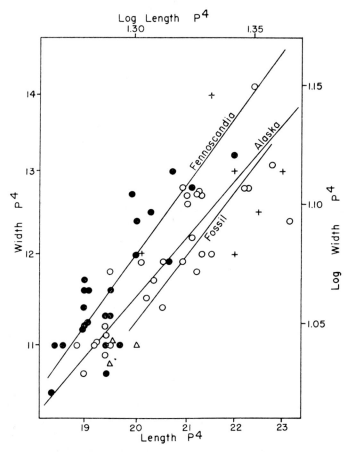

Fig. 4 Allometric relationship of width to length of the upper carnassial tooth in the glutton, from several localities. Fennoscandia (black circle); Alaska (open circle); Arctic Canada (triangle); late Pleistocene of Europe (cross). (Reproduced with permission from Kurtén and Rausch, 1959.)

Fossil members of a species often differ in size from their living representatives and, on the basis of dental and other skeletal evidence, Kurtén (1968) has shown that in Europe since the last glaciation the brown bear, glutton and pine marten have all become smaller, the lynx and polecat have stayed the same size and the otter and badger have become larger. One of the major problems in interpreting the evolutionary significance of these changes is that many mammals show geographical clines in size. For example, Kurtén (1957a), using the mean length of the lower carnassial as a measure of size, showed that spotted hyaenas living in Africa are smallest near the equator and larger in higher latitudes, and hence it is not unexpected that fossil material from even higher northern latitudes is even larger (Fig. 6).

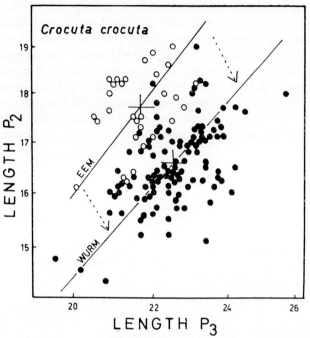

Fig. 5 Allometric relationship of the lengths of the second and third lower premolars in two samples of cave hyaena, which lived at different times in England. Eemian, Tornewton Cave (black circle); Wurmian, Kent's Cavern (open circle). Arrows indicate transposition of the allometry axis. (Reproduced with permission from Kurtén, 1967.)

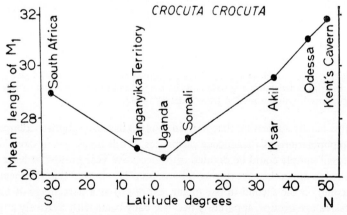

Fig. 6 Cline (character gradient) in the living African and late Pleistocene Syrian and European populations of the spotted hyaena (*Crocuta crocuta*); the mean length of the lower carnassial tooth in each local population is plotted against latitude. The gradual transition is interpreted as demonstrating the specific unity of the cave hyaena (*C. crocuta spelaea*) and the living form. (Reproduced with permission from Kurtén, 1957.)

In several publications Kurtén (1960, 1968) has drawn attention to fluctuations in the size of the brown bear during the Pleistocene of Europe, it being larger in the glacials and smaller in the interglacials (Fig. 7). The present geographical range of the brown bear in Europe is too restricted to determine whether a size cline exists, and so it is difficult to judge whether these fluctuations in size represent evolutionary changes in a more or less static population (as a result of fluctuating selection sometimes favouring larger bears and sometimes smaller bears), or whether they represent movements into the area of bears of different sizes (as a result of the changing climatic regime). One cannot rule out the possibility that the fluctuations in size are phenotypic, for it is well known that some mammals attain different sizes under different environmental conditions (as has often been demonstrated inadvertently in both domestic and wild stock one generation after transportation to new areas). The situation in brown bears is further complicated by the observation that in Palestine during the late Pleistocene the changes in size appear to be inverse to those in Europe (Kurtén, 1965).

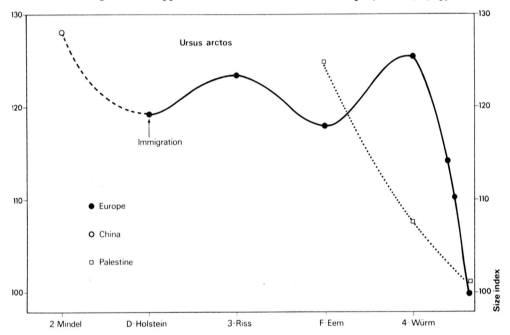

Fig. 7 Oscillation in size of Pleistocene and Recent brown bear, based on the relative lengths of the lower molars, the size index 100 being based on the living Scandinavian form. The time scale is not absolute. (Reproduced with permission from Kurtén, 1968.)

In this context it is hardly surprising that Kurtén (1957b) has also sought evidence for the action of natural selection on populations of Pleistocene carnivores. While investigating cave bears from Odessa he found that the fossil sample could be divided into successive year groups, at least when young, because the bears tended to die in the cave during hibernation. He showed that the mean length of a prominent cusp (paracone) on the second upper molar, expressed as a percentage of total tooth length, decreased in successive age groups, apparently because baby bears with relatively large cusps had a far greater mortality than those with relatively small cusps. Indeed, the differential selection was so strong that few animals which possessed an above average cusp size reached sexual maturity, but despite this there was no evidence of any evolutionary change in the population in the thousands of years during which the cave was inhabited. It is, of course, easy to rationalise several possible genetic explanations which could account for the cave bears retaining their large paracone despite apparent selection against it, but it is tantalising that one of the few good examples of directional selection in the fossil record should have failed to evolve.

It is, therefore, reassuring that the fossil record does provide a good documentation of evolution under the effects of domestic selection. Degerbøl (1963) has described the history of cattle in Denmark and, using dimensions of the lower hindmost molar (M3) which have been shown to be correlated with overall size, he has demonstrated a decrease in size through successive samples drawn from populations representing the wild *Bos primigenius*, the semi-domesticated Ertebølle culture, and the domestic stock in Neolithic, Bronze age, Iron age and Medieval times. In this character there is hardly any overlap in the range of variation present in the wild form and the domestic Neolithic cattle, but the gap is bridged by the Ertebølle sample. Further selection for smaller cattle between Neolithic and Medieval times carried the range of variation right outside that which had been present in the wild ancestral stock. Medieval cattle in Denmark were diminutive and it is only in post-Medieval times that domestic selection has begun to favour larger size, and some modern cattle are now almost as beefy as those in the Neolithic.

I am convinced that studies of Pleistocene mammals, such as those described in the previous paragraphs, are beginning to fill in details of the pattern of evolution and will, in due course, help to close the gap between palaeontological and genetical studies. The results obtained are completely consistent with the theoretical considerations discussed in this essay and as might be expected, with the incomplete data available, it *has* proved difficult to distinguish between evolutionary change and the effects of changes in distribution.

The intensive study of particular, carefully chosen species, with the aim of documenting the distribution of their successive populations in both space and time, combined with detailed comparisons between those populations, may in due course provide a better understanding of the pattern of evolution at the species level. (At present, man himself is the only species which has been studied at this sort of level, but it seems likely that the history of our species is one of atypical complexity.)

At the present time some palaeontologists are advocating that the Linnaean system of nomenclature should be abandoned for fossils on the grounds that it is too rigid a system, pre-evolutionary in concept and inadequate for the description of evolutionary processes. In my view, the concepts of geographical and chronological subspecies provide the necessary flexibility. They have been found useful in the discussion of Pleistocene to Recent evolutionary phenomena and have in no way blocked progress in this field. It has been argued that the problems involved in earlier parts of the geological succession are quite different because of the different time spans involved, but in my own experience the phenomena which I have encountered while working on Carboniferous blastoids, Mesozoic, Tertiary and Recent echinoids, Jurassic oysters, and Pleistocene mammals seem to be completely analogous.

REFERENCES

DEGERBØL, M. 1963. Prehistoric cattle in Denmark and adjacent areas. *Royal Anth. Inst.* Occasional paper No. 18, London.
—— 1964. Some remarks on late- and post-glacial vertebrate fauna and its ecological relations in Northern Europe. *J. Anim. Ecol.*, **33** (Suppl.), 71-85.
FORBES, C. L., JOYSEY, K. A. and WEST, R. G. 1958. On Post-glacial pelicans in Britain. *Geol. Mag.*, **95**, 153-160.
GOULD, S. J. 1966. Allometry and size in ontogeny and phylogeny. *Biol. Rev.*, **41**, 587-640.
HATTING, T. 1963. On subfossil finds of Dalmatian Pelican (*Pelecanus crispus* Bruch) from Denmark. *Vidensk. Medd. fra Dansk naturh. Foren.*, **125**, 337-351.
HECHT, M. K. 1952. Natural selection in the lizard genus *Aristelliger*. *Evolution*, **6**, 112-124.

HUXLEY, J. S. 1932. *Problems of relative growth*. Methuen. London.
—— 1945. (ed.) *The new systematics*. Oxford University Press.
JOYSEY, K. A. 1952. Fossil lineages and environmental change. *Geol. Mag.*, **89**, 357-360.
—— 1953. A study of the type-species of the blastoid *Codaster* McCoy. *Geol. Mag.*, **90**, 208-218.
—— 1956. The nomenclature and comparison of fossil communities. In *The species concept in palaeontology*, pp. 83-94, edited by P. C. Sylvester-Bradley. Systematics Assoc. Publ. No. 2, London.
—— 1959a. A study of variation and relative growth in the blastoid *Orbitremites*. *Phil. Trans. R. Soc.* B **242**, 99-125.
—— 1959b. The evolution of the Liassic oysters *Ostrea-Gryphaea*. *Biol. Rev.*, **34**, 297-332.
—— 1963. A scrap of bone. In *Science in Archaeology* pp. 197-203, edited by D. Brothwell and E. S. Higgs, Thames and Hudson, London.
KERMACK, K. A. and HALDANE, J. B. S. 1950. Organic correlation and allometry. *Biometrika*, **37**, 30-41.
KURTÉN, B. 1957a. The bears and hyenas of the interglacials. *Quaternaria*, **4**, 69-81.
—— 1957b. A case of Darwinian selection in bears. *Evolution*, **11**, 412-416.
—— 1960. Rates of evolution in fossil mammals. *Cold Spring Harbor Symp. Quant. Biol.*, **24**, 205-215.
—— 1965. The Carnivora of the Palestine caves. *Acta. Zool. Fennica.*, No. 107, 1-74.
—— 1967. Some quantitative approaches to dental microevolution. *J. Dental Res.*, **46**, 817-828.
—— 1968. *Pleistocene mammals of Europe*. Weidenfeld and Nicolson, London.
KURTÉN, B. and RAUSCH, R. 1959. Biometric comparisons between North American and European mammals. *Acta Arctica*, **11**, 5-44.
MAYR, E. 1942. *Systematics and the origin of species*. Columbia University Press, New York.
MORRIS, P. A. 1970. A method for determining absolute age in the hedgehog. *J. Zool.*, **161**, 277-281.
NICHOLS, D. 1962. Differential selection in populations of a heart-urchin. In *Taxonomy and Geography*. pp. 105-118, edited by D. Nichols. Systematics Assoc. Publ. No. 4, London.
SYLVESTER-BRADLEY, P. C. 1956 (ed.). *The species concept in palaeontology*. Systematics Assoc. Publ. No. 2, London.
WEST, R. G. 1968. *Pleistocene geology and biology, with especial reference to the British Isles*. Longmans, London.

W. G. KÜHNE

Progress in biological evolution

Very often a contribution to a *Festschrift* consists of a 'minor matter' of one's research. A self-contained marginal subject is treated in less than 20 pages; such a paper is a "contribution to science" but it defies its object. A contribution to a *Festschrift* is, I think, the ideal means to vent ideas not yet fully accepted, to leave for a moment the properly acknowledged subject matter and to take stock of methods and ideas.

Some philosophers, and many politicians, underwrite the philosophy of dialectical materialism. In the so-called western civilisation this philosophy is still taboo for good reasons. It is the philosophy of change, and, for a ruling class, change means a decrease in power or at least a questioning of its claims to power.

Palaeontology, however, is the science of evolutionary change during more than 500 million years. To discuss theorems of dialectical materialism in palaeontological terms may be profitable. The attempt to do so is new. To politicians it would be a revelation if some of their theorems were to find support in the venerable and respectable science of palaeontology. To palaeontologists it would provide their subject with a new attraction; and they might find in books of philosophy answers to the problems they like to solve. To the author it is an essay on a subject he hopes to write more about in the near future; even a meagre echo he might receive may encourage him to do so. To the honoured recipient of the *Festschrift* it may be satisfying and consoling that intellectual change, in the minds of his pupils, has its dialectical counterpart in his own mind and that his words, now turned into their dialectical antithesis, are still his very words and that such intellectual evolution can be the best a worker can experience.

Dialectical materialism proposes, as one of its theorems, the progressiveness of social evolution. A sequence leads from the earliest manifestation of society through savage, feudal, capitalist and socialist society and this sequence is obviously one in chronological order as well as in increasing complexity; it is also partly a sequence of size in which the earliest society is also the smallest. That social evolution can skip one of the mentioned stages is not impossible, but it is unlikely. My task is now to elaborate on the theory of progressive biological evolution, exemplified in palaeontology.

Is it possible to find in biological evolution evidence either of progress or of regress, or neither or both?

To come to the subject matter, we have to state: the progressiveness, or its contrary, in biological evolution cannot be observed in isolated phenomena but only in related phenomena. An institution is not progressive *per se*, but only when measured against another related institution. Ritualized cannibalism is progressive in relation to the custom of devouring one's victim to satisfy hunger, and symbolized cannibalism is progressive compared with ordinary ritual cannibalism.

After the Middle Devonian no bactritid descendant possesses the straight phragmocone characteristic of the cephalopods of the Ordovician and Silurian. After the zone of *Gattendorfia* no Lower Carboniferous goniatite possesses a concave nautiloid septum, but they have a convex one instead which was present only rarely in the Devonian. After the Upper Cretaceous no cephalopod (except *Nautilus*), possesses an external shell.

Without much reflection we would regard the straight phragmocone as being the simpler, less elaborate condition, compared with the planispirally coiled shell of goniatites. Our hydrodynamical experience would favour a disc-like shell, compared with a straight conical one. We could easily rationalize the selective advantage and would, in passing, point to the vulnerability of the straight cone of the Nautiloidea (of the Palaeozoic).

Is the more elaborate goniatite shell progressive compared with the phragmocone of straight-shelled nautiloids? The more elaborate structure, the goniatite shell, is more recent than the straight shell. The waning of the orthoceratids in the Triassic is the corollary to the acme of the goniatites and the descendants of the prolecanitids. Elaboration of structure and, as a consequence, success in geological time would, in my opinion, be progressive.

Our second observation regarding change in ammonoids is a more difficult case. As a nautiloid heritage most Devonian goniatites possess a concave septum; from the beginning of the Carboniferous a convex septum predominates. Is this progressive change? I would not dare to say! The feature is concomitant with the existence of the ammonites of the Upper Palaeozoic and the Mesozoic. The feature is not linked chronologically with an acme. It may be linked with sutural elaboration, but rudimentary elaboration of the contact-line between outer shell and septum is already present in Devonian goniatites with a concave septum. From a hydrostatic point of view a concave septum is more stable than a straight septum, which is unknown among cephalopods. Our second example could be summarized under the headings: 'indeterminable', 'indifferent', or 'still unknown'. My choice is with the last alternative.

Another stock of cephalopods, probably stemming from a sister-group of the Belemnoidea, jettisoned its outer shell and is represented by the living cephalopods. But for *Nautilus*, all other cephalopods bearing an outer shell have been extinct since the beginning of the Tertiary; the survivors are shell-less. The 'obvious' anthropomorphic conclusion drives us to accept the progressiveness of this evolutionary phenomenon. Evolved cephalopods could cope with their enemies without an organ of protection and have proved their superiority by their existence today. The fossil record of cephalopods devoid of a shell is a poor one. Whether they ever experienced an acme cannot be established. Their first occurrence quite certainly antedates the Jurassic and Cretaceous acme of the Ammonoidea. Neither their post-Cretaceous survival, nor the extinction, in the Upper Cretaceous, of ammonites and belemnites could have been predicted by a Jurassic palaeontologist, if such had existed. The size range, a rather good measure of evolutionary activity, of present day cephalopods exceeds that of all earlier ones. It is improbable that replacement of ammonoids and belemnites by teuthoids ever happened, in particular as a displacement of the former two by the latter, and it certainly cannot be proved. As the teuthoids are practically non-fossilizable, nothing useful can be said about their number of taxa and their diversity in any geological time.

As soon as extinction enters into our consideration of progressive evolution we can recognize it as a corollary of progressive evolution. But progressive evolution, and the future fate of the respective taxon experiencing it, are two matters not intimately, or even causally, connected. It has taken a long time to see that there is no aim in the process of evolution; (though Spinoza fought bravely against teleology). If there is no pinnacle, no summit, no aim in biological evolution, then no evolution can 'overshoot its aim', as once was said of *Megaloceras*' antlers or *Machairodus*' canines.

Any live taxon coping with, or reacting to, the everchanging Umwelt it finds itself in, experiences progressive evolution. As long as it results in the maintenance of life, the experienced change cannot be but progressive. Regressive change would mean the negation of the taxon's life.

Loss of organs is still sometimes regarded as regressive change; but far from it. The Pekinese prognathous lower incisors may not be regarded as beautiful, but they are part and parcel of the Chinese Palace dog with a venerable span of life. When the last Chinese palace was sacked by the infuriated populace, or by an invader, the Pekinese even found a new biotope in western civilization.

And so for *Loris* and *Poto*. Being slow and tailless is part and parcel of a life in a biotope practically devoid of tetrapod enemies, *viz.*, life in the upper region of tropical forests. The same applies to *Chamaeleo* with its prehensile tail. To call a *Sacculina* degenerate is simply naïve if degenerate means unfit or unable to maintain life. A jockey weighing 75 kg, or a miner measuring 189 cm is degenerate, but a Dean and Nobel prizewinner who suffers from flatfoot, hernia and premature baldness, and who is lefthanded and shortsighted is *not* degenerate.

At one time there were authors who wrote about the senescence of the Permian trilobites or the Cretaceous Saurischia. Simpson, *inter alia*, has been instrumental in stemming loose thinking of this kind. Regression and degeneration are concepts which, in connection with evolution, are objectionable and have to be eliminated from evolutionary theory. Both, like senescence, apply to the individual but not to the lineage or to the cladus. They imply unfitness of the individual in its specific biotope, or unfitness because of old age, but nothing more. Life of the individual and life of the cladus are unidirectional processes in time, and regression is a concept which is inapplicable to both life processes. The individual progresses from birth to death but does not undergo a kind of 'regression' by which the senile becomes 'childish'.

To compare individual unfitness such as shortsightedness in a hunter, with the waning of vision in trilobites and cave animals, or the taillessness of *Poto* and *Homo* is irrelevant and leads to nothing. Progressive change is as much a shedding of useless structures as a building of new ones; their balance conditions the future of the cladus experienceing these changes, but this future is determined by the relation of the cladus to its *Umwelt* (which implies more than its biotope).

From its beginning stratigraphy made it its task to predict the relative geological age of a stratum below or above that of a stratum whose stratigraphical age was known. In the early decades of the 19th century the index fossil was the means to accomplish this task. Later accumulation of knowledge, further sampling and subsequent systematic work led to a voluminous body of observations which allowed a far better stratigraphy than the one gained by means of index fossils; it is the stratigraphy by means of evolutionary trends.

The progressiveness of a whole fauna is assessed in respect to another fauna, or a single lineage is analysed, and different stages in the manifestation of trends are understood and put into the stratigraphical statement. What has happened is the substitution of 'timeless' objects, index fossils, by a sequence of phylogenetically related faunas undergoing evolutionary change; the change being manifest in a trend.

I now wish to consider the question: are trends progressive, regressive or neutral?

Simpson stated at some length that evolutionary trends are adaptive. To disprove this statement is well-nigh impossible, and the statement "evolution is adaptive" might be acceptable by Simpson. The apparent conflict between those who deny progressiveness of evolution and those who understand evolution as progressive in all instances is resolved if we understand "progressive" as simply *pro-gressive*, that is, advancing through time. Progressive is not better or worse than something else, but newer than something older. And, the maintenance of life is progressive. It may sound cynical to state that slavery, colonialism and monopoly capitalism are progressive, when measured against the fate of slaves, colonial people or powerless workers subjected to do work devoid of sense, but when measured against human conditions without these institutions, they are indeed progressive and are the real prerequisite for their dialectical antithesis. The loss of an organ in an acme-forming taxon would hardly be called regressive, *viz.*, toothlessness of birds, or limblessness of snakes, or loss of the hind pair of wings in Diptera. The inauguration of electrically-driven railborne traffic in a town is certainly progressive. Progressive too is the substitution of trams by buses. Progressive is the erection of fortified walls round towns in the Middle Ages, progressive too is the demolition of such walls and the filling in of the moats and their substitution by parks. There are many social institutions which have come into existence and which are today obsolete. Both instances are progressive.

Life, being fitted with such minimal requirements as reproduction, metabolism, the power to repair a damaged organ, the power to react to stimuli also has the faculty to evolve, to change. To evolve is historic; and in history no previous phase is repeated, regress is unthinkable. Progress does not only make life possible where there was no life before. Progress results in evading selection pressure where life is already established. Regress, to my mind, means undergoing a change which leads its bearer into a situation where he experiences, not less, but more selection pressure; hence this route is barren. In other words, without progressive change, an individual can only maintain itself in a biotope already inhabited by its taxon, and by living in this biotope it increases the selection pressure. By regressive change an individual is less fit than the members of the taxon to which it belongs and this is lethal; by progressive change it can live in a biotope not inhabited previously, or only partly inhabited, by its taxon, and hence it can live under decreased selection pressure and thrive.

Stratigraphical prediction and stratigraphy itself would be impossible if evolution were to be random. Stratigraphy is possible because evolution is historic, is directional and is progressive.

SOMERVILLE COLLEGE
LIBRARY